ORGANIC SYNTHESES

ORGANIC SYNTHESES

AN ANNUAL PUBLICATION OF SATISFACTORY METHODS FOR THE PREPARATION OF ORGANIC CHEMICALS

VOLUME 93

2016

HUW M. L. DAVIES AND ERICK M. CARREIRA

VOLUME EDITORS

WILEY

Published by John Wiley & Sons, Inc., Hoboken, New Jersey.

Published simultaneously in Canada.

Library of Congress Catalog Card Number: 21-17747

ISBN: 978-1-119-34902-0

Printed in the United States of America

10 9 8 7 6 5 4 3 2 1

ORGANIC SYNTHESES

VOLUME	VOLUME EDITOR	PAGES
I*	† Roger Adams	84
II*	† James Bryant Conant	100
III*	† Hans Thacher Clark	105
IV*	† Oliver Kamm	89
V*	† Carl Shipp Marvel	110
VI*	† Henry Gilman	120
VII*	† Frank C. Whitmore	105
VIII*	† Roger Adams	139
IX*	† James Bryant Conant	108
Collective Vol. I	A revised edition of Annual Volumes I-IX †Henry Gilman, *Editor-in-Chief* 2nd Edition revised by † A. H. Blatt	580
X*	† Hans Thacher Clarke	119
XI*	† Carl Shipp Marvel	106
XII*	† Frank C. Whitmore	96
XIII*	† Wallace H. Carothers	119
XIV	† William W. Hartman	100
XV*	† Carl R. Noller	104
XVI*	† John R. Johnson	104
XVII*	† L. F. Fieser	112
XVIII*	† Reynold C. Fuson	103
XIX*	† John R. Johnson	105
Collective Vol. II	A revised edition of Annual volumes X-XIX † A. H. Blatt, *Editor-in-Chief*	654
20*	† Charles F. H. Allen	113
21*	† Nathan L. Drake	120
22*	† Lee Irvin Smith	114
23*	† Lee Irvin Smith	124
24*	† Nathan L. Drake	119
25*	† Werner E. Bachmann	120
26*	† Homer Adkins	124
27*	† R. L. Shriner	121
28*	† H. R. Snyder	121
29*	† Cliff S. Hamilton	119
Collective Vol. III	A revised edition of Annual Volumes 20-29 † E. C. Horning, *Editor-in-Chief*	890
30*	† Arthur C. Cope	115
31*	† R. S. Schreiber	122
32*	† Richard Arnold	119
33*	† Charles Price	115
34*	† William S. Johnson	121
35*	† T. L. Cairns	122

**Out of print.*
†Deceased.

VOLUME	VOLUME EDITOR	PAGES
36*	† N. J. LEONARD	120
37*	† JAMES CASON	109
38*	† JAMES C. SHEEHAN	120
39*	† MAX TISHLER	114
Collective Vol. IV	A revised edition of Annual Volumes 30-39 † NORMAN RABJOHN, *Editor-in-Chief*	1036
40*	† MELVIN S. NEWMAN	114
41*	† JOHN D. ROBERTS	118
42*	† VIRGIL BOEKELHEIDE	118
43*	† B. C. MCKUSICK	124
44*	† WILLIAM E. PARHAM	131
45*	† WILLIAM G. DAUBEN	118
46*	E. J. COREY	146
47*	† WILLIAM D. EMMONS	140
48*	† PETER YATES	164
49*	KENNETH B. WIBERG	124
Collective Vol. V	A revised edition of Annual Volumes 40-49 † HENRY E. BAUMGARTEN, *Editor-in-Chief*	1234
Cumulative Indices to Collective Volumes, I, II, III, IV, V	† RALPH L. AND † RACHEL H. SHRINER, *Editors*	
50*	RONALD BRESLOW	136
51*	† RICHARD E. BENSON	209
52*	† HERBERT O. HOUSE	192
53*	† ARNOLD BROSSI	193
54*	† ROBERT E. IRELAND	155
55*	† SATORU MASAMUNE	150
56*	† GEORGE H. BÜCHI	144
57*	CARL R. JOHNSON	135
58*	† WILLIAM A. SHEPPARD	216
59*	ROBERT M. COATES	267
Collective Vol. VI	A revised edition of Annual Volumes 50-59 WAYLAND E. NOLAND, *Editor-in-Chief*	1208
60*	† ORVILLE L. CHAPMAN	140
61*	† ROBERT V. STEVENS	165
62*	MARTIN F. SEMMELHACK	269
63*	† GABRIEL SAUCY	291
64*	ANDREW S. KENDE	308
Collective Vol. VII	A revised edition of Annual Volumes 60-64 † JEREMIAH P. FREEMAN, *Editor-in-Chief*	602
65*	EDWIN VEDEJS	278
66*	CLAYTON H. HEATHCOCK	265
67*	BRUCE E. SMART	289
68*	JAMES D. WHITE	318
69*	LEO A. PAQUETTE	328

**Out of print.*
†Deceased.

VOLUME	VOLUME EDITOR	PAGES
Reaction Guide to Collective Volumes I-VII and Annual Volumes 65-68	DENNIS C. LIOTTA AND MARK VOLMER, *Editors*	854
Collective Vol. VIII	A revised edition of Annual Volumes 65-69 † JEREMIAH P. FREEMAN, *Editor-in-Chief*	696
Cumulative Indices to Collective Volumes, I, II, III, IV, V, VI, VII, VIII		
70*	† ALBERT I. MEYERS	305
71*	LARRY E. OVERMAN	285
72*	† DAVID L. COFFEN	333
73*	ROBERT K. BOECKMAN, JR.	352
74*	ICHIRO SHINKAI	341
Collective Vol. IX	A revised edition of Annual Volumes 70-74 † JEREMIAH P. FREEMAN, *Editor-in-Chief*	840
75	AMOS B. SMITH, III	257
76	STEPHEN MARTIN	340
77	DAVID S. HART	312
78	WILLIAM R. ROUSH	326
79	LOUIS S. HEGEDUS	328
Collective Vol. X	A revised edition of Annual Volumes 75-79 † JEREMIAH P. FREEMAN, *Editor-in-Chief*	810
80	STEVEN WOLFF	259
81	RICK L. DANHEISER	296
82	EDWARD J. J. GRABOWSKI	195
83	DENNIS P. CURRAN	221
84	MARVIN J. MILLER	410
Collective Vol. XI	A revised edition of Annual Volumes 80-84 CHARLES K. ZERCHER, *Editor-in-Chief*	1138
85	SCOTT E. DENMARK	321
86	JOHN A. RAGAN	403
87	PETER WIPF	403
88	JONATHAN A. ELLMAN	472
89	MARK LAUTENS	561
90	DAVID L. HUGHES	376
Collective Vol. XII	A revised edition of Annual Volumes 85-89 CHARLES K. ZERCHER, *Editor-in-Chief*	1570
91	KAY M. BRUMMOND	365
92	VIRESH H. RAWAL	386
93	HUW M. L. DAVIES AND ERICK CARREIRA	421

**Out of print.*
†Deceased.

NOTICE

Beginning with Volume 84, the Editors of *Organic Syntheses* initiated a new publication protocol, which is intended to shorten the time between submission of a procedure and its appearance as a publication. Immediately upon completion of the successful checking process, procedures are assigned volume and page numbers and are then posted on the Organic Syntheses website (www.orgsyn.org). The accumulated procedures from a single volume are assembled once a year and submitted for publication. The annual volume is published by John Wiley and Sons, Inc., and includes an index. The hard cover edition is available for purchase through the publisher. Incorporation of graphical abstracts into the Table of Contents began with Volume 77. Annual volumes 70–74, 75–, 80–84 and 85–89 have been incorporated into five-year versions of the collective volumes of *Organic Syntheses.* Collective Volumes IX, X, XI and XII are available for purchase in the traditional hard cover format from the publishers.

Beginning with Volume 88, a new type of article, referred to as Discussion Addenda, appeared. In these articles submitters are provided the opportunity to include updated discussion sections in which new understanding, further development, and additional application of the original method are described. Organic Syntheses intends for Discussion Addenda to become a regular feature of future volumes.

Organic Syntheses, Inc., joined the age of electronic publication in 2001 with the release of its free web site (www.orgsyn.org). The site is accessible through internet browsers using Macintosh and Windows operating systems, and the database can be searched by key words and sub-structure. John Wiley & Sons, Inc., and Accelrys, Inc., partnered with Organic Syntheses, Inc., to develop a database (onlinelibrary.wiley.com/book/10.1002/0471264229) that is available for license with internet solutions from John Wiley & Sons, Inc. and intranet solutions from Accelrys, Inc.

Both the commercial database and the free website contain all annual and collective volumes and indices of *Organic Syntheses*. Chemists can draw structural queries and combine structural or reaction transformation queries with full-text and bibliographic search terms, such as chemical name, reagents, molecular formula, apparatus, or even hazard warnings or phrases. The contents of individual or collective volumes can be browsed by lists of titles, submitters' names, and volume and page references, with or without structures.

The commercial database at onlinelibrary.wiley.com/book/10.1002/0471264229 also enables the user to choose his/her preferred chemical drawing package, or to utilize several freely available plug-ins for entering queries. The user is also able to cut and paste existing structures and reactions directly into the structure search query or their preferred chemistry editor, streamlining workflow. Additionally, this database contains links to the full text of primary literature references via CrossRef, ChemPort, Medline, and ISI Web of Science. Links to local holdings for institutions using open url technology can also be enabled. The database user can limit his/her search to, or order the search results by, such factors as reaction type, percentage yield, temperature, and publication date, and can create a customized table of reactions for comparison. Connections to other Wiley references are currently made via text search, with cross-product structure and reaction searching to be added in the near future. Incorporations of new preparations will occur as new material becomes available.

INFORMATION FOR AUTHORS OF PROCEDURES

Organic Syntheses welcomes and encourages submissions of experimental procedures that lead to compounds of wide interest or that illustrate important new developments in methodology. Proposals for *Organic Syntheses* procedures will be considered by the Editorial Board upon receipt of an outline proposal as described below. A full procedure will then be invited for those proposals determined to be of sufficient interest. These full procedures will be evaluated by the Editorial Board, and if approved, assigned to a member of the Board for checking. In order for a procedure to be accepted for publication, each reaction must be successfully repeated in the laboratory of a member of the Editorial Board at least twice, with similar yields (generally ±5%) and selectivity to that reported by the submitters.

Organic Syntheses Proposals

A cover sheet should be included providing full contact information for the principal author and including a scheme outlining the proposed reactions (an *Organic Syntheses* Proposal Cover Sheet can be downloaded at orgsyn.org). Attach an outline proposal describing the utility of the methodology and/or the usefulness of the product. Identify and reference the best current alternatives. For each step, indicate the proposed scale, yield, method of isolation and purification, and how the purity of the product is determined. Describe any unusual apparatus or techniques required, and any special hazards associated with the procedure. Identify the source of starting materials. Enclose copies of relevant publications (attach pdf files if an electronic submission is used).

Submit proposals by mail or as e-mail attachments to:

Professor Charles K. Zercher
Associate Editor, *Organic Syntheses*
Department of Chemistry
University of New Hampshire
23 Academic Way, Parsons Hall
Durham, NH 03824

Electronic submission through the website (www.orgsyn.org) is strongly encouraged.

Submission of Procedures

Authors invited by the Editorial Board to submit full procedures should prepare their manuscripts in accord with the Instructions to Authors which are described below or may be downloaded at orgsyn.org. Submitters are also encouraged to consult this volume of *Organic Syntheses* for models with regard to style, format, and the level of experimental detail expected in *Organic Syntheses* procedures. Manuscripts should be submitted to the Associate Editor. Electronic submissions are encouraged; procedures will be accepted as e-mail attachments in the form of Microsoft Word files with all schemes and graphics also sent separately as ChemDraw files.

Procedures that do not conform to the Instructions to Authors with regard to experimental style and detail will be returned to authors for correction. Authors will be notified when their manuscript is approved for checking by the Editorial Board, and it is the goal of the Board to complete the checking of procedures within a period of no more than six months.

Additions, corrections, and improvements to the preparations previously published are welcomed; these should be directed to the Associate Editor. However, checking of such improvements will only be undertaken when new methodology is involved.

NOMENCLATURE

Both common and systematic names of compounds are used throughout this volume, depending on which the Volume Editor felt was more appropriate. The Chemical Abstracts indexing name for each title compound, if it differs from the title name, is given as a subtitle. Systematic Chemical Abstracts nomenclature, used in the Collective Indexes for the title compound and a selection of other compounds mentioned in the procedure, is provided in an appendix at the end of each preparation. Chemical Abstracts Registry numbers, which are useful in computer searching and identification, are also provided in these appendices.

ACKNOWLEDGMENT

Organic Syntheses wishes to acknowledge the contributions of Amgen, Inc. and Boehringer Ingelheim to the success of this enterprise through their support, in the form of time and expenses, of members of the Board of Editors.

INSTRUCTIONS FOR AUTHORS

All organic chemists have experienced frustration at one time or another when attempting to repeat reactions based on experimental procedures found in journal articles. To ensure reproducibility, *Organic Syntheses* requires experimental procedures written with considerably more detail as compared to the typical procedures found in other journals and in the "Supporting Information" sections of papers. In addition, each *Organic Syntheses* procedure is carefully "checked" for reproducibility in the laboratory of a member of the Board of Editors.

Even with these more detailed procedures, the experience of *Organic Syntheses* editors is that difficulties often arise in obtaining the results and yields reported by the submitters of procedures. To expedite the checking process and ensure success, we have prepared the following "Instructions for Authors" as well as a **Checklist for Authors** and **Characterization Checklist** to assist you in confirming that your procedure conforms to these requirements. Please include a completed Checklist together with your procedure at the time of submission. Procedures submitted to *Organic Syntheses* will be carefully reviewed upon receipt and procedures lacking any of the required information will be returned to the submitters for revision.

Scale and Optimization

The appropriate scale for procedures will vary widely depending on the nature of the chemistry and the compounds synthesized in the procedure. However, some general guidelines are possible. For procedures in which the principal goal is to illustrate a synthetic method or strategy, it is expected, in general, that the procedure should result in at least 5 g and no more than 50 g of the final product. In cases where the point of the procedure is to provide an efficient method for the preparation of a useful reagent or synthetic building block, the appropriate scale also should not exceed 50 g of final product. Exceptions to these guidelines may be granted in special circumstances. For example, procedures describing the preparation of reagents employed as catalysts will often be acceptable on a scale of less than 5 g.

In considering the scale for an *Organic Syntheses* procedure, authors should also take into account the cost of reagents and starting materials. In general, the Editors will not accept procedures for checking in which the cost of any one of the reactants exceeds **$500** for a single full-scale run. Authors are requested to identify the most expensive reagent or starting material on the procedure submission checklist and to estimate its cost per run of the procedure.

It is expected that all aspects of the procedure will have been optimized by the authors prior to submission, and it is required that each reaction will have been carried out at least twice on exactly the scale described in the procedure, and with the results reported in the manuscript.

It is appropriate to report the weight, yield, and purity of the product of each step in the procedure as a range. In any case where a reagent is employed in significant excess, a Note should be included explaining why an excess of that reagent is necessary. If possible, the Note should indicate the effect of using amounts of reagent less than that specified in the procedure.

The Checking Process

A unique feature of papers published in *Organic Syntheses* is that each procedure and all characterization data is carefully checked for reproducibility in the laboratory of a member of the Board of Editors. In the event that an editor finds it necessary to make any modifications in an experimental procedure, then the published article incorporates the modified procedure, with an explanation and mention of the original protocol often included in a Note. The yields reported in the published article are always those obtained by the checkers. In general, the characterization data in the published article also is that of the checkers, unless there are significant differences with the data obtained by the authors, in which case the author's data will also be reported in a Note.

Reaction Apparatus

Describe the size and type of flask (number of necks) and indicate how *every* neck is equipped.

> "A 500-mL, three-necked, round-bottomed flask equipped with an 3-cm Teflon-coated magnetic stirbar, a 250-mL pressure-equalizing addition funnel fitted with an argon inlet, and a rubber septum is charged with . . . ."

Indicate how the reaction apparatus is dried and whether the reaction is conducted under an inert atmosphere. Note that in general balloons are not acceptable as a means of maintaining an inert atmosphere unless warranted by special circumstances. The description of the reaction apparatus can be incorporated in the text of the procedure or included in a Note.

> "The apparatus is flame-dried and maintained under an atmosphere of argon during the course of the reaction."

In the case of procedures involving unusual glassware or especially complicated reaction setups, authors are encouraged to include a photograph or drawing of the apparatus in the text or in a Note (for examples, see *Org. Syn.,* Vol. 82, 99 and Coll. Vol. X, pp 2, 3, 136, 201, 208, and 669).

Use of Gloveboxes

When a glovebox is employed in a procedure, justification must be provided in a Note and the consequences of carrying out the operation without using a glovebox should be discussed.

Reagents and Starting Materials

All chemicals employed in the procedure must be commercially available or described in an earlier *Organic Syntheses* or *Inorganic Syntheses* procedure. For other compounds, a procedure should be included either as one or more steps in the text or, in the case of relatively straightforward preparations of reagents, as a Note. In the latter case, all requirements with regard to characterization, style, and detail also apply. Authors are encouraged to consult with the Associate Editor if they have any question as to whether to include such steps as part of the text or as a Note.

Authors are encouraged to consider the use of "substitute solvents" in place of more hazardous alternatives. For example, the use of *t*-butyl methyl ether (MTBE) should be considered as a substitute for diethyl ether, particularly in large scale work. Authors are referred to the articles "Sanofi's Solvent Selection Guide: A Step Toward More Sustainable Processes" (Prat, D.; Pardigon, O.; Flemming, H.-W.; Letestu, S.; Ducandas, V.; Isnard, P.; Guntrum, E.; Senac, T.; Ruisseau, S.; Cruciani, P. Hosek, P. O*rg. Process Res. Dev.* **2013**, *17*, 1517-1525) and "Solvent Replacement for Green Processing" (Sherman, J.; Chin, B.; Huibers, P. D. T.; Garcia-Valis, R.; Hatton, T. A. *Environ. Health Perspect.* **1998**, *106* (Supplement I, 253-271) as well as the references cited therein for discussions of this subject. In addition, a link to a "solvent selection guide" can be accessed via the American Chemical Society Green Chemistry website at http://www.acs.org/content/acs/en/greenchemistry/research-innovation/tools-for-green-chemistry.html.

In one or more Notes, indicate the purity or grade of each reagent, solvent, etc. It is highly desirable to also indicate the source (company the chemical was purchased from), particularly in the case of chemicals where it is suspected that the composition (trace impurities, etc.) may vary from one supplier to another. In cases where reagents are purified, dried, "activated" (e.g., Zn dust), etc., a detailed description of the procedure used should be included in a Note. In other cases, indicate that the chemical was "used as received".

"Diisopropylamine (99.5%) was obtained from Aldrich Chemical Co., Inc. and distilled under argon from calcium hydride before use. THF (99+%) was obtained from Mallinckrodt, Inc. and distilled from sodium benzophenone ketyl. Diethyl ether (99.9%) was purchased from Aldrich Chemical Co., Inc. and purified by pressure filtration under argon through activated alumina. Methyl iodide (99%) was obtained from Aldrich Chemical Co., Inc. and used as received."

The amount of each reactant must be provided in parentheses in the order mL, g, mmol, and equivalents with careful consideration to the correct number of **significant figures**. Avoid indicating amounts of reactants with more significant figures than makes sense. For example, "437 mL of THF" implies that the amount of solvent must be measured with a level of precision that is unlikely to affect the outcome of the reaction. Likewise, "5.00 equiv" implies that an amount of excess reagent must be controlled to a precision of 0.01 equiv.

The concentration of solutions should be expressed in terms of molarity or normality, and not percent (e.g., 1 N HCl, 6 M NaOH, not "10% HCl").

Reaction Procedure

Describe every aspect of the procedure clearly and explicitly. Indicate the order of addition and time for addition of all reagents and how each is added (via syringe, addition funnel, etc.).

Indicate the temperature of the reaction mixture (preferably internal temperature). Describe the type of cooling (e.g., "dry ice-acetone bath") and heating (e.g., oil bath, heating mantle) methods employed. Be careful to describe clearly all cooling and warming cycles, including initial and final temperatures and the time interval involved.

Describe the appearance of the reaction mixture (color, homogeneous or not, etc.) and describe all significant changes in appearance during the course of the reaction (color changes, gas evolution, appearance of solids, exotherms, etc.).

Indicate how the reaction can be monitored to determine the extent of conversion of reactants to products. In the case of reactions monitored by TLC, provide details in a Note, including eluent, R_f values, and method of visualization. For reactions followed by GC, HPLC, or NMR analysis, provide details on analysis conditions and relevant diagnostic peaks.

"The progress of the reaction was followed by TLC analysis on silica gel with 20% EtOAc-hexane as eluent and visualization with *p*-anisaldehyde. The ketone starting material has $R_f = 0.40$ (green) and the alcohol product has $R_f = 0.25$ (blue)."

Reaction Workup

Details should be provided for reactions in which a "quenching" process is involved. Describe the composition and volume of quenching agent, and time and temperature for addition. In cases where reaction mixtures are added to a quenching solution, be sure to also describe the setup employed.

> "The resulting mixture was stirred at room temperature for 15 h, and then carefully poured over 10 min into a rapidly stirred, ice-cold aqueous solution of 1 N HCl in a 500-mL Erlenmeyer flask equipped with a magnetic stirbar."

For extractions, the number of washes and the volume of each should be indicated as well as the size of the separatory funnel.

For concentration of solutions after workup, indicate the method and pressure and temperature used.

> "The reaction mixture is diluted with 200 mL of water and transferred to a 500-mL separatory funnel, and the aqueous phase is separated and extracted with three 100-mL portions of ether. The combined organic layers are washed with 75 mL of water and 75 mL of saturated NaCl solution, dried over 25 g of $MgSO_4$, filtered through a 250-mL medium porosity sintered glass funnel, and concentrated by rotary evaporation (25 °C, 20 mmHg) to afford 3.25 g of a yellow oil."

> "The solution is transferred to a 250-mL, round-bottomed flask equipped with a magnetic stirbar and a 15-cm Vigreux column fitted with a short path distillation head, and then concentrated by careful distillation at 50 mmHg (bath temperature gradually increased from 25 to 75 °C)."

In cases where solid products are filtered, describe the type of filter funnel used and the amount and composition of solvents used for washes.

> " … and the resulting pale yellow solid is collected by filtration on a Büchner funnel and washed with 100 mL of cold (0 °C) hexane."

When solid or liquid compounds are dried under vacuum, indicate the pressure employed (rather than stating "reduced pressure" or "dried *in vacuo*").

> ". . . . and concentrated at room temperature by rotary evaporation (20 mmHg) and then at 0.01 mmHg to provide . . . . "

> "The resulting colorless crystals are transferred to a 50-mL, round-bottomed flask and dried overnight in a 100 °C oil bath at 0.01 mmHg."

Purification: Distillation

Describe distillation apparatus including the size and type of distillation column. Indicate temperature (and pressure) at which all significant fractions are collected.

"…. and transferred to a 100-mL, round-bottomed flask equipped with a magnetic stirbar. The product is distilled under vacuum through a 12-cm, vacuum-jacketed column of glass helices (Note 16) topped with a Perkin triangle. A forerun (ca. 2 mL) is collected and discarded, and the desired product is then obtained, distilling at 50-55 °C (0.04-0.07 mmHg) …. "

Purification: Column Chromatography

Provide information on TLC analysis in a Note, including eluent, R_f values, and method of visualization.

Provide dimensions of column and amount of silica gel used; in a Note indicate source and type of silica gel.

Provide details on eluents used, and number and size of fractions.

"The product is charged on a column (5 x 10 cm) of 200 g of silica gel (Note 15) and eluted with 250 mL of hexane. At that point, fraction collection (25-mL fractions) is begun, and elution is continued with 300 mL of 2% EtOAc-hexane (49:1 hexanes:EtOAc) and then 500 mL of 5% EtOAc-hexane (19:1 hexanes:EtOAc). The desired product is obtained in fractions 24-30, which are concentrated by rotary evaporation (25 °C, 15 mmHg) …."

Purification: Recrystallization

Describe procedure in detail. Indicate solvents used (and ratio of mixed solvent systems), amount of recrystallization solvents, and temperature protocol. Describe how crystals are isolated and what they are washed with. A photograph of the crystalline product is often valuable to indicate the form and color of the crystals.

"The solid is dissolved in 100 mL of hot diethyl ether (30 °C) and filtered through a Buchner funnel. The filtrate is allowed to cool to room temperature, and 20 mL of hexane is added. The solution is cooled at -20 °C overnight and the resulting crystals are collected by suction filtration on a Buchner funnel, washed with 50 mL of ice-cold hexane, and then transferred to a 50-mL, round-bottomed flask and dried overnight at 0.01 mmHg to provide …."

Characterization

Physical properties of the product such as color, appearance, crystal forms, melting point, etc. should be included in the text of the procedure. Comments on the stability of the product to storage, etc. should be provided in a Note.

In a Note, provide data establishing the identity of the product. This will generally include IR, MS, ^{1}H-NMR, and ^{13}C-NMR data, and in some cases UV data. Copies of the proton and carbon NMR spectra for the products of each step in the procedure should be submitted showing integration for all resonances. Submission of copies of the NMR spectra for other nuclei are encouraged as appropriate.

In the same Note, provide analytical data establishing that the purity of the **isolated** product is at least 97%. **Note that this data should be obtained for the material on which the yield of the reaction is based**, not for a sample that has been subjected to additional purification by chromatography, distillation, or crystallization. Elemental analysis for carbon and hydrogen (and nitrogen if present) agreeing with calculated values within 0.4% is preferred. However, **quantitative** NMR, GC, or HPLC analyses involving measurements versus an internal standard will also be accepted. See *Instructions for Authors* at orgsyn.org for procedures for quantitative analysis of purity by NMR and chromatographic methods. Provide details on equipment and conditions for GC and HPLC analyses.

In procedures involving non-racemic, enantiomerically enriched products, optical rotations should generally be provided, but **enantiomeric purity must be determined by another method** such as chiral HPLC or GC analysis.

In cases where the product of one step is used without purification in the next step, a Note should be included describing how a sample of the product can be purified and providing characterization data for the pure material. Copies of the proton NMR spectra of both the product both *before* and *after* purification should be submitted.

Hazard Warnings

Particularly significant hazards should be indicated in a statement within a box at the beginning of the procedure in italicized, red type. Hazard warnings should only be included in the case of procedures that involve unusual hazards such as the use of pyrophoric or explosive substances, and substances with a high degree of acute or chronic toxicity. Instructions are provided in the Article Template. For other procedures, it is not necessary to include a special caution note that refers to standard operating procedures such as working in a hood, avoiding skin contact, etc., since this is referenced in the "Working with Hazardous Chemicals" statement within each article. Efforts should be made to avoid the use of toxic and hazardous solvents and reagents when less hazardous alternatives are available.

Discussion Section

The style and content of the discussion section will depend on the nature of the procedure.

For procedures that provide an improved method for the preparation of an important reagent or synthetic building block, the discussion should focus on the advantages of the new approach and should describe and reference all of the earlier methods used to prepare the title compound.

In the case of procedures that illustrate an important synthetic method or strategy, the discussion section should provide a mini-review on the new methodology. The scope and limitations of the method should be discussed, and it is generally desirable to include a table of examples. Please be sure each table is numbered and has a title. Competing methods for accomplishing the same overall transformation should be described and referenced. A brief discussion of mechanism may be included if this is useful for understanding the scope and limitations of the method.

Titles of Articles

In cases where the main thrust of the article is the illustration of a synthetic method of general utility, the title of the article should incorporate reference to that method. Inclusion of the name of the final product is acceptable but not required. In the case of articles where the objective is the preparation of a specific compound of importance (such as a chiral ligand), then the name of that compound should be part of the title.

Examples

Title without name of product:

"Stereoselective Synthesis of 3-Arylacrylates by Copper-Catalyzed Syn Hydroarylation" (*Org. Synth.* **2010**, *87*, 53).

Title including name of final product (note name of product is not required):

"Catalytic Enantioselective Borane Reduction of Benzyl Oximes: Preparation of (S)-1-Pyridin-3-yl-ethylamine Bis Hydrochloride" (*Org. Synth.* **2010**, *87*, 36).

Title where preparation of specific compound is the subject:

"Preparation of (S)-3,3'-Bis-Morpholinomethyl-5,5',6,6',7,7',8,8'-octahydro-1,1'-bi-2-naphthol" (*Org. Synth.* **2010**, *87*, 59).

Style and Format for Text

Articles should follow the style guidelines used for organic chemistry articles published in the ACS journals such as *J. Am. Chem. Soc.*, *J. Org. Chem.*, *Org. Lett.*, etc. as described in the ACS Style Guide (3rd Ed.). The text of the procedure should be created using the Word template available on the *Organic Syntheses* website. Specific instructions with regard to the manuscript format (font, spacing, margins) is available on the website in the "Instructions for Article Template" and embedded within the Article Template itself.

Style and Format for Tables and Schemes

Chemical structures and schemes should be drawn using the standard ACS drawing parameters (in ChemDraw, the parameters are found in the "ACS Document 1996" option) with a maximum full size width of 15 cm (5.9 inches). The graphics files should then be pasted into the Word document at the correct location and the size reduced to 75% using "Format Picture" (Mac) or "Size and Position" (Windows). Graphics files must also be submitted separately. All Tables that include structures should be entirely prepared in the graphics (ChemDraw) program and inserted into the word processing file at the appropriate location. Tables that include multiple, separate graphics files prepared in the word processing program will require modification.

Tables and schemes should be numbered and should have titles. The title for a Table should be included within the ChemDraw graphic and placed immediately above the table. The title for a scheme should be included within the ChemDraw graphic and placed immediately below the scheme. Use 12 point Palatino Bold font in the ChemDraw file for all titles. For footnotes in Tables use Helvetica (or Arial) 9 point font and place these immediately below the Table.

Acknowledgments and Author's Contact Information

Contact information (institution where the work was carried out and mailing address for the principal author) should be included as footnote 1. This footnote should also include the email address for the principal author. Acknowledgment of financial support should be included in footnote 1.

Biographies and Photographs of Authors

Photographs and 100-word biographies of all authors should be submitted as separate files at the time of the submission of the procedure. The format of the biographies should be similar to those in the Volume 84 procedures found at the orgsyn.org website. Photographs can be accepted in a number of electronic formats, including tiff and jpeg formats.

DISPOSAL OF CHEMICAL WASTE

General Reference: *Prudent Practices in the Laboratory* National Academy Press, Washington, D.C. 2011.

Effluents from synthetic organic chemistry fall into the following categories:

1. **Gases**
   - 1a. Gaseous materials either used or generated in an organic reaction.
   - 1b. Solvent vapors generated in reactions swept with an inert gas and during solvent stripping operations.
   - 1c. Vapors from volatile reagents, intermediates and products.
2. **Liquids**
   - 2a. Waste solvents and solvent solutions of organic solids (see item 3b).
   - 2b. Aqueous layers from reaction work-up containing volatile organic solvents.
   - 2c. Aqueous waste containing non-volatile organic materials.
   - 2d. Aqueous waste containing inorganic materials.
3. **Solids**
   - 3a. Metal salts and other inorganic materials.
   - 3b. Organic residues (tars) and other unwanted organic materials.
   - 3c. Used silica gel, charcoal, filter aids, spent catalysts and the like.

The operation of industrial scale synthetic organic chemistry in an environmentally acceptable manner* requires that all these effluent categories be dealt with properly. In small scale operations in a research or academic setting, provision should be made for dealing with the more environmentally offensive categories.

- 1a. Gaseous materials that are toxic or noxious, e.g., halogens, hydrogen halides, hydrogen sulfide, ammonia, hydrogen cyanide, phosphine, nitrogen oxides, metal carbonyls, and the like.
- 1c. Vapors from noxious volatile organic compounds, e.g., mercaptans, sulfides, volatile amines, acrolein, acrylates, and the like.

*An environmentally acceptable manner may be defined as being both in compliance with all relevant state and federal environmental regulations *and* in accord with the common sense and good judgment of an environmentally aware professional.

2a. All waste solvents and solvent solutions of organic waste.
2c. Aqueous waste containing dissolved organic material known to be toxic.
2d. Aqueous waste containing dissolved inorganic material known to be toxic, particularly compounds of metals such as arsenic, beryllium, chromium, lead, manganese, mercury, nickel, and selenium.
3. All types of solid chemical waste.

Statutory procedures for waste and effluent management take precedence over any other methods. However, for operations in which compliance with statutory regulations is exempt or inapplicable because of scale or other circumstances, the following suggestions may be helpful.

Gases

Noxious gases and vapors from volatile compounds are best dealt with at the point of generation by "scrubbing" the effluent gas. The gas being swept from a reaction set-up is led through tubing to a large trap to prevent suck-back and into a sintered glass gas dispersion tube immersed in the scrubbing fluid. A bleach container can be conveniently used as a vessel for the scrubbing fluid. The nature of the effluent determines which of four common fluids should be used: dilute sulfuric acid, dilute alkali or sodium carbonate solution, laundry bleach when an oxidizing scrubber is needed, and sodium thiosulfate solution or diluted alkaline sodium borohydride when a reducing scrubber is needed. Ice should be added if an exotherm is anticipated.

Larger scale operations may require the use of a pH meter or starch/iodide test paper to ensure that the scrubbing capacity is not being exceeded.

When the operation is complete, the contents of the scrubber can be poured down the laboratory sink with a large excess (10-100 volumes) of water. If the solution is a large volume of dilute acid or base, it should be neutralized before being poured down the sink.

Liquids

Every laboratory should be equipped with a waste solvent container in which *all* waste organic solvents and solutions are collected. The contents of these containers should be periodically transferred to properly labeled waste solvent drums and arrangements made for contracted disposal in a regulated and licensed incineration facility.**

**If arrangements for incineration of waste solvent and disposal of solid chemical waste by licensed contract disposal services are not in place, a list of providers of such services should be available from a state or local office of environmental protection.

Aqueous waste containing dissolved toxic organic material should be decomposed *in situ*, when feasible, by adding acid, base, oxidant, or reductant. Otherwise, the material should be concentrated to a minimum volume and added to the contents of a waste solvent drum.

Aqueous waste containing dissolved toxic inorganic material should be evaporated to dryness and the residue handled as a solid chemical waste.

Solids

Soluble organic solid waste can usually be transferred into a waste solvent drum, provided near-term incineration of the contents is assured.

Inorganic solid wastes, particularly those containing toxic metals and toxic metal compounds, used Raney nickel, manganese dioxide, etc. should be placed in glass bottles or lined fiber drums, sealed, properly labeled, and arrangements made for disposal in a secure landfill.** Used mercury is particularly pernicious and small amounts should first be amalgamated with zinc or combined with excess sulfur to solidify the material.

Other types of solid laboratory waste including used silica gel and charcoal should also be packed, labeled, and sent for disposal in a secure landfill.

Special Note

Since local ordinances may vary widely from one locale to another, one should always check with appropriate authorities. Also, professional disposal services differ in their requirements for segregating and packaging waste.

PREFACE

With the publication of Volume 93, *Organic Syntheses* remains true to its mission of providing to the community enabling synthetic methods that form the basic tools of modern chemical synthesis. The volume covers a wide range of procedures that will find broad utility with the assurance that comes with an experimental protocol that has been independently checked, optimized where necessary, and deemed reproducible. The collection is also notable because of the procedures that are included, which span the range of modern transformations. The interested practitioner can find detailed preparation of ligands that are used in a wide range of processes involving Au, Pd, Ni and Cu-catalyzed reactions, enjoying wide usage in academic and industrial research laboratories. The timeliness of the collection is also evident in its inclusion of procedures utilizing chiral Brønsted acids from the recent literature. In line with current interests several procedures describe the synthesis of fluorinated building blocks and a versatile, practical sulfinyl imine auxiliary for asymmetric synthesis; additionally, a protocol is included for the Cu-catalyzed trifluoromethylation of arenes. One also finds procedures for cyanation, formylation, cycloborylation as well as approaches to benzofurans, alkyne semireduction, and chlorination. The preparation of several building blocks, encompassing carbolines, 1,1-disubstituted cyclopropanes, and protected 1,3-dihydroxyacetone, will surely find applications in discovery. The latter has found use in a range of organocatalytic reactions. In line with a vision for *Organic Syntheses* by the Board that aims to identify, critically examine, and incorporate new technologies and approaches, the Volume provides experimental details for the execution of photoaddition reactions in a flow apparatus and as well as metal-mediated visible-light photocatalytic coupling reactions. In summary, Volume 93 represents an addition to the family of high-quality, checked procedures for which *Organic Syntheses* has become well known and highly appreciated by the science community.

Erick M. Carreira

TABLE OF CONTENTS

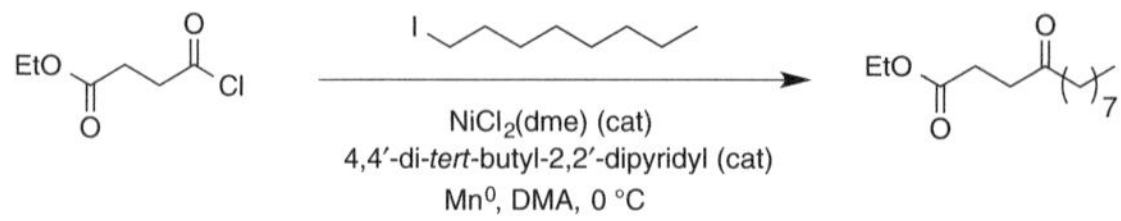

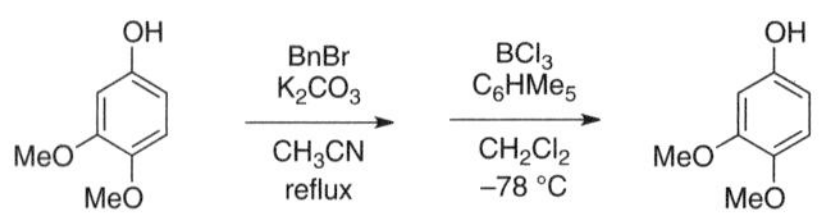

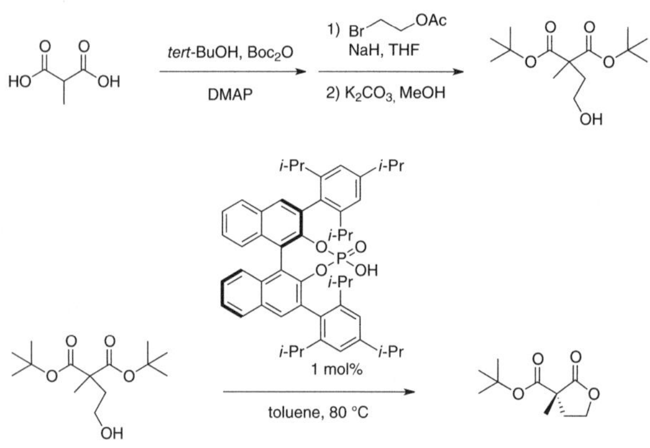

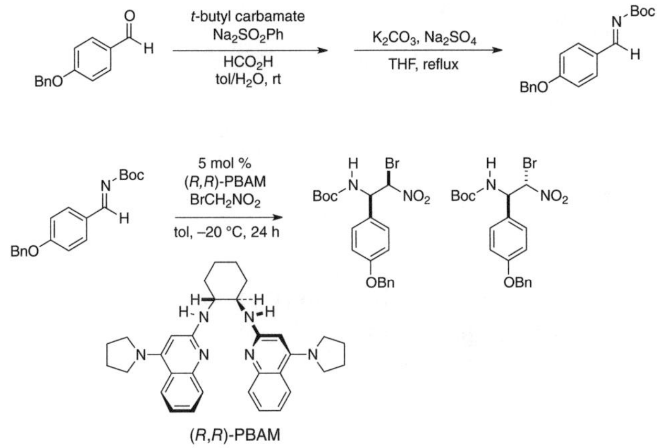
t-butyl carbamate
Na_2SO_2Ph
HCO_2H
tol/H_2O, rt
K_2CO_3, Na_2SO_4
THF, reflux
5 mol %
(*R*,*R*)-PBAM
$BrCH_2NO_2$
tol, –20 °C, 24 h
(*R*,*R*)-PBAM

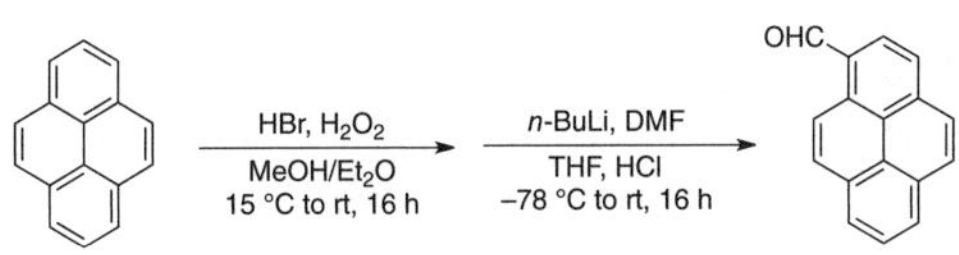
HBr, H_2O_2
MeOH/Et_2O
15 °C to rt, 16 h
n-BuLi, DMF
THF, HCl
–78 °C to rt, 16 h

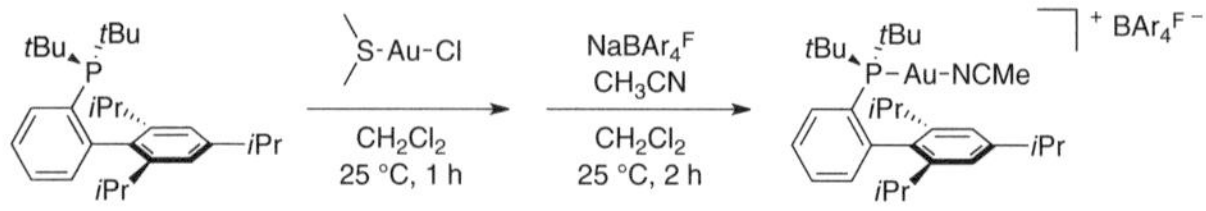

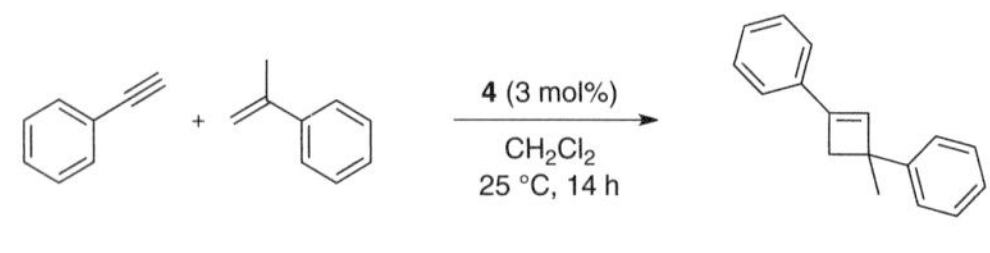

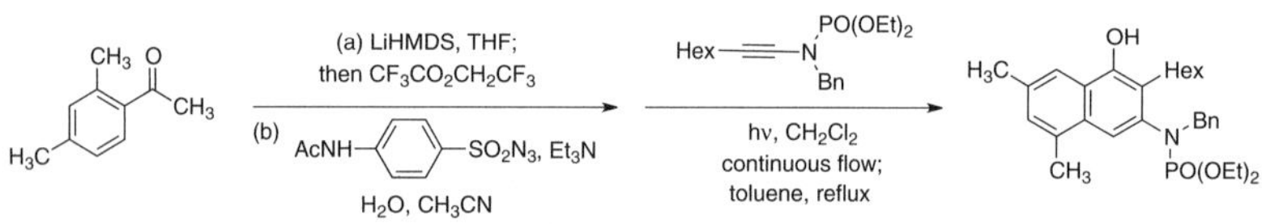

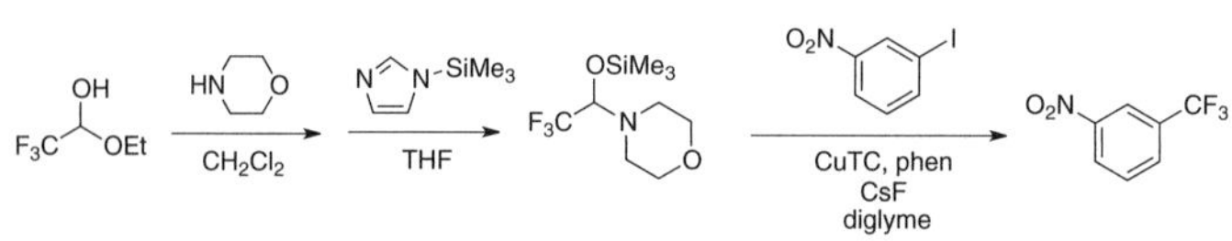

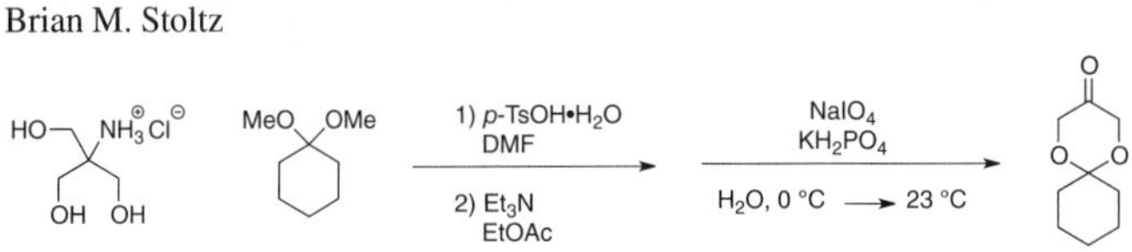
HO
NH3 Cl
OH OH
MeO OMe
1) p-TsOH•H2O
DMF
2) Et3N
EtOAc
NaIO4
KH2PO4
H2O, 0 °C → 23 °C

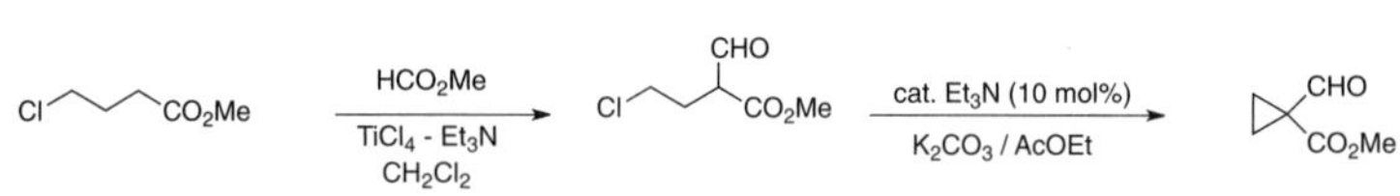
Cl
CO_2Me
HCO_2Me
$TiCl_4$ - Et_3N
CH_2Cl_2
CHO
Cl
CO_2Me
cat. Et_3N (10 mol%)
K_2CO_3 / AcOEt
CHO
CO_2Me

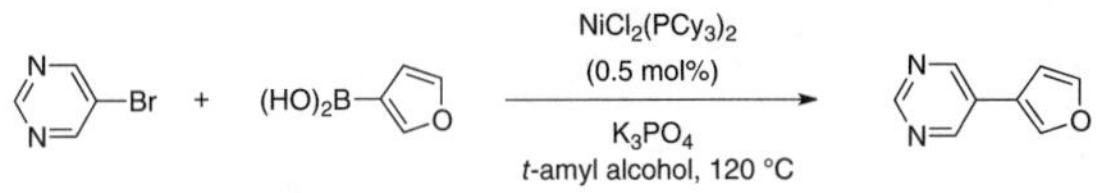
NiCl2(PCy3)2
(0.5 mol%)
K3PO4
t-amyl alcohol, 120 °C

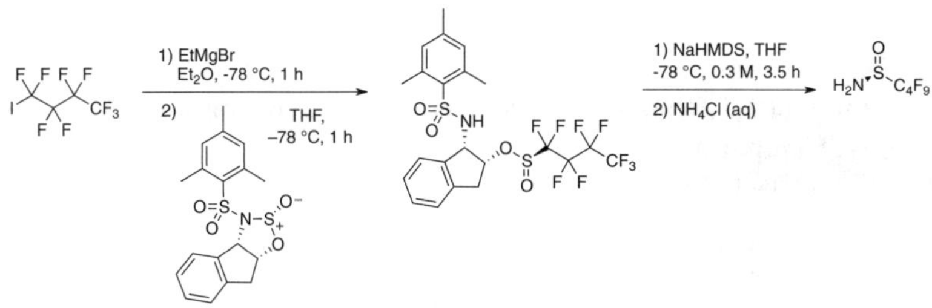
1) EtMgBr
Et2O, -78 °C, 1 h
2)
THF,
−78 °C, 1 h
1) NaHMDS, THF
-78 °C, 0.3 M, 3.5 h
2) NH4Cl (aq)

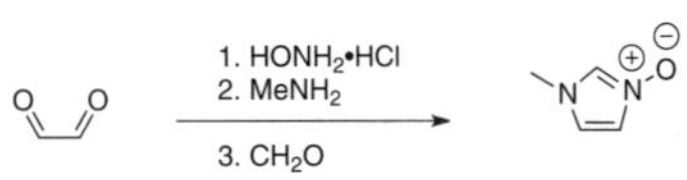
1. HONH2•HCl
2. MeNH2
3. CH2O

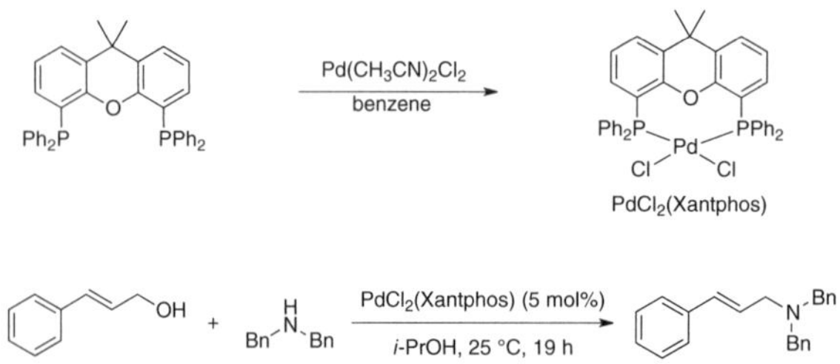

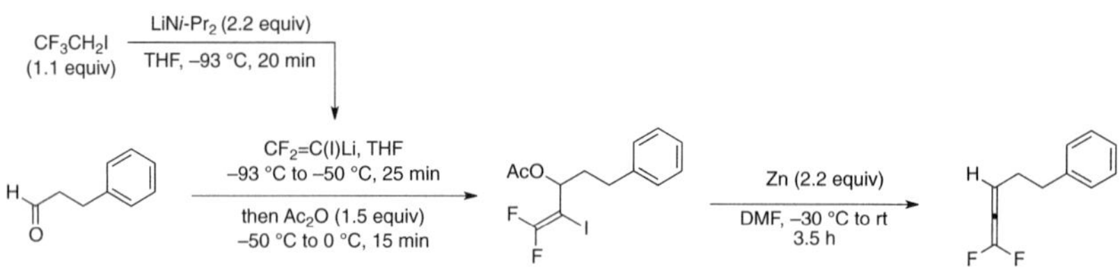

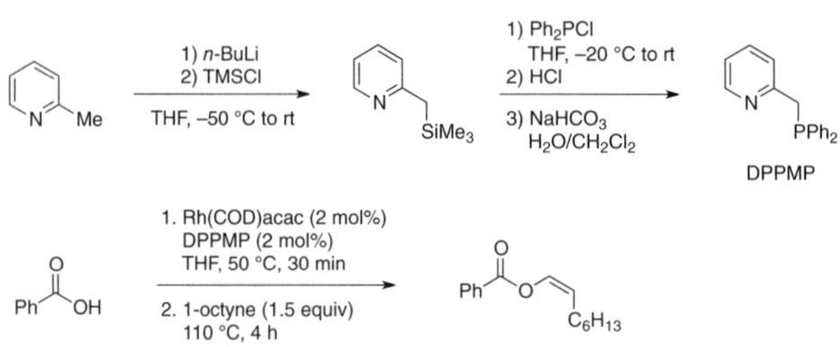

Organic Syntheses

DIBALH-Mediated Reductive Ring-Expansion Reaction of Cyclic Ketoxime

Taku Imaizumi, Kentaro Okano, and Hidetoshi Tokuyama*[1]

Graduate School of Pharmaceutical Sciences, Tohoku University, Aoba 6-3, Aramaki, Aoba-ku, Sendai 980-8578, Japan

Checked by Lang-Tin Xu and Dawei Ma

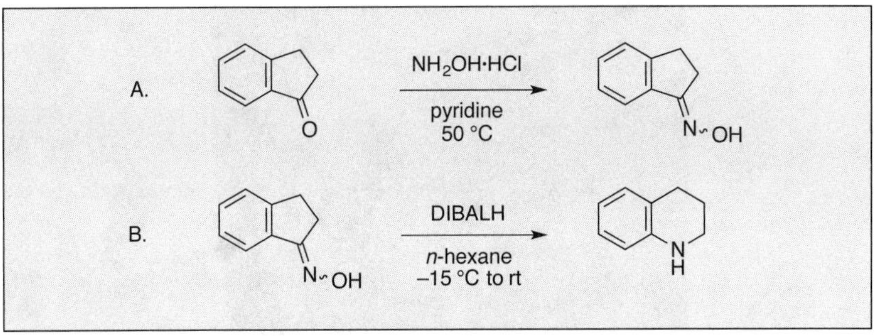

Procedure

A. *1-Indanone oxime.* A 500-mL one-necked round-bottomed flask equipped with a Teflon-coated magnetic stir bar (3.5 x 1.5 cm) is charged with 1-indanone (13.2 g, 100 mmol) (Note 1), pyridine (50 mL) (Note 2), and hydroxylamine hydrochloride (7.30 g, 105 mmol, 1.05 equiv) (Note 3). After stirring for 20 min (500 rpm) at 50 °C (Note 4), the reaction mixture is cooled to ambient temperature and concentrated on a rotary evaporator under reduced pressure (20 °C, 1 mmHg) to remove pyridine. To the residue are added ethyl acetate (300 mL) and 1 M aqueous HCl (50 mL). After the mixture is partitioned, the aqueous layer is extracted with ethyl acetate (50 mL), and the combined organic extracts are washed with 1 M aqueous HCl (2 x 100 mL) and brine (100 mL). The organic layer is dried over Na_2SO_4 (40 g) and filtered through cotton wool. The filtrate is concentrated on a

DOI: 10.15227/orgsyn.093.0001

rotary evaporator under reduced pressure (20 °C, 16 mmHg) to afford 1-indanone oxime (Notes 5, 6 and 7) as a pale yellow solid (14.5 g, 99%).

B. *Tetrahydroquinoline.* A 1-L oven-dried three-necked round-bottomed flask equipped with an overhead mechanical stirrer (paddle size: w = 8 cm, h = 1.5 cm), a 200-mL pressure-equalizing dropping funnel with a rubber septum, and a straight inlet adapter with a side arm 3-way stopcock fitted with an argon gas inlet and a thermometer (Figure 1) is charged with DIBALH (1.00 M in n-hexane, 199 mL, 199 mmol, 4.50 equiv) (Note 8) through the pressure-equalizing dropping funnel (Note 9). After the

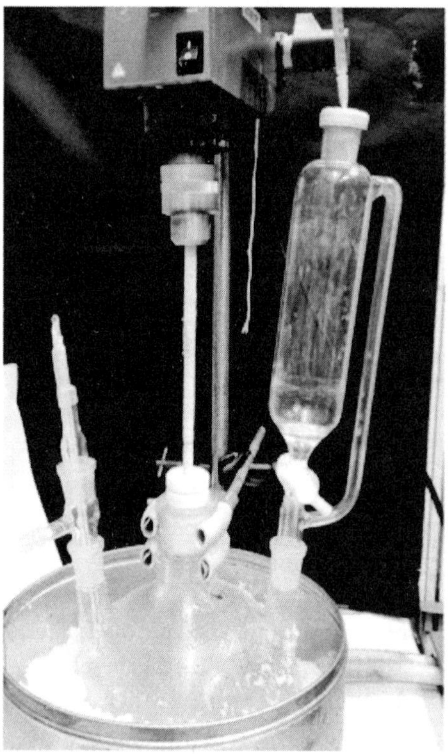

Figure 1. Reaction Assembly for Step B

dropping funnel is replaced with a rubber septum, the solution is cooled to –16 ~ –11 °C (internal temperature) using a brine/ice bath. The rubber septum is removed, and 1-indanone oxime (6.50 g, 44.2 mmol) is added portionwise over 3 min with stirring (300 rpm) *via* a powder addition funnel, while keeping the internal temperature below 22 °C. The funnel is washed with distilled hexanes (5 mL), and the septum is fitted to the flask. After stirring for 5 min, the reaction mixture is warmed to ambient temperature and stirred for 2.5 h (Note 10). The paddle of the overhead mechanical stirrer is washed with distilled hexanes (10 mL) and removed (Note 11). The mixture is cooled to 2–3 °C (internal temperature) with an ice bath and gradually poured into saturated aqueous Rochelle salt (400 mL, precooled to 3 °C using an ice bath) (Note 12) in a 2-L three-necked round-bottomed flask with stirring (300 rpm) using an overhead mechanical stirrer (paddle size: w = 8 cm, h = 1.5 cm), while keeping the internal temperature below 30 °C. After the mixture is transferred to the flask with the aid of ethyl acetate (50 mL), the mixture is warmed to ambient temperature and stirred for 3.5 h. The resulting two-phase solution is partitioned, and the aqueous phase is extracted with ethyl acetate (2 x 100 mL). The combined organic extracts are washed with brine (200 mL), dried over Na_2SO_4 (40 g), and filtered through cotton wool. The filtrate is concentrated on a rotary evaporator under reduced pressure (20 °C, 16 mmHg), and the resulting residual oil is dried *in vacuo* for 2 h in a desiccator (20 °C) (Note 13) to afford 5.95–5.96 g of a pale yellow liquid, which is purified by silica gel column chromatography (hexanes-ethyl acetate = 13:1) (Notes 14 and 15). The fractions containing the product are collected and concentrated on a rotary evaporator under reduced pressure (40 °C, 16 mmHg), and the residual oil is dried *in vacuo* for 3.5 h in a desiccator (20 °C) to yield tetrahydroquinoline as a pale yellow oil (5.08 g, 86%) (Note 16 and 17).

Notes

1. 1-Indanone (>99%) was purchased from Sigma-Aldrich Co. and used as received without further purification.
2. Pyridine (99.5%) was purchased from J&K Scientific and used as received without further purification.
3. Hydroxylamine hydrochloride (98%) was purchased from Sigma-Aldrich Co. and used as received without further purification.
4. The reaction typically requires 20 min to consume all the 1-indanone and is monitored by TLC analysis on Merck silica gel 60 F_{254} plates developing with hexanes/ethyl acetate (3:1). The R_f values of 1-indanone and 1-indanone oxime are 0.39 and 0.33, respectively (visualized with 254 nm UV lamp and stained with an ethanol solution

of $Ce_2(SO_4)_3$ and phosphomolybdic acid (Ce-PMA). After dipping the TLC plate to the Ce-PMA solution, the chromatogram is stained by heating).

5. The mixture of (*E*)- and (*Z*)-1-indanone oxime (*E*/*Z* = 7:1) exhibits the following physicochemical properties: R_f = 0.33 and 0.17 (hexanes/ethyl acetate = 3:1); Merck silica gel 60 F_{254} plates (visualized with 254 nm UV lamp and stained with an ethanol solution of $Ce_2(SO_4)_3$ and phosphomolybdic acid (Ce-PMA). After dipping the TLC plate to the Ce-PMA solution, the chromatogram is stained by heating); ^{1}H NMR (400 MHz, $CDCl_3$) δ: 2.86–2.89 (m, 0.26 H), 2.97–3.00 (m, 2 H), 3.05–3.08 (m, 2.29 H), 7.23–7.34 (m, 3.55 H), 7.69 (d, *J* = 7.6 Hz, 1.0 H), 8.46 (d, *J* = 7.6 Hz, 0.13 H), 9.05 (br s, 0.16 H), 9.45 (br s, 0.87 H); ^{13}C NMR (100 MHz, $CDCl_3$) δ: 26.20, 28.70, 28.87, 29.04, 121.82, 125.67, 125.80, 127.02, 127.21, 129.81, 130.61, 131.27, 133.89, 136.10, 148.66, 149.64, 160.77, 164.26; Chemical shifts were given relative to $CDCl_3$ (7.26 ppm for ^{1}H NMR, 77.16 ppm for ^{13}C NMR). IR (neat, cm^{-1}): 3157, 3124, 3096, 3064, 2849, 1654, 1479, 1458, 1432, 1411, 1334, 1070, 986, 956, 833, 818, 753; Anal. calcd. for C_9H_9NO: C, 73.45; H, 6.16; N, 9.52. Found: C, 72. 67; H, 6.19; N, 9.45.

6. The E and Z isomers the 1-indanone oxime can be isolated in pure form by preparative TLC (hexanes/ethyl acetate = 1:1). (*E*)-1-Indanone oxime exhibits the following physicochemical properties: R_f = 0.33 (hexanes/ethyl acetate = 3:1); mp = 146.5–150.0 °C (ethyl acetate); IR (neat, cm^{-1}): 3178, 3063, 2850, 1655, 1480, 1456, 1432, 1410, 1334, 1070, 987, 957, 833, 819, 775, 753, 580, 569, 544, 526, 514; ^{1}H NMR (400 MHz, $CDCl_3$) δ: 2.96–3.00 (m, 2 H), 3.05–3.08 (m, 2 H), 7.22–7.36 (m, 3 H), 7.67 (d, *J* = 7.7 Hz, 1 H), 8.79 (br s, 1H); ^{13}C NMR (100 MHz, $CDCl_3$) δ: 26.13, 28.73, 121.76, 125.83, 127.22, 130.63, 136.16, 148.63, 164.26; Chemical shifts were given relative to $CDCl_3$ (7.26 ppm for ^{1}H NMR, 77.16 ppm for ^{13}C NMR). Anal. calcd. for C_9H_9NO: C, 73.45; H, 6.16; N, 9.52. Found: C, 73.46; H, 6.16; N, 9.50. These spectral data are identical with those reported in the literature.[2] (Z)-1-Indanone oxime exhibits the following physicochemical properties: R_f = 0.17 (hexanes/ethyl acetate = 3:1); mp = 137.6–140.9 °C (ethyl acetate); IR (neat, cm^{-1}): 3200, 3075, 2884, 1662, 1462, 1442, 994, 954, 808, 750; ^{1}H NMR (400 MHz, $CDCl_3$) δ: 2.83–2.87 (m, 2 H), 3.05–3.09 (m, 2 H), 7.25–7.39 (m, 3 H), 8.35 (br s, 1H), 8.43 (d, *J* = 7.7 Hz, 1H); ^{13}C NMR (100 MHz, $CDCl_3$) δ: 28.88, 29.08, 125.70, 127.02, 129.80, 131.24, 133.87, 149.60, 160.74; Chemical shifts were given relative to $CDCl_3$ (7.26 ppm for ^{1}H NMR, 77.16 ppm for ^{13}C NMR).

Anal. calcd. for C_9H_9NO: C, 73.45; H, 6.16; N, 9.52. Found: C, 73.43; H, 6.14; N, 9.43.

7. A second reaction provided 14.6 g (99%) of the mixture of E and Z isomers.
8. DIBALH (1.00 M in *n*-hexane) was purchased from J&K Scientific and used as received without further purification.
9. A pressure-equalizing dropping funnel is used to handle DIBALH safely. DIBALH is transferred via disposable syringe from the bottle into the pressure-equalizing dropping funnel. The volume of 199 mL is marked on the dropping funnel prior to oven drying.
10. The reaction process is monitored by TLC analysis on silica gel 60 F_{254} plate. The R_f values of 1,2,3,4-tetrahydroquinoline, (*E*)-1-indanone oxime, (Z)-1-indanone oxime, and 2-isobutyl-1,2,3,4-tetrahydroquinoline are 0.48, 0.33, 0.17, and 0.67, respectively (hexanes/ethyl acetate = 3:1) (visualized with 254 nm UV lamp and stained with an ethanol solution of $Ce_2(SO_4)_3$ and phosphomolybdic acid (Ce-PMA). After dipping the TLC plate to the Ce-PMA solution, the chromatogram is stained by heating).
11. When this reaction is carried out in ca. >10 g scale, the following procedure is recommended to avoid large volume extraction: the reaction mixture is cooled to 2 °C (internal temperature). After portionwise addition of sodium fluoride (28.0 equiv) for 2 min (internal temperature should be kept below 40 °C), water (20.0 equiv) is carefully added dropwise for 8 min, causing a temperature rise to ca. 44 °C. The mixture is stirred (300 rpm) using an overhead mechanical stirrer (paddle size: w = 8 cm, h = 1.5 cm) for 30 min followed by stirring for 1 h at ambient temperature. After the paddle of the overhead mechanical stirrer is washed with ethyl acetate (30 mL), the contents of the flask are passed through a pad of Celite (40 g) fitted to a Kiriyama funnel® (diameter = 14 cm, height = 8 cm, fitted with a filter paper No. 4, diameter = 9.5 cm) with ethyl acetate (3 × 150 mL). The filtrate is concentrated on a rotary evaporator under reduced pressure (20 °C, 16 mmHg), and the residual oil is dried *in vacuo* for 2 h in a desiccator (20 °C) to afford a pale yellow liquid, which is purified by column chromatography (Note 13) and dried to give tetrahydroquinoline as a pale yellow oil.
12. Potassium sodium (+)-tartrate tetrahydrate (Rochelle salt) was purchased from Sinopharm Chemical Reagent Co., Ltd and used as received without further purification.

13. Drying *in vacuo* more than 2 h results in loss of the tetrahydroquinoline due to volatilization.
14. The crude material is dissolved in eluent (5 mL) and then charged onto a column (diameter = 10 cm, height = 40 cm) of 380-gram silica gel. The column is sequentially eluted with hexanes/ethyl acetate = 13:1 (4 L), hexanes/ethyl acetate = 6:1. During chromatography, 100-mL fractions are collected. Fractions 12 − 17 and 21 − 42 were combined and concentrated on a rotary evaporator under reduced pressure (20 °C, 16 mmHg) and dried *in vacuo* to provide undesired 2-isobutyl-1,2,3,4-tetrahydroquinoline (fractions 12−17, 128–132 mg, 2%) and desired 1,2,3,4-tetrahydroquinoline (fractions 21 − 42, 5.04–5.12 g, 86–87%), respectively.
15. 2-Isobutyl-1,2,3,4-tetrahydroquinoline would be generated through the migration of the isobutyl group of DIBALH to the imine intermediate. Physicochemical properties of 2-isobutyl-1,2,3,4-tetrahydroquinoline: pale yellow oil; R_f = 0.67 (hexanes/ethyl acetate = 3:1); Merck silica gel 60 F_{254} plates (visualized with 254-nm UV lamp and stained with an ethanol solution of $Ce_2(SO_4)_3$ and phosphomolybdic acid (Ce-PMA). After dipping the TLC plate to the Ce-PMA solution, the chromatogram is stained by heating); ^{1}H NMR (400 MHz, $CDCl_3$) δ: 0.94 (d, *J* = 6.6 Hz, 6 H), 1.28–1.43 (m, 2 H), 1.52–1.62 (m, 1 H), 1.70–1.80 (m, 1 H), 1.90–1.96 (m, 1 H), 2.68–2.87 (m, 2 H), 3.28–3.34 (m, 1 H), 6.47 (d, *J* = 8.2 Hz, 1 H), 6.60 (dd, *J* = 7.3, 7.3 Hz, 1 H), 6.93–6.97 (m, 2 H); ^{13}C NMR (100 MHz, $CDCl_3$) δ: 22.64, 23.36, 24.61, 26.58, 28.72, 46.05, 49.40, 114.25, 117.07, 121.51, 126.84, 129.43, 144.84; Chemical shifts were given relative to $CDCl_3$ (7.26 ppm for ^{1}H NMR, 77.16 ppm for ^{13}C NMR). IR (neat, cm^{-1}): 3408, 3051, 3015, 2954, 2867, 2843, 1607, 1585, 1484, 1434, 1384, 1366, 1354, 1309, 1273, 1258, 1153, 745, 716; Anal. calcd. for $C_{13}H_{19}N$: C, 82.48; H, 10.12; N, 7.40. Found: C, 82.20; H, 9.98; N, 7.39. These spectral data are identical with those reported in the literature.[3]
16. Physicochemical properties of 1,2,3,4-tetrahydroquinoline: pale yellow oil; R_f = 0.48 (hexanes/ethyl acetate = 3:1); Merck silica gel 60 F_{254} plates (visualized with 254-nm UV lamp and stained with an ethanol solution of $Ce_2(SO_4)_3$ and phosphomolybdic acid (Ce-PMA). After dipping the TLC plate to the Ce-PMA solution, the chromatogram is stained by heating); ^{1}H NMR (400 MHz, $CDCl_3$) δ: 1.91–1.97 (m, 2 H), 2.75 (t, *J* = 6.4 Hz, 2 H), 3.28 (t, *J* = 5.5 Hz, 2 H), 6.46 (d, *J* = 7.8 Hz, 1 H), 6.58 (dd, *J* = 7.4, 7.4 Hz, 1 H), 6.94–6.98 (m, 2 H); ^{13}C NMR (100 MHz, $CDCl_3$) δ: 22.29, 27.10, 42.09, 114.29, 117.01, 121.51, 126.83, 129.62, 144.90; Chemical shifts

were given relative to $CDCl_3$ (7.26 ppm for ^1H NMR, 77.16 ppm for ^{13}C NMR). IR (neat, cm^{-1}): 3406, 2926, 2840, 1606, 1584, 1504, 1468, 1355, 1310, 1267, 746; Anal. calcd. for $C_9H_{11}N$: C, 81.16; H, 8.32; N, 10.52. Found: C, 80.86; H, 8.33; N, 10.53. These spectral data are identical with those reported in the literature.[4]

17. A second reaction provided 5.10 g (87%) of the product.

Working with Hazardous Chemicals

The procedures in *Organic Syntheses* are intended for use only by persons with proper training in experimental organic chemistry. All hazardous materials should be handled using the standard procedures for work with chemicals described in references such as "Prudent Practices in the Laboratory" (The National Academies Press, Washington, D.C., 2011; the full text can be accessed free of charge at http://www.nap.edu/catalog.php?record_id=12654). All chemical waste should be disposed of in accordance with local regulations. For general guidelines for the management of chemical waste, see Chapter 8 of Prudent Practices.

In some articles in *Organic Syntheses*, chemical-specific hazards are highlighted in red "Caution Notes" within a procedure. It is important to recognize that the absence of a caution note does not imply that no significant hazards are associated with the chemicals involved in that procedure. Prior to performing a reaction, a thorough risk assessment should be carried out that includes a review of the potential hazards associated with each chemical and experimental operation on the scale that is planned for the procedure. Guidelines for carrying out a risk assessment and for analyzing the hazards associated with chemicals can be found in Chapter 4 of Prudent Practices.

The procedures described in *Organic Syntheses* are provided as published and are conducted at one's own risk. *Organic Syntheses, Inc.*, its Editors, and its Board of Directors do not warrant or guarantee the safety of individuals using these procedures and hereby disclaim any liability for any injuries or damages claimed to have resulted from or related in any way to the procedures herein.

Discussion

Since benzene ring-fused nitrogen heterocycles are often found in pharmaceuticals and other bioactive compounds, considerable efforts have been devoted toward the development of effective synthetic methodologies for these compounds. In addition to the classical Beckmann rearrangement, ring-expansion reactions of oxime derivatives have also been investigated so far,[5] owing to the wide availability of the corresponding ketones (Scheme 1). The silylation of oxime **1** and subsequent treatment with $BH_3{\cdot}SMe_2/BF_3{\cdot}OEt_2$ gives tetrahydroquinoline in excellent yields.[4] Oxime ether **2** can also undergo a similar reaction.[6] Unlike the reactions that utilize *O*-functionalized oximes, the presented method provides a more facile and economical access to the cyclic amines because the hydroxyl group does not require protection.

In contrast to the Beckmann rearrangement, the two possible geometries of the oxime C–N double bond can give exclusively an aromatic secondary amine in the presented method (Scheme 2).[7] Cho and co-workers reported that the separated (*E*)-oxime **3** was converted to the N-benzoyl-furo[3,2-*b*]azepine **4** in 48% yield over 2 steps. (Z)-Oxime **3** was also converted to the same compound, and its regioisomer **5** was not detected. The reaction is known to proceed via a hydroxylamine intermediate.[5a,5f] This was further supported by control experiments, where the synthesis of hydroxylamine **6** from a 1,2-reduction of oxime **7** gave the corresponding aromatic secondary amine **8** under the same reaction conditions (Scheme 3).[8a]

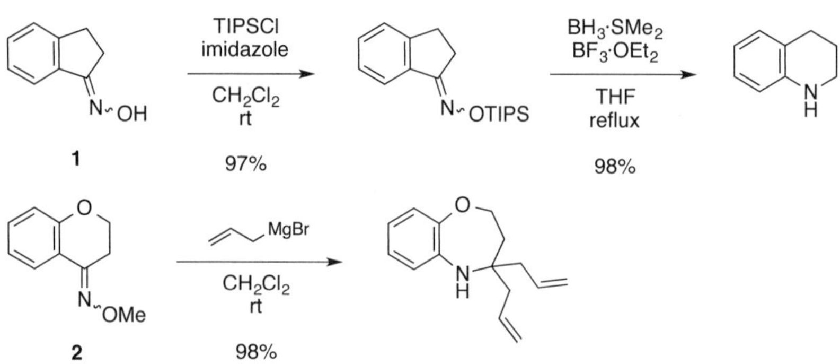

Scheme 1. Ring-expansion reaction of protected O-silyl or O-alkyloxime substrates

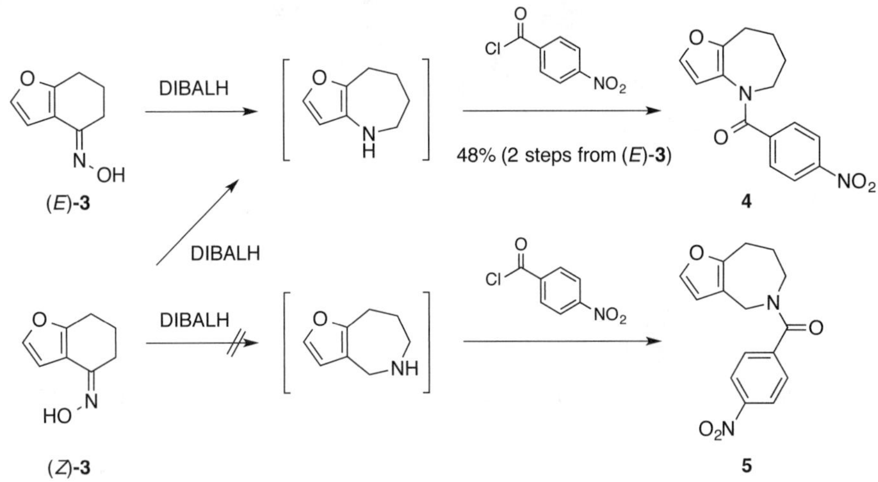

Scheme 2. Selective migration of the aryl group

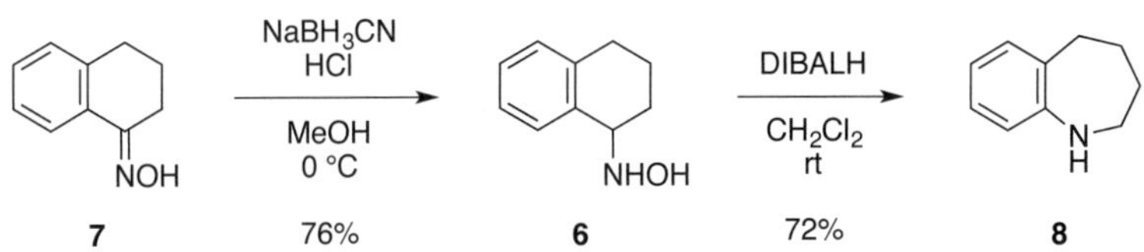

Scheme 3. Control experiments using hydroxylamine as a substrate

This reaction has a broad scope for the synthesis of cyclic secondary anilines **9**.[8] It should be noted that seven and eight-membered structures, which have previously been difficult to construct through the palladium- or copper-catalyzed intramolecular aryl amination, are easily accessible with this reaction (Scheme 4). In addition, heterocyclic ketoximes **10** containing an oxygen or a sulfur atom are also compatible. In addition, the reaction is applicable to the industrial-scale synthesis of 2,3,4,5-tetrahydrobenzo[*b*][1,4] oxazepine (25.6 kg oxime, 91%).[9]

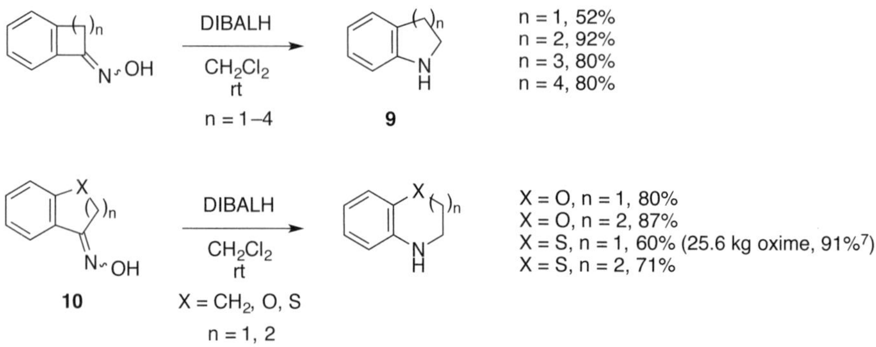

Scheme 4. Scope of this reductive ring-expansion reaction

Tokuyama and co-workers applied this ring-expansion reaction to construct an azepinoindole skeleton and accomplished a total synthesis of (–)-mersicarpine (Scheme 5).[10]

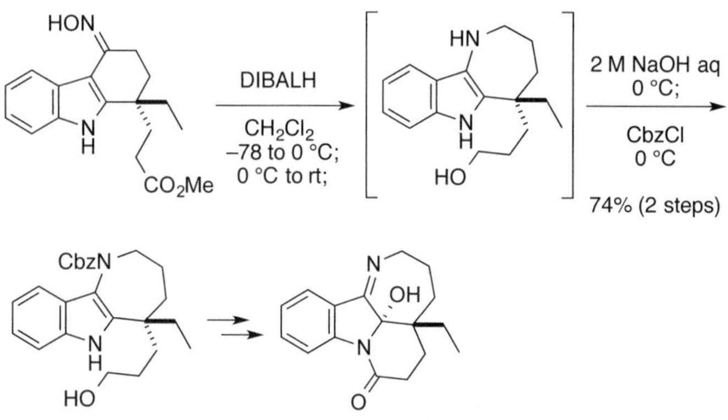

Scheme 5. Application to total synthesis of (–)-mersicarpine

Unlike the previous protocol that uses DIBALH in a dichloromethane solution, this modified protocol allows us to an environment-friendly alternative of DIBALH in *n*-hexane solution.

References

1. Graduate School of Pharmaceutical Sciences, Tohoku University, Aramaki, Aoba-ku, Sendai 980-8578, Japan.
2. Clive, D. L. J.; Pham, M. P.; Subedi, R. *J. Am. Chem. Soc.* **2007**, *129*, 2713.
3. Wang, C.; Li, C.; Wu, X.; Pettman, A.; Xiao, J. *Angew. Chem. Int. Ed.* **2009**, *48*, 6524.
4. Ortiz-Marciales, M.; Rivera, L. D.; De Jesús, M.; Espinosa, S.; Benjamin, J. A.; Casanova, O. E.; Figueroa, I. G.; Rodríguez, S.; Correa, W. *J. Org. Chem.* **2005**, *70*, 10132.
5. (a) Rerick, M. N.; Trottier, C. H.; Daignault, R. A.; DeFoe, J. D. *Tetrahedron Lett.* **1963**, 629. (b) Girault, Y.; Decouzon, M.; Azzaro, M. *Bull. Soc. Chim. Fr.* **1975**, 385. (c) Hattori, K.; Matsumura, Y.; Miyazaki, T.; Maruoka, K.; Yamamoto, H. *J. Am. Chem. Soc.* **1981**, *103*, 7368. (d) Sasatani, S.; Miyazaki, T.; Maruoka, K.; Yamamoto, H. *Tetrahedron Lett.* **1983**, *24*, 4711. (e) Fujiwara, J.; Sano, H.; Maruoka, K.; Yamamoto, H. *Tetrahedron Lett.* **1984**, *25*, 2367. (f) Maruoka, K.; Yamamoto, H. *Angew. Chem. Int. Ed.* **1985**, *24*, 668.
6. Mukhopadhyay, P. P.; Miyata, O.; Naito, T. *Synlett* **2007**, 1403.
7. Cho, H.; Murakami, K.; Nakanishi, H.; Isoshima, H.; Hayakawa, K.; Uchida, I. *Heterocycles* **1998**, *48*, 919.
8. (a) Cho, H.; Iwama, Y.; Sugimoto, K.; Mori, S.; Hidetoshi, T. *J. Org. Chem.* **2010**, *75*, 627. (b) Cho, H.; Iwama, Y.; Sugimoto, K.; Kwon, E.; Tokuyama, H. *Heterocycles* **2009**, *78*, 1183. (c) Cho, H.; Sugimoto, K.; Iwama, Y.; Mitsuhashi, N.; Okano, K.; Tokuyama, H. *Heterocycles* **2011**, *82*, 1633. (d) Cho, H.; Iwama, Y.; Mitsuhashi, N.; Sugimoto, K.; Okano, K.; Tokuyama, H. *Molecules* **2012**, *17*, 7348. (e) Cho, H.; Iwama, Y.; Okano, K.; Tokuyama, H. *Synlett* **2013**, 813. (f) Cho, H.; Iwama, Y.; Okano, K.; Tokuyama, H. *Chem. Pharm. Bull.* **2014**, *62*, 354.
9. Hobson, L. A.; Nugent, W. A.; Anderson, S. R.; Deshmukh, S. S.; Haley, J. J., III; Liu, P.; Magnus, N. A.; Sheeran, P.; Sherbine, J. P.; Stone, B. R. P.; Zhu, J. *Org. Proc. Res. Dev.* **2007**, *11*, 985.
10. (a) Iwama, Y.; Okano, K.; Sugimoto, K.; Tokuyama, H. *Org. Lett.* **2012**, *14*, 2320. (b) Iwama, Y.; Okano, K.; Sugimoto, K.; Tokuyama, H. *Chem. Eur. J.* **2013**, *19*, 9325. (c) Iwama, Y.; Okano, K.; Tokuyama, H. *J. Synth. Org. Chem., Jpn.* **2013**, *71*, 926.

Appendix
Chemical Abstracts Nomenclature (Registry Number)

1-Indanone: 1*H*-Inden-1-one, 2,3-dihydro-; (83-33-0)
Hydroxylamine, hydrochloride: Hydroxylamine, hydrochloride (1:1); (5470-11-1)
Pyridine: Pyridine; (110-86-1)
Aluminum, hydrodiisobutyl-: Aluminum, hydrobis(2-methylpropyl)-; (1191-15-7)

Hidetoshi Tokuyama was born in Yokohama in 1967. He received his Ph.D. in 1994 from Tokyo Institute of Technology under the direction of Professor Ei-ichi Nakamura. He spent one year (1994-1995) at the University of Pennsylvania as a postdoc with Professor Amos B. Smith, III. He joined the group of Professor Tohru Fukuyama at the University of Tokyo in 1995 and was appointed Associate Professor in 2003. In 2006, he was move to Tohoku University, where he is currently Professor of Pharmaceutical Sciences His research interest is on the development of synthetic methodologies and total synthesis of natural products.

Kentaro Okano was born in Tokyo in 1979. He received his B.S. in 2003 from Kyoto University under the supervision of Professor Tamejiro Hiyama. He then moved to the laboratories of Professor Tohru Fukuyama at the University of Tokyo. In 2007, he joined the faculty at Tohoku University, where he is currently an assistant professor in Professor Hidetoshi Tokuyama's group. In 2014, he visited Professor Amir Hoveyda's laboratories at Boston College as a visiting researcher. His current research interest is natural product synthesis based on the development of new synthetic methodologies.

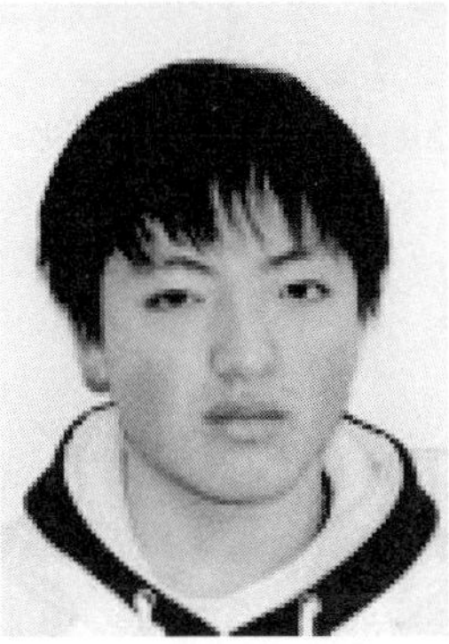

Taku Imaizumi was born in Ibaraki in 1992. He is an undergraduate student in the laboratories of Professor Tokuyama at the Faculty of Pharmaceutical Sciences, Tohoku University. His current research interest is synthesis of indole via novel rearrangement reaction.

Dr. Lan-Ting Xu received her B.S. degree from West China School of Pharmacy, Sichuan University in 2008, and her Ph.D. degree from Fudan University in 2013, under the supervision of Dawei Ma. She is now a MSD China R&D Postdoc Research Fellow in Shanghai Institute of Organic Chemistry. Her research interests include copper-catalyzed coupling reactions, metal-catalyzed direct C-H functionalization and heterocycle synthesis.

Organic Syntheses

Preparation of 2-(2-(Dicyclohexylphosphino)phenyl)-1-methyl-1*H*-indole (CM-phos)

Shun Man Wong, On Ying Yuen, Pui Ying Choy, Chau Ming So,* and Fuk Yee Kwong*[1]

State Key Laboratory of Chirosciences and Department of Applied Biology and Chemical Technology, The Hong Kong Polytechnic University, Hung Hom, Kowloon, Hong Kong

Checked by Takumi Fukazawa and Keisuke Suzuki

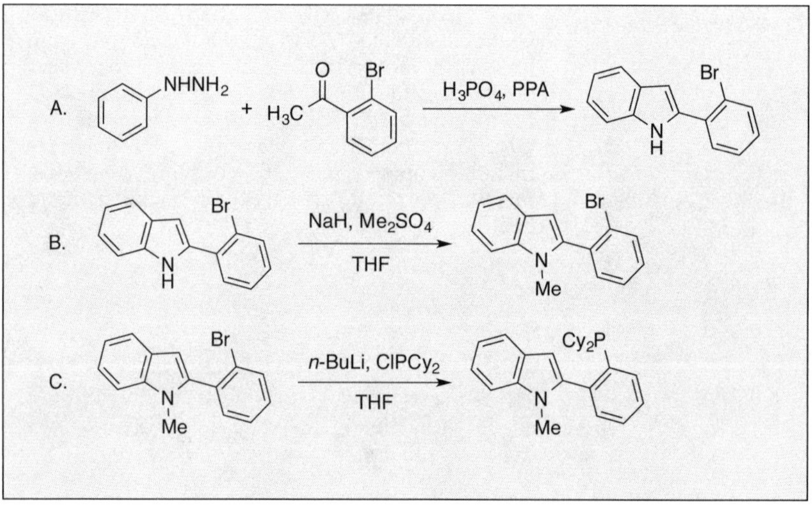

Procedure

A. *2-(2-Bromophenyl)-1H-indole.* An oven-dried 100-mL single-necked round-bottomed flask equipped with a Teflon-coated magnetic stir bar (oval, 25 mm × 15 mm) is charged with 2-bromoacetophenone (2.7 mL, 4.0 g, 20 mmol) (Note 1) and phenylhydrazine (2.4 mL, 2.6 g, 24 mmol) (Note 2) *via* syringe, and stirring is started. Phosphoric acid (10.0 mL) (Note 3) is added slowly over 1 min, and the mixture is stirred for 50 min (Note 4).

Polyphosphoric acid (56.7 g, Note 5) is added to the reaction mixture slowly over 30 min with continuous stirring (Note 6). The reaction flask is equipped with a Dimroth condenser, and is placed in a pre-heated oil bath (85 °C). The bath temperature is raised to 120 °C within 1 h under air atmosphere (Note 7). The resulting mixture is stirred at 120 °C for an additional 2 h, and then carefully poured into ice water (90 g ice, 20 mL water) in a 300 mL beaker (Note 8) with stirring by means of a glass rod (Note 9). The remaining crude product in the reaction flask is rinsed with water (3 × 10 mL) (Note 10) and dichloromethane (3 × 10 mL). All rinses are transferred to the beaker. The resulting mixture is stirred by means of a glass rod for 30 min (Note 11), and then poured into 500-mL separatory funnel. The beaker is rinsed with dichloromethane (3 × 10 mL), water (3 × 10 mL), and again dichloromethane (100 mL). All rinses are transferred to the separatory funnel. The funnel is shaken, and the organic layer is separated. The aqueous layer is further extracted with dichloromethane (3 × 100 mL). The combined organic extracts are concentrated by rotary evaporation (37 °C, 52 mmHg). The resulting dark-brown residues are purified by column chromatography (4.5-cm diameter × 16-cm packed height) with 81 g of silica gel (Note 12) in 10% ethyl acetate–hexane (1:9 ethyl acetate:hexane). At this point, 1400 mL is collected with elution of 10% ethyl acetate–hexane (1:9 ethyl acetate:hexane). The eluent containing the product is concentrated by rotary evaporation (37 °C, 20 mmHg) to afford an orange-yellow oil and dried at 7 mmHg for 1 h. This oil is cooled in an ice-water bath, and hexane (20 mL) is added with stirring by means of a Teflon-coated magnetic stir bar (cylindrical, 25 mm × 8 mm) for 1 h to afford an off-white powder. The resulting powder is collected by suction filtration on a funnel, washed with ice-cold hexane (3 × 5 mL), dried for 5 h at 0.2 mmHg to provide the crude material as a pale yellow powder (3.5–3.7 g, 64–68%) (Note 13).

The crude powder (3.6 g) is suspended in hexane (20 mL) and vigorously stirred for 1 h with a Teflon-coated magnetic stir bar (cylindrical, 25 mm × 8 mm). The resulting white powder is collected by suction filtration on a funnel, washed with ice-cold hexane (3 × 5 mL), and dried for 7 h at 0.2 mmHg to provide 2-(2-bromophenyl)-1*H*-indole (2.3 g, 42%) (Note 14).

B. *2-(2-Bromophenyl)-1-methyl-1H-indole.* An oven-dried 100-mL three-necked round-bottomed flask is equipped with a Teflon-coated magnetic stir bar (cylindrical, 25 mm × 8 mm), is charged with sodium hydride (63 % dispersion in mineral oil) (0.46 g of the sodium hydride in oil,

12 mmol) (Note 15). The necks are fitted with a glass stopper, a rubber septum and a nitrogen stopcock inlet. The flask is evacuated and backfilled with nitrogen three times. Sodium hydride is washed free of mineral oil by stirring with distilled hexane (10 mL), and the supernatant liquid is decanted. This operation is repeated for three times. The remaining solvent is removed by placing the flask under vacuum (0.2 mmHg) for 15 min, after which the flask is filled with nitrogen. The glass stopper is removed and 100-mL pressure-equalizing dropping funnel attached (Figure 1). The flask is evacuated and backfilled with nitrogen three times. Anhydrous tetrahydrofuran (5 mL) (Note 16) is added to the reaction flask *via* syringe to form a suspension of sodium hydride, and stirring is started. The suspension is cooled to 0 °C in an ice-water bath.

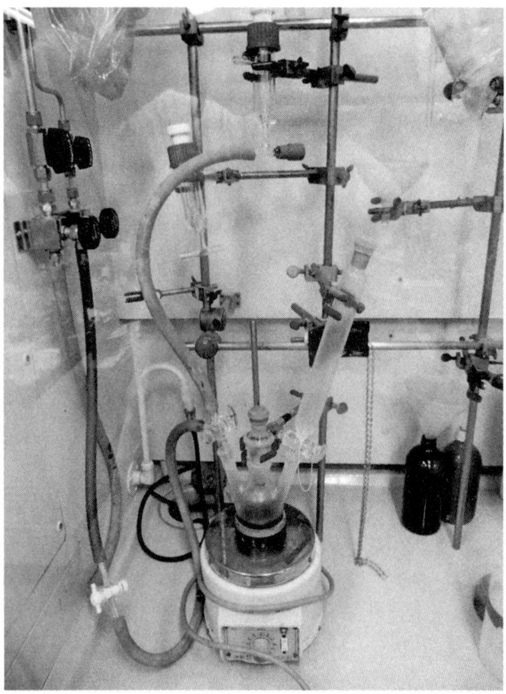

Figure 1. Apparatus Assembly for Step B

A separate oven-dried 100 mL, two-necked, round-bottomed flask is equipped with a Teflon-coated magnetic stir bar (cylindrical, 25 mm × 8 mm) and charged with 2-(2-bromophenyl)-1*H*-indole (2.7 g, 10 mmol). The necks are fitted with a rubber septum and a nitrogen stopcock inlet. The flask is evacuated and backfilled with nitrogen three times. Anhydrous tetrahydrofuran (20 mL) (Note 16) is added *via* syringe, and stirring is started, to form a solution of 2-(2-bromophenyl)-1*H*-indole. The solution of 2-(2-bromophenyl)-1*H*-indole is transferred to the dropping funnel *via* cannula and additional anhydrous tetrahydrofuran (5 mL) (Note 16) is added to the two-necked flask to rinse all the 2-(2-bromophenyl)-1*H*-indole solution to the dropping funnel. The solution of 2-(2-bromophenyl)-1*H*-indole is added dropwise *via* the dropping funnel to the reaction mixture over 5 min and additional anhydrous tetrahydrofuran (5 mL) (Note 16) is added to the dropping funnel to rinse all the 2-(2-bromophenyl)-1*H*-indole solution to the reaction mixture. Upon completion of the addition, the reaction mixture is allowed to warm to room temperature and stirred for 30 min. Dimethyl sulfate (0.99 mL, 10.3 mmol) (Note 17) is added *via* syringe to the reaction mixture. The mixture is stirred overnight at room temperature, at which time TLC analysis indicates the reaction has been completed (Notes 18 and 19). After adding methanol (5 mL) to quench the excess sodium hydride, the reaction mixture is concentrated by rotary evaporation (36 °C, 20 mmHg) to afford a brown oil. This oil is diluted with ethyl acetate (20 mL), and then transferred to a 300-mL separatory funnel. The reaction flask is rinsed with ethyl acetate (2 × 20 mL), deionized water (2 × 20 mL), and again ethyl acetate (20 mL). All rinses are transferred to the separatory funnel. The funnel is shaken, and the layers are separated. The aqueous phase is separated out and the organic phase is washed with water (2 × 20 mL) and brine (20 mL). The organic extract is concentrated by rotary evaporation (36 °C, 20 mmHg) to afford a brown oil. The oil is purified by column chromatography (3.5-cm diameter × 15-cm packed height) on 50 g of silica gel (Note 12) in 5% ethyl acetate-hexane (1:19 ethyl acetate:hexane). The oil is loaded as a solution in ethyl acetate (2 × 1 mL). At this point, 250 mL of fraction is collected with elution of 5% ethyl acetate-hexane (1:19 ethyl acetate:hexane). The eluent containing the product is concentrated by rotary evaporation (36 °C, 20 mmHg) and dried at 0.2 mmHg for 4 h to afford yellow oil (2.8–2.9 g). The oil contains a small amount of ethyl acetate, which cannot be removed under vacuum (Note 20). The material can be used for the next step after co-evaporation with toluene, as described in Step C.

Alternatively, crystallization of the oil is facilitated by the following process that involves treatment in ethanol. The yellow oil is transferred to a 100 mL, single-necked, round-bottomed flask equipped with a Teflon-coated magnetic stir bar (cylindrical, 25 mm × 8 mm). Ethanol (10 mL) is added to the flask, and stirring is initiated. The mixture is refluxed until a homogenous solution results. The solution is allowed to cool to room temperature, and stirred to afford a white powder (Note 21). The mixture is cooled to 0 °C in an ice-water bath, and stirring is maintained for 10 min. The precipitate is collected by suction filtration on a Büchner funnel, washed with ice-cold ethanol (3 × 5 mL), and dried for 3 h at 0.2 mmHg (Note 22) to provide 2-(2-bromophenyl)-1-methyl-1*H*-indole (2.7 g, 96%) (Note 23) as a white powder.

C. *2-(2-(Dicyclohexylphosphino)phenyl)-1-methyl-1H-indole* (CM-phos). An oven-dried 100-mL three-necked round-bottomed flask is charged with 2-(2-bromophenyl)-1-methyl-1*H*-indole (2.3 g, 8.0 mmol) in the form of the oily material described in Step B. The necks are fitted with two glass stoppers. Anhydrous toluene (2 mL) (Note 24) is added to the flask. The solution is concentrated by rotary evaporation (50 °C, 20 mmHg). This operation is repeated two additional times (Note 25). The flask is equipped with a Teflon-coated magnetic stir bar (cylindrical, 25 mm × 8 mm), glass stoppers are changed to 50-mL pressure-equalizing dropping funnel, rubber septum, and nitrogen stopcock inlet. The flask is evacuated and backfilled with nitrogen three times. Anhydrous tetrahydrofuran (40 mL) (Note 16) is added to the reaction flask *via* syringe to form a solution of 2-(2-bromophenyl)-1-methyl-1*H*-indole and stirring is started. The solution is cooled to –78 °C in a dry ice-acetone bath. *n*-Butyllithium (5.4 mL, 1.63 M in hexane, 8.8 mmol) (Note 26) is added dropwise *via* dropping funnel over 3 min and additional anhydrous tetrahydrofuran (2 mL) (Note 16) is added to the dropping funnel to rinse all the *n*-butyllithium solution into the reaction mixture. The resulting mixture is stirred for an additional 30 min at –78 °C. Chlorodicyclohexylphosphine (2.2 mL, 9.6 mmol) (Note 27) and anhydrous tetrahydrofuran (5 mL) (Note 16) are added to the dropping funnel to form a solution of chlorodicyclohexylphosphine. The solution is added dropwise *via* dropping funnel at –78 °C over 3 min, and additional anhydrous tetrahydrofuran (2 mL) (Note 16) is added to the dropping funnel to rinse all phosphine solution to the reaction mixture. The resulting mixture is allowed to warm to room temperature within 2 h and stirred for overnight at 22 °C, at which time TLC analysis indicates completion of the reaction (Notes 18 and 28). After adding methanol (5 mL) to quench the excess *n*-butyllithium, the reaction mixture is concentrated by rotary evaporation (36 °C, 20 mmHg) and dried at 0.2 mmHg for 15 min to afford an off-white solid. This solid is cooled in an ice-water bath, and methanol (20 mL) is added with stirring for 30 min to afford an off-white powder.

The resulting powder is collected by suction filtration on a Büchner funnel, washed with ice-cold methanol (2 × 5 mL) and ice-cold hexane (5 mL), and dried for 3 h at 0.2 mmHg to give 2-(2-(dicyclohexylphosphino)phenyl)-1-methyl-1*H*-indole (CM-phos) (2.2–2.4 g, 69–75%) (Notes 29 and 30) as a white powder.

Notes

1. 2-Bromoacetophenone (>98%) was obtained from Tokyo Chemical Industry Co., Ltd. (TCI) (checkers). 2-Bromoacetophenone (99%) was obtained from Aldrich Co., Inc. and used as received (submitters).
2. Phenylhydrazine (>98%) was obtained from Tokyo Chemical Industry Co., Ltd. (TCI) (checkers). Phenylhydrazine (97%) was obtained from Acros Organics and used as received (submitters).
3. Phosphoric acid (ACS reagent, ≥85% H_3PO_4) was obtained from Aldrich Co., Inc. and used as received.
4. Upon addition of phosphoric acid, the reaction mixture solidified, and the color changed from clear pale-yellow to turbid yellow. After heat was released, the solid started to disappear and the reaction mixture was allowed to stir for 30 min until the reaction mixture reached room temperature.
5. Polyphosphoric acid (reagent grade, 115% H_3PO_4 basis) was obtained from Aldrich Co., Inc. and used as received.
6. Since polyphosphoric acid is a viscous liquid, it was first weighed in a beaker, and then transferred to the reaction flask slowly over 10 min. Notably, not all polyphosphoric acid in the beaker could be transferred to the reaction flask, the amount of polyphosphoric acid could be determined by weight difference of the beaker before and after addition. If the polyphosphoric acid is added too quickly, the reaction mixture becomes very viscous and is unable to be stirred with the stir bar. If needed, the mixture can be mixed thoroughly by glass rod or overhead stirring.
7. The color of the reaction mixture was changed from yellow to dark-green to dark-brown. The reaction mixture should be efficiently stirred during heating, otherwise uneven-heating of the reaction mixture will occur, resulting in lower yield.
8. Precaution: Heat insulated gloves should be put on when pouring the hot reaction mixture into ice water.

9. Overhead stirring could also be employed.
10. Grey precipitates were formed when the crude product was added to ice water.
11. To make the mixture less viscous, thorough stirring by a glass rod is necessary, otherwise a huge amount of product will be lost. Overhead stirring could also be employed.
12. Silica gel 60N (Spherical, neutral, 63–210 μm) was obtained from KANTO CHEMICAL Co., Inc. (checkers). Silica gel 60 (0.040–0.063 mm, 230–400 mesh ASTM) was obtained from Merck Millipore and used as received (submitter).
13. Checkers noted that the product is contaminated by the remaining starting material (2–8%).
14. Melting point: 76.7–77.3 °C; IR (cm^{-1}) 3382, 3053, 1614, 1559, 1463, 1447, 1435, 1401, 1355, 1301, 1229, 1178, 1115, 1073, 1022, 958, 942, 931, 849, 795, 745, 721, 695, 676, 644, 609; ^1H NMR (600 MHz, $CDCl_3$) δ: 6.82 (s, 1H), 7.14 (dd, 1H, J = 7.8, 7.2 Hz), 7.22 (dd, 1H, J = 8.4, 7.8 Hz), 7.23 (dd, 1H, J = 7.8, 7.2 Hz), 7.39 (dd, 1H, J = 7.8, 7.2 Hz), 7.43 (d, 1H, J = 8.4 Hz), 7.62 (d, 1H, J = 7.8 Hz), 7.67 (d, 1H, J = 7.8 Hz), 7.69 (d, 1H, J = 7.8 Hz), 8.65 (s, 1H); ^{13}C NMR (150 MHz, $CDCl_3$) δ: 103.7, 111.0, 120.2, 120.8, 121.3, 122.6, 127.7, 128.2, 129.2, 131.5, 133.5, 134.0, 136.3; HRMS calcd. for $C_{14}H_{10}NBrH^+$: 272.0069, found 272.0065; Anal. calcd. for $C_{14}H_{10}NBr$: C, 61.79; H, 3.70; N, 5.15. found: C, 62.01; H, 3.56; N, 5.17; R_f = 0.23 in 10% ethyl acetate–hexane (1:9 ethyl acetate:hexane) solvent system.
15. Sodium hydride (63 % dispersion in mineral oil) was obtained from Yoneyama Yakuhin Kogyo. Co., Ltd., (checkers). Sodium hydride (60 % dispersion in mineral oil) was obtained from Aldrich Co., Inc. (submitters).
16. Tetrahydrofuran (anhydrous) was obtained from KANTO CHEMICAL Co., Inc. (checkers). Tetrahydrofuran (ACS grade) was obtained from Tedia and distilled from sodium benzophenone ketyl under nitrogen (submitters).
17. Dimethyl sulfate (>98%) was obtained from Tokyo Chemical Industry Co., Ltd. (TCI) (checkers) and Aldrich Co., Inc. (submitters), and used as received.
18. Thin layer chromatography was performed on pre-coated TLC-plates (Merck Co., Inc. TLC silica gel 60 F_{254}, Art 5715, 0.25 mm) (checkers). Thin layer chromatography was performed on pre-coated TLC-sheets ALUGRAM® SIL G/UV254 and obtained from MACHEREY-NAGEL GmbH & Co. KG (submitters).

19. The R_f value of 2-(2-bromophenyl)-1-methyl-1*H*-indole: 0.56 (1:9 ethyl acetate:hexane).
20. Checkers noted that the oil contained ethyl acetate (4–6%), which cannot be removed even after prolonged standing under vacuum (0.2 mmHg, 4 h).
21. Crystallization of the oil is facilitated in ethanol. Ethanol (99.5 %) was obtained from Nacalai Tesque Co. (checkers) and VWR (submitters), and used as received. The time required for the precipitation varies from 10 min (submitters) to 2–3 days (checkers).
22. In preparation for use in the next step, the precipitate should be well dried (0.2 mmHg, 3 h). By this protocol, ethanol was no longer detected by ^{1}H NMR.
23. Melting point: 92.2-92.8 °C; ^{1}H NMR (600 MHz, $CDCl_3$) δ: 3.57 (s, 3H), 6.51 (s, 1H), 7.15 (dd, 1H, *J* = 7.8, 7.2 Hz), 7.27 (dd, 1H, *J* = 7.8, 7.2 Hz), 7.29–7.33 (m, 1H), 7.35–7.43 (m, 3H), 7.66 (d, 1H, *J* = 7.8 Hz), 7.71 (d, 1H, *J* = 7.8 Hz); ^{13}C NMR (150 MHz, $CDCl_3$) δ: 30.6, 102.1, 109.5, 119.8, 120.7, 121.8, 125.2, 127.2, 127.6, 130.1, 132.8, 132.9, 134.3, 137.3, 139.7; IR (cm^{-1}) 3053, 2937, 1924, 1560, 1542, 1463, 1431, 1386, 1360, 1339, 1311, 1235, 1204, 1165, 1145, 1130, 1098, 1058, 1024, 1004, 947, 923, 896, 783, 762, 749, 738, 689, 665, 643, 579, 536; HRMS calcd. for $C_{15}H_{12}NBrH^+$: 286.0231, found 286.0221.
24. Toluene was obtained from KANTO CHEMICAL Co., Inc. (checkers).
25. Treatment with toluene and evaporation of the toluene is not necessary if crystalline material is used.
26. *n*-Butyllithium (1.6 M solution in hexane) was obtained from KANTO CHEMICAL Co., Inc. and used as received (checkers). *n*-Butyllithium (1.6 M solution in hexane) was obtained from J&K Chemical Co. and titrated with 1,3-diphenylacetone *p*-tosylhydrazone in anhydrous tetrahydrofuran under nitrogen prior to use, the *n*-butyllithium was titrated[2] to be 1.32 M (submitters).
27. Chlorodicyclohexylphosphine (97%) was obtained from Aldrich Co., Inc. and used as received.
28. The R_f value of CM-phos is 0.59 (ethyl acetate:hexane ,1:9).
29. Melting point: 181.2–182.4 °C; ^{1}H NMR (600 MHz, CD_2Cl_2) δ: 1.05–1.36 (m, 11H), 1.62–1.85 (m, 11H), 3.50 (s, 3H), 6.39 (s, 1H), 7.14 (dd, 1H, *J* = 7.8, 7.2 Hz), 7.24 (dd, 1H, *J* = 8.4, 7.2 Hz), 7.35–7.42 (m, 2H), 7.47 (dd, 1H, *J* = 7.8, 7.2 Hz), 7.52 (t, 1H, *J* = 7.2 Hz), 7.62 (d, 1H, *J* = 7.8 Hz), 7.70 (brd, 1H, *J* = 5.4 Hz); ^{13}C NMR (150 MHz, CD_2Cl_2) δ: 26.4, 27.2, 30.66, 30.69, 103.0, 103.1, 109.3, 119.3, 120.0, 120.9, 127.7, 128.0, 128.1, 131.6, 131.7,

132.77, 132.80, 136.6, 137.5, 137.6, 140.8, 141.0, 141.5, 141.6 (unresolved complex C-P splittings were observed); ^{31}P NMR (243 MHz, CD_2Cl_2) δ: –9.96; IR (cm^{-1}) 3054, 2923, 2847, 1539, 1467, 1445, 1419, 1385, 1363, 1339, 1310, 1265, 1234, 1199, 1179, 1167, 1145, 1122, 1099, 1000, 920, 886, 849, 778, 768, 745, 731, 670; HRMS calcd. for $C_{27}H_{34}NPH^+$: 404.2501, found 404.2513; Anal. calcd. for $C_{27}H_{34}NP$: C, 80.36; H, 8.49; N, 3.47. found: C, 80.53; H, 8.15; N, 3.47.

30. CM-phos was stored with an argon atmosphere in the cool and dark space.

Working with Hazardous Chemicals

The procedures in *Organic Syntheses* are intended for use only by persons with proper training in experimental organic chemistry. All hazardous materials should be handled using the standard procedures for work with chemicals described in references such as "Prudent Practices in the Laboratory" (The National Academies Press, Washington, D.C., 2011; the full text can be accessed free of charge at http://www.nap.edu/catalog.php?record_id=12654). All chemical waste should be disposed of in accordance with local regulations. For general guidelines for the management of chemical waste, see Chapter 8 of Prudent Practices.

In some articles in *Organic Syntheses*, chemical-specific hazards are highlighted in red "Caution Notes" within a procedure. It is important to recognize that the absence of a caution note does not imply that no significant hazards are associated with the chemicals involved in that procedure. Prior to performing a reaction, a thorough risk assessment should be carried out that includes a review of the potential hazards associated with each chemical and experimental operation on the scale that is planned for the procedure. Guidelines for carrying out a risk assessment and for analyzing the hazards associated with chemicals can be found in Chapter 4 of Prudent Practices.

The procedures described in *Organic Syntheses* are provided as published and are conducted at one's own risk. *Organic Syntheses, Inc.*, its Editors, and its Board of Directors do not warrant or guarantee the safety of individuals using these procedures and hereby disclaim any liability for any injuries or damages claimed to have resulted from or related in any way to the procedures herein.

Discussion

It is well known that ligand design and synthesis are very important for the metal-catalyzed coupling reactions. After recognizing the preliminary relationship between the ligand structure and the outcome of the coupling reaction, various research groups have designed and synthesized countless supporting ligands for the coupling reactions, such as P*t*-Bu_3,[3] Buchwald ligand kit,[4] and Josiphoskit.[5] Although a variety of ligands have been developed, rapid assembly of structurally diverse ligand systems via simple synthetic methods is still important for the development of versatile catalysts for more widespread applications of coupling reactions.

We have the following strategic considerations for the design of effective ligands for the coupling reactions: (1) the starting materials should be inexpensive and readily available; (2) the ligand synthesis should be simple and straightforward (an elimination of metal/halogen exchange (from ArBr or ArI) would be even more advantageous); (3) the ligand diversity should be easily accessible, and should conveniently provide high level of steric and electronic fine-tunings. Based on these three strategies, indole was selected as template to design and explore a new class of phosphine ligands. There were several reasons for the choice of indole: (1) indole and starting materials to synthesize indole derivatives, such as phenylhydrazine and acetophenone are generally inexpensive; (2) indole and most of its derivatives are readily and commercially available; (3) many synthetic methods of indoles have been reported for over hundreds of years,[6] such as Fischer indole synthesis and metal-catalyzed indole synthesis.

Notably, the chemical properties of indole scaffold can provide potential for diversification of the new ligands. One of the modern concepts for the phosphine ligand design is the rapid assembly of the two major compartments. One compartment is the phosphorus atom that attaches two of the simple equivalent groups that can be an aryl or alkyl group, such as diphenyl or dicyclohexyl group. Another compartment is the ligand template that provides steric and electronic fine-tuning for the phosphine ligand. Herein, CM-phos that derived from indole ligand template was highly tunable in both steric and electronic effects (Figure 2).

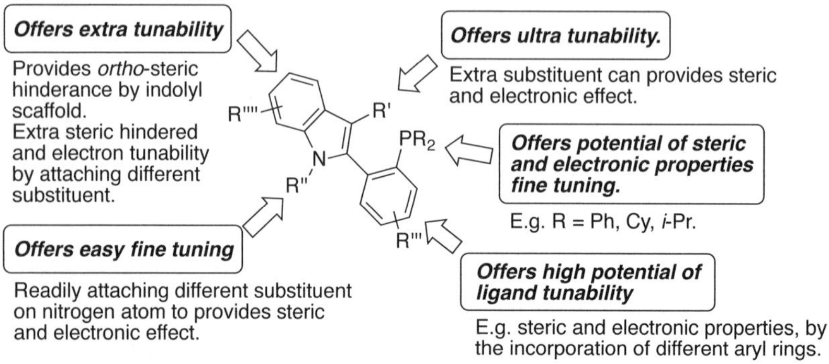

Figure 1. Ligand design and diversity

To sum up, the indole ligand template 1) is inexpensive; 2) can be readily available from commercially accessible starting materials; 3) can be synthesized from well-developed synthesis methods; and 4) has high potential for diversification and easy steric and electronic fine-tuning.

We finally selected the Fischer indole synthesis[6b] as our primary tool to synthesize the indole templates from phenylhydrazine and 2'-bromoacetophenone. With the methylated ligand precursor, CM-phos could be afforded easily by lithiation and subsequently trapping with chlorodicyclohexylphosphine. It should be noted that CM-phos in combination with the precatalyst palladium(II) acetate is highly active and effective in amination and Suzuki cross-coupling reaction of aryl mesylates.[7]

References

1. State Key Laboratory of Chirosciences and Department of Applied Biology and Chemical Technology, The Hong Kong Polytechnic University, Hung Hom, Kowloon, Hong Kong. E-mail: chau.ming.so@polyu.edu.hk, fuk-yee.kwong@polyu.edu.hk; Fax: +852-2364-9932. **Acknowledgement**: We thank the Research Grants Council of Hong Kong (PolyU5010/11P) and Collaborative Research Fund (C5023-14G) for financial support. F.Y.K. is grateful to The Croucher Foundation for the Croucher Senior Research Fellowship 2013.

2. F. Lipton, M.; M. Sorensen, C.; C. Sadler, A.; H. Shapiro, R. *J. Organomet. Chem.* **1980**, *186*, 155.

3. (a) Littke, A. F.; Dai, C.; Fu, G. C. *J. Am. Chem. Soc.* **2000**, *122*, 4020. (b) Netherton, M. R.; Dai, C.; Neuschütz, K.; Fu, G. C. *J. Am. Chem. Soc.* **2001**, *123*, 10099. (c) Kirchhoff, J. H.; Netherton, M. R.; Hills, I. D.; Fu, G. C. *J. Am. Chem. Soc.* **2002**, *124*, 13662. (d) Kudo, N.; Perseghini, M.; Fu, G. C. *Angew. Chem. Int. Ed.* **2006**, *45*, 1282.
4. (a) Nguyen, H. N.; Huang, X.; Buchwald, S. L. *J. Am. Chem. Soc.* **2003**, *125*, 11818. (b) Barder, T. E.; Walker, S. D.; Martinelli, J. R.; Buchwald, S. L. *J. Am. Chem. Soc.* **2005**, *127*, 4685. (c) Wolfe, J. P.; Tomori, H.; Sadighi, J. P.; Yin, J.; Buchwald, S. L. *J. Org. Chem.* **2000**, *65*, 1158. (d) Yin, J.; Rainka, M. P.; Zhang, X.-X.; Buchwald, S. L. *J. Am. Chem. Soc.* **2002**, *124*, 1162. (e) Walker, S. D.; Barder, T. E.; Martinelli, J. R.; Buchwald, S. L. *Angew. Chem. Int. Ed.* **2004**, *43*, 1871. (f) Billingsley, K. L.; Anderson, K. W.; Buchwald, S. L. *Angew. Chem. Int. Ed.* **2006**, *45*, 3484. (g) Billingsley, K.; Buchwald, S. L. *J. Am. Chem. Soc.* **2007**, *129*, 3358.
5. (a) Stambuli, J. P.; Kuwano, R.; Hartwig, J. F. *Angew. Chem. Int. Ed.* **2002**, *41*, 4746. (b) Kataoka, N.; Shelby, Q.; Stambuli, J. P.; Hartwig, J. F. *J. Org. Chem.* **2002**, *67*, 5553.
6. (a) Gräbe, C.; Liebermann, C. *Ber. Dtsch. Chem. Ges.* **1869**, *2*, 678. (b) Fischer, E.; Jourdan, F. *Ber. Dtsch. Chem. Ges.* **1883**, *16*, 2241. (c) Dalton, L.; Humphrey, G. L.; Cooper, M. M.; Joule, J. A. *J. Chem. Soc., Perkin Trans. 1* **1983**, 2417. (d) Reissert, A. *Ber. Dtsch. Chem. Ges.* **1897**, *30*, 1030. (e) Madelung, W. *Ber. Dtsch. Chem. Ges.* **1912**, *45*, 1128. (f) Allen, G. R.; Pidacks, C.; Weiss, M. J. *J. Am. Chem. Soc.* **1966**, *88*, 2536. (g) Hemetsberger, H.; Knittel, D. *Monatsh. Chem.* **1972**, *103*, 194. (h) Gassman, P. G.; Van Bergen, T. J.; Gruetzmacher, G. *J. Am. Chem. Soc.* **1973**, *95*, 6508. (i) Batcho, A. D.; Leimgruber, W. *Org. Synth.* **1985**, *63*, 214; (j) Bartoli, G.; Palmieri, G.; Bosco, M.; Dalpozzo, R. *Tetrahedron Lett.* **1989**, *30*, 2129. (k) Larock, R. C.; Yum, E. K.; Refvik, M. D. *J. Org. Chem.* **1998**, *63*, 7652. (l) Tokuyama, H.; Yamashita, T.; Reding, M. T.; Kaburagi, Y.; Fukuyama, T. *J. Am. Chem. Soc.* **1999**, *121*, 3791.
7. (a) So, C. M.; Zhou, Z.; Lau, C. P.; Kwong, F. Y. *Angew. Chem. Int. Ed.* **2008**, *47*, 6402. (b) So, C. M.; Lau, C. P.; Kwong, F. Y. *Angew. Chem. Int. Ed.* **2008**, *47*, 8059. (c) So, C. M.; Kwong, F. Y. *Chem. Soc. Rev.* **2011**, *40*, 4963. (d) Wong, S. M.; Choy, P. Y.; Yuen, O. Y.; So, C. M.; Kwong, F. Y. *Org. Synth.* **2015**, *92*, 195.

Appendix
Chemical Abstracts Nomenclature (Registry Number)

2′-Bromoacetophenone; (2142-69-0)
Phenylhydrazine; (100-63-0)
Phosphoric acid; (7664-38-2)
Polyphosphoric acid; (8017-16-1)
Sodium hydride; (7646-69-7)
Dimethyl sulfate; (77-78-1)
n-Butyllithium; (109-72-8)
Chlorodicyclohexylphosphine; (16523-54-9)

Shun Man Wong received his B.Sc. in chemical technology from The Hong Kong Polytechnic University in 2010. He pursued his postgraduate study at the same university and obtained his Ph.D. degree in 2014. He is currently a postdoctoral fellow under the supervision of Prof. Fuk Yee Kwong, researching the synthesis of new heterocyclic phosphine ligands and their potential applications.

On Ying Yuen received her B.Sc. (1st class honor) in Chemical Technology from the Hong Kong Polytechnic University in 2011. Currently, she is pursuing her Ph.D. under the guidance of Prof. Fuk Yee Kwong. Her main research focuses on palladium-catalyzed direct functionalization of aromatics: process and catalyst design.

Pui Ying Choy received her B.Sc. in Chemical Technology in The Hong Kong Polytechnic University in 2010. She pursued her postgraduate study at the same university and obtained her Ph.D. degree in 2014. She is currently a research associate under the supervision of Prof. Fuk Yee Kwong, researching the synthesis of new heterocyclic phosphine ligands and their potential applications in transition-metal catalysis.

Chau Ming So is currently a Visiting Assistant Professor in the Department of Applied Biology and Chemical Technology of The Hong Kong Polytechnic University. He received his B.Sc. (1st class honor) from PolyU in 2006. He pursued his postgraduate study at the same university and obtained his Ph.D. degree in 2010. He received the Hong Kong Young Scientist Award in the same year. Moreover, he was the winner of Eli Lilly the Best Thesis Award (1st Prize). In 2012-2013, he moved to Institute of Materials Research and Engineering (IMRE) as postdoctoral fellow in Prof. Tamio Hayashi's research group.

Fuk Yee (Michael) Kwong is currently a professor of the Department of Applied Biology and Chemical Technology at The Hong Kong Polytechnic University. He received his B.Sc. in 1996, and completed his Ph.D. at The Chinese University of Hong Kong in 2000 under the supervision of Professor Kin Shing Chan. In 2001–2003, he was at the Massachusetts Institute of Technology (MIT), USA, as a Croucher Foundation postdoctoral fellow. Kwong's research interests are in the areas of new crosscoupling methodologies, carbon–hydrogen bond functionalization, and catalytic enantioselective transformations.

Takumi Fukazawa was born in Chiba, Japan. He received his B. Sc. degree in 2012 at Waseda University under the supervision of Prof. Kuniaki Tatsuta. In the same year, he joined the research group of Prof. Keisuke Suzuki at Tokyo Institute of Technology. In 2014, he received his M. Sc., and currently is pursuing his Ph.D. degree.

Organic Syntheses

Synthesis of 5-(Hydroxymethyl)furfural (HMF)

Svilen P. Simeonov, Jaime A. S. Coelho, and Carlos A. M. Afonso*[1]

Research Institute for Medicines (iMed.ULisboa), Faculty of Pharmacy, University of Lisbon, Av. Prof. Gama Pinto, 1649-003, Lisboa, Portugal and Institute of Organic Chemistry with Centre of Phytochemistry, Bulgarian Academy of Sciences, Acad. G. Bonchev str., bl. 9, 1113, Sofia, Bulgaria

Checked by Mengyang Fan and Dawei Ma

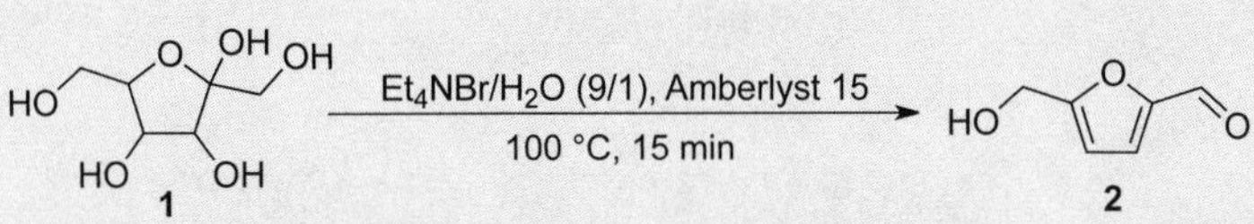

Procedure

5-(Hydroxymethyl)furfural (**2**). A 500-mL single-necked, round-bottomed flask equipped with a 3.5-cm ellipsoidal magnetic stir bar is charged with tetraethylammonium bromide (45 g, 214 mmol) (Note 1), distilled water (5 mL) and fructose (**1**) (10 g, 0.056 mol) (Note 2). The flask is placed in an oil bath at 100 °C. The reaction temperature is monitored by a thermometer placed inside the reaction mixture. When the internal temperature reaches 90 °C (Note 3), Amberlyst® 15 (1 g, 10%w/w) (Note 4) is added. The stirring is continued for 15 min, during which time the internal temperature reaches 96–98 °C and the color of the reaction mixture turns brown (Note 5). After this period the flask is removed from the heat source and the solution separated from the Amberlyst® 15 by decantation of the hot solution into a 2-L single-necked round-bottomed flask (Note 6). Absolute ethanol (20–30 mL) is used to wash the catalyst after the transfer, and the ethanol is added to the aqueous solution. The reaction mixture is placed on a rotary evaporator for 1 h (15 mmHg, 50 °C) (Note 7). A 3.5-cm ellipsoidal magnetic stir bar is added, and the residue is dissolved in hot absolute anhydrous ethanol (25 mL), which is followed by immediate addition of ethyl acetate

DOI: 10.15227/orgsyn.093.0029
Published on the Web 1/5/2016

(1.5 L) under vigorous stirring. Immediate precipitation of the tetraethylammonium bromide is observed, and the mixture is cooled down to room temperature. The precipitate is vacuum-filtered through a fritted filter and the cake rinsed with ethyl acetate (100 mL) (Note 8). The collected filtrate is vacuum-filtered through a pad of silica gel (30 g) using a fritted filter, and the silica gel is rinsed with ethyl acetate (100 mL) (Notes 9 and 10). The filtrate is concentrated by rotary evaporation (32 °C, 15 mmHg). The oily residue is transferred into a 50 mL single-necked round-bottomed flask and the residual organic solvent is removed *in vacuo* (0.5 mmHg). The title compound is obtained as deeply orange liquid (5.8–6.1 g, 83–87%) and 90-92% purity (Note 11). Flash column chromatography (Note 12) was applied to give the title compound as an orange liquid (5.5 g, 79%) with 97% purity (Notes 13, 14 and 15).

Figure 1. Reaction Assembly

Notes

1. Tetraethylammonium bromide (TEAB) (98%) was purchased from Alfa-Aesar and used directly. The submitter used tetraethylammonium bromide (98%) purchased from Sigma Aldrich.
2. Fructose (>99%) was purchased from TCI and used directly. The submitter used fructose (100%) that was obtained from a supermarket.
3. The temperature is reached within 15-20 min. The submitter reports that the reaction mixture became homogeneous at this point, but the checkers observed insoluble particles in the solution.
4. Amberlyst® 15 hydrogen form wet was purchased from Sigma Aldrich and used as received (~48% moisture according to product specification). An old batch (apparently dried) afforded less pure isolated HMF (80%).
5. The reaction temperature should not exceed 98–100 °C, because this can result in formation of impurities.
6. If the catalyst is not removed, formation of impurities in the final product is observed. The tetraethylammonium bromide may precipitate from the solution during cooling, which may make filtering of the solution difficult. Consequently, it is recommended that the solution remains hot during the decantation.
7. The water should be completely removed, because the presence of water can inhibit precipitation in the subsequent operation.
8. The collected tetraethylammonium bromide (TEAB) precipitate is dried using high vacuum rotary pump for 3 h to give 44 g of TEAB as a pale yellow powder. New TEAB (1 g) is added to provide 45 g. The procedure can be repeated using this source of TEAB to provide HMF (6.1 g) as a deep orange liquid with purity of 80%. The submitters report that the purity of the second run can reach 97%.
9. This step is required in order to remove any traces of dissolved TEAB.
10. 230-400 mesh SiliaFlash® P60 (Silicycle Inc.) was used.
11. The submitters report that the deeply orange liquid would solidify in a freezer at –12 °C, but the checkers found the impure product would not solidify completely. Only a semisolid was obtained.
12. The inner diameter of the column is 8 cm and 320 g (700 mL) silica gel (Note 10) was used. The eluent was comprised of hexane/EtOAc, ranging from 3/1 (300 mL/100 mL) to 2/1 (600 mL/300 mL). TLC on silica gel 60 F254 plate (EtOAc:Hexane, 1:1), product $R_f = 0.4$.

13. 5-(Hydroxymethyl)furfural is known to be light and temperature unstable. It should be kept in a freezer for long term storage. No observable degradation occurred when stored in a freezer for up to 8 months.
14. TLC on silica gel 60 F254 plate (EtOAc:Hexane, 1:1) R_f = 0.4. ^{1}H NMR (400 MHz, $CDCl_3$) δ: 2.60 (br s, 1H), 4.71 (s, 2H), 6.51 (d, J = 3.6 Hz, 1H), 7.21 (d, J = 3.6 Hz, 1H), 9.58 (s, 1H); ^{13}C NMR (100 MHz, $CDCl_3$) δ: 57.1, 110.0, 123.8, 152.0, 161.2, 177.9. IR (liquid film, cm^{-1}): 3465 (O-H), 3121, 2929, 2840, 1690 (C=O), 1528 (C=C), 1398, 1280, 1194, 1016. The product was obtained as an orange liquid that solidifies in a freezer at –12 °C. Due to the low mp, it was not possible for the submitters to determine the mp.
15. The purity of the product was determined by HPLC. The analyses were performed on Dionex Ultimate 3000RS, detection at 220 nm, Kinetex, 2.6 μ, XB-C18, 100 x 4.6mm. R_t (HMF) = 1.01 min. Mobile phase gradient from 10:90 to 100:0 for 5 min acetonitrile:water, flow 1.8 mL/min. The purity of HMF was determined by comparing the obtained integration area of HMF with other observed minor peaks. The obtained values were in the range of 85-92% from the different reactions before further purification. After flash column chromatography, the purity reached 97%.

Working with Hazardous Chemicals

The procedures in *Organic Syntheses* are intended for use only by persons with proper training in experimental organic chemistry. All hazardous materials should be handled using the standard procedures for work with chemicals described in references such as "Prudent Practices in the Laboratory" (The National Academies Press, Washington, D.C., 2011; the full text can be accessed free of charge at http://www.nap.edu/catalog.php?record_id=12654). All chemical waste should be disposed of in accordance with local regulations. For general guidelines for the management of chemical waste, see Chapter 8 of Prudent Practices.

In some articles in *Organic Syntheses*, chemical-specific hazards are highlighted in red "Caution Notes" within a procedure. It is important to recognize that the absence of a caution note does not imply that no significant hazards are associated with the chemicals involved in that procedure. Prior to performing a reaction, a thorough risk assessment should be carried out that includes a review of the potential hazards associated with each chemical and experimental operation on the scale that is planned for the procedure. Guidelines for carrying out a risk assessment and for analyzing the hazards associated with chemicals can be found in Chapter 4 of Prudent Practices.

The procedures described in *Organic Syntheses* are provided as published and are conducted at one's own risk. *Organic Syntheses, Inc.*, its Editors, and its Board of Directors do not warrant or guarantee the safety of individuals using these procedures and hereby disclaim any liability for any injuries or damages claimed to have resulted from or related in any way to the procedures herein.

Discussion

During the recent years 5-(hydroxymethyl)furfural (HMF) collected considerable interest from the researchers worldwide as a very important chemical building block derived from biorenewable resources.[2] Under investigations are broad range of HMF conversions to useful chemicals for the industry such as polymer industry,[3] pharmacy,[4] fuel chemistry.[5-7] Derived from HMF Furan-2.5-dicarboxylic acid is described as one of the building blocks of the future.[8]

5-(Hydroxymethyl)furfural (HMF) is synthesized by dehydration of carbohydrates under acid catalysis and fructose is the best starting material for its lab scale synthesis. Because of the low solubility of fructose in organic solvents the dehydration is typically performed in water, ionic liquids, DMSO, eutectic mixtures or biphasic systems.[9,10] Although high yields up to 100% are reported in some cases the purification of HMF remains challenging. HMF is not easy to extract from aqueous phase, since the distribution coefficient between the organic and the aqueous phase is not favorable. Polar organic solvents such as DMSO and DMF have high boiling points but HMF is not stable at high temperatures and the distillation is undesirable.

The procedure described herein is originated on our previous communication[11] and is based on simple precipitation of the reaction media (tetraethyl ammonium bromide) allowing complete separation of HMF in high yields and purity in a single step without extraction or column chromatography. Although HMF is known to be unstable, no formation of side products or decomposition was observed during the work up procedure and it was isolated with above 95% purity by HPLC and can be directly used in further reactions or if required stored in the freezer.

References

1. Research Institute for Medicines (iMed.ULisboa), Faculty of Pharmacy, University of Lisbon, Av. Prof. Gama Pinto, 1649-003, Lisboa, Portugal and Institute of Organic Chemistry with Centre of Phytochemistry, Bulgarian Academy of Sciences, Acad. G. Bonchev str., bl. 9, 1113, Sofia, Bulgaria (Simeonov, S.P.). E-mail: carlosafonso@ff.ul.pt. The authors are grateful to Fundação para a Ciência e a Tecnologia (Ref. SFRH/BD/73971/ 2010, SFRH/BD/67025/2009, PTDC/QUI-QUI/119823/2010, PEst-OE/SAU/ UI4013/2011), European Research Area Network; ERANet LAC (ref. ELAC2014/BEE-0341) and Fundação Calouste Gulbenkian for financial support, REDE/1518/REM/2005 and Portuguese NMR Network (IST-UTL Center) for providing access to the MS and NMR facilities).
2. van Putten, R.-J.; van der Waal, J. C.; de Jong, E.; Rasrendra, C. B.; Heeres, H. J.; de Vries, J. G. *Chem. Rev.* **2013**, *113*, 1499.
3. Lewkowski, J. *Arkivoc* **2001**, 17.
4. Mascal, M.; Dutta, S. *Green Chem.* **2011**, *13*, 3101.
5. Thananatthanachon, T.; Rauchfuss, T. B. *Ang. Chem. Int. Ed.* **2010**, *49*, 6616.
6. Daniel, R.; Xu, H.; Wang, C.; Richardson, D.; Shuai, S. *Appl. Energy* **2012**, *98*, 59.
7. Roman-Leshkov, Y.; Barrett, C. J.; Liu, Z. Y.; Dumesic, J. A. *Nature* **2007**, *447*, 982.
8. Werpy, T. P., G., *US Dept. of Energy* **2004**, *1*, 26.
9. Rosatella, A. A.; Simeonov, S. P.; Frade, R. F. M.; Afonso, C. A. M. *Green Chem.* **2011**, *13*, 754.
10. Zakrzewska, M. E.; Bogel-Łukasik, E.; Bogel-Łukasik, R. *Chem. Rev.* **2010**, *111*, 397.
11. Simeonov, S. P.; Coelho, J. A. S.; Afonso, C. A. M. *ChemSusChem*, **2012**, *5*, 1388.

Appendix
Chemical Abstracts Nomenclature (Registry Number)

Tetraethylammonium bromide (71-91-0)
Fructose (57-48-7)
Amberlyst 15 (39389-20-3)
5-(Hydroxymethyl)furfural (HMF) (67-47-0)

Svilen P. Simeonov was born in 1980 in Ruse (Bulgaria). He graduated from Sofia University in 2004 and joined the Institute of Organic Chemistry with Centre of Phytochemistry, Bulgarian Academy of sciences as a Research Fellow. In 2010 he moved to University of Lisbon and received his PhD in 2014 under the supervision of Prof. Carlos A. M. Afonso. After, he returned back to Sofia and is currently working as a Researcher at the the Institute of Organic Chemistry with Centre of Phytochemistry, Bulgarian Academy of Sciences. His research interested are mainly focused on green chemistry methodologies, flow chemistry, valorisation of natural resources and asymmetric catalysis.

Jaime A. S. Coelho was born in 1986 in Lisbon, Portugal. He received his B.Sc and M.Sc degrees in Chemistry from Instituto Superior Técnico, Portugal. He then moved to Faculty of Pharmacy, University of Lisbon, where he completed a Ph.D. (2014) under the supervision of Prof. Carlos A. M. Afonso. During his doctoral studies he also joined Prof. Nuno Maulide group at Max-Planck-Institut für Kohlenforschung, Mülheim an der Ruhr, Germany, for six months. After a one year postdoctoral experience at Faculty of Pharmacy, University of Lisbon, Portugal, he joined Prof. Dean Toste group, University of California, Berkeley, where he is a visiting researcher. His research interests include catalysis, development of new methodologies and valorization of natural resources.

Carlos A. M. Afonso (born in 1962) is currently Full Professor at Faculty of Pharmacy, University of Lisbon. He received his Ph.D. in 1990 under the supervision of Professor Christopher Maycock, subsequently became Assistant Professor at New University of Lisbon and Associate Professor at the Instituto Superior Técnico in Lisbon (2004-2010), and took a postdoctoral fellow (1990) at the Imperial College of Science Technology and Medicine (W. B. Motherwell) and sabbatical leave (1997/1998) at University of Bath, UK (Jonathan Williams) and at the University of Toronto (Robert Batey). His main interests are new catalyzed synthetic methodologies, green chemistry (reactions in water, ionic liquids, PEG, solvent free, catalyst reuse, membrane technology), synthetic valorization of national natural resources, development of new materials (ionic liquids, probes), and biotransformations.

Mengyang Fan was born in 1989 in Xuzhou, China. He obtained his B.Sc. degree from the Department of Chemistry and Chemical Engineering, Southeast University in 2011. He then joined the Ph.D. program in the laboratory of Professor Dawei Ma at Shanghai Institute of Organic Chemistry and is studying transition metal catalyzed C-H bond activation.

Organic Syntheses

Erratum for: Titanium-Mediated Addition of Silyl Dienol Ethers to Electrophilic Glycine: 4-Ketopipecolic Acid Hydrochloride[1]

Clarisse Mühlemann, Peter Hartmann, and Jean-Pierre Obrecht

Dr.R.Maag AG (CRD), from 1990 Ciba-Geigy Plant Protection Division (Disease Control), Ueberlandstrasse 138, CH-8600 Duebendorf, Switzerland.

Checked by Eugene Ho and David L. Coffen
Rechecked by Yasuhiro Shimamoto, Heemal Dhanjee and John L. Wood

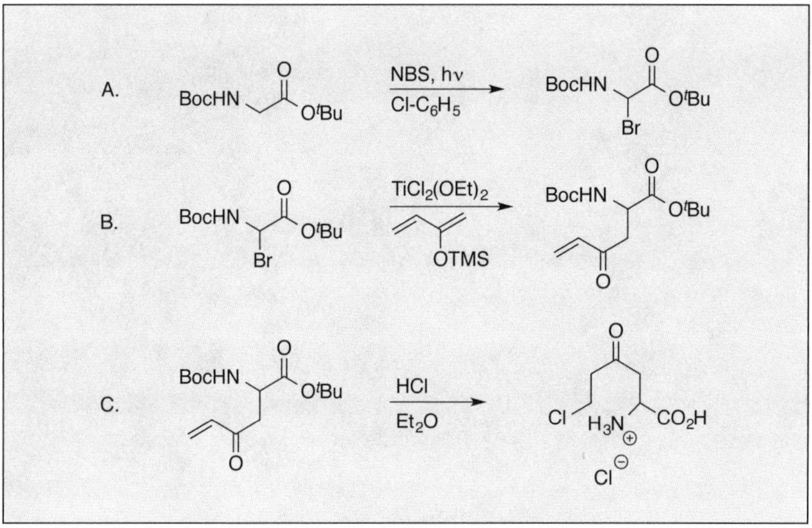

Procedure

A. 2-Bromo-N-Boc-glycine tert-butyl ester. A 500 mL round-bottomed flask equipped with an egg-shaped, teflon-coated magnetic stir bar (3.0 cm x 1.5 cm) is capped with a rubber septum. The flask is placed under nitrogen by way of a needle through the rubber septum (Note 1). The septum is removed temporarily, *N*-Boc-glycine *tert*-butyl ester (10.0 g, 43.2 mmol,

DOI: 10.15227/orgsyn.93.0037

1.0 equiv) (Note 2) and *N*-bromosuccinimide (8.10 g, 45.5 mmol, 1.05 equiv) (Note 3) are added. The flask is placed under vacuum and backfilled with nitrogen three times after which chlorobenzene (175 mL, 0.25 M) (Note 4) is added. The flask is then cooled in an ice water bath, and irradiated with a 300 W tungsten lamp for 1 h (Note 5) (Figure 1) during which the colorless solution becomes a dark red solution. The suspension is filtered using a fritted filter funnel (4.5 cm diameter, Fine) and chlorobenzene is removed under reduced pressure using a rotary evaporator (50 mmHg, 40 °C). The residue is further dried under high vacuum (1.0 mmHg, 25 °C) for 2 h. The yellow crude oil is used in the next step without further purification (Note 6).

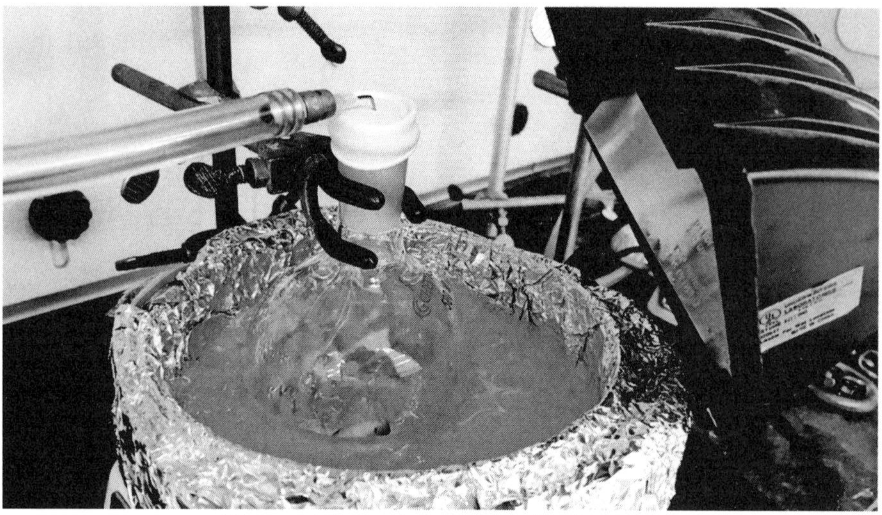

Figure 1. Reaction Set-up for Step A.

B. *tert-Butyl [1-(tert-butoxycarbonyl)-3-oxo-4-pentenyl]carbamate.* The crude bromination product from the previous step is taken up in dry tetrahydrofuran (120 mL) (Note 7) and transferred to a 500 mL 3-necked round-bottomed flask equipped with an egg-shaped, teflon-coated magnetic stir bar (3.0 cm x 1.5 cm), a rubber septum with thermocouple temperature probe (Neck 1), a 100-mL dropping funnel capped with a rubber septum (Neck 2), and a nitrogen inlet (Neck 3) (Figure 2). The solution is cooled to –78 °C in a dry ice-acetone bath and a solution of dichlorodiethoxytitanium [$TiCl_2(OEt)_2$] (21.0 g, 100 mmol, 2.3 equiv) in THF (40 mL, 2.5 M) (Note 8) is added slowly to ensure the internal temperature does not exceed –72 °C. When the addition is complete, the reaction mixture is stirred for 10 min at which point a solution of 2-trimethylsiloxybutadiene (12.0 g, 85 mmol, 2.0 equiv) (Note 9) in THF (50 mL, 1.7 M) is added dropwise, resulting in a slight increase in temperature (to –72 °C). The reaction mixture is allowed to warm to room temperature. After 12 h, the mixture is poured into 350 mL of ice-cooled, saturated sodium bicarbonate solution and is filtered through a pad of Celite (60 g) using a fritted filter funnel (10 cm diameter, Fine). The celite is washed with diethyl ether (2 x 100 mL). The filtrate is transferred to a 1 L separatory funnel. The layers are separated, and the aqueous phase is extracted with diethyl ether (2 x 100 mL). The combined organic layers are washed with brine (1 x 100 mL), dried over anhydrous sodium sulfate (Na_2SO_4), filtered through a cotton plug and collected in a 1-L round-bottomed flask. The Na_2SO_4 is rinsed with diethyl ether (2 x 20 mL). The solvent is removed under reduced pressure by rotary evaporation (50 mmHg, 30 °C). The dark crude oil is purified by column chromatography. A fritted chromatography column (3.0 cm diameter) is dry-packed with silica gel (20 cm height, 50 g) (Note 10) and then wetted with hexanes under air pressure. The crude oil is loaded directly onto the column and eluted with 1 L of 10:1 hexane/ethyl acetate. The eluent is collected in 25 mL fractions and monitored by TLC (Product R_f = 0.11, hexane/ethyl acetate: 10/1) (Notes 6 and 11). The fractions are combined in a 1-L round-bottomed flask and concentrated under reduced pressure by rotary evaporation (50 mmHg, 30 °C) to give *tert*-butyl [1-(*tert*-butoxycarbonyl)-3-oxo-4-pentenyl] carbamate as a yellowish oil (2.82 g, 22%) (Note 12).

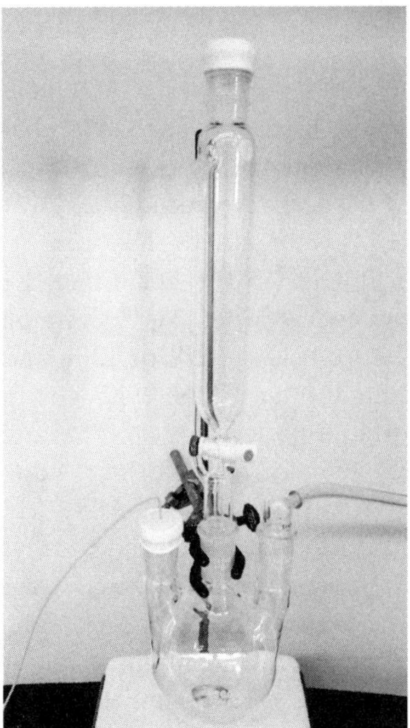

Figure 2. Reaction Set-up for Step B

C. 1-Carboxy-5-chloro-3-oxopentan-1-aminium chloride. In a 500 mL, one-necked round-bottomed flask, *tert*-butyl [1-(*tert*-butoxycarbonyl)-3-oxo-4-pentenyl] carbamate (8.72 g, 29.1 mmol, 1.0 equiv) is dissolved in an ice-cooled, saturated solution of hydrogen chloride in diethyl ether (280 mL) (Note 14). The solution is kept, without stirring, at room temperature overnight. The solvent is removed with a pipette and the solid sequentially washed with diethyl ether (4 x 100 mL) (Note 13). The resulting solid is dried under reduced pressure (1.0 mmHg, 25 °C) to afford 1-carboxy-5-chloro-3-oxopentan-1-aminium chloride as a pale cream solid (5.56 g, 88%) (Notes 15 and 16).

Notes

1. Unless otherwise noted, all glassware and stir bars were flame dried under vacuum, cooled to room temperature under vacuum, and placed under nitrogen atmosphere.
2. *N*-Boc-glycine *tert*-butyl ester was prepared as follows: A 1 L one-necked round-bottomed flask was equipped with an egg shaped stir bar (3.0 cm x 1.5 cm) and capped with a rubber septum. Boc-Gly-OH (99%, purchased from Oakwood Chemical and used as received) (30.2 g, 171 mmol, 1.0 equiv) and DCM (600 mL, 0.29 M) were added. The solution was cooled in ice-water bath, and then *t*BuOH (15.3 g, 202 mmol, 1.2 equiv), DMAP (1.5 g 12 mmol, 0.07 equiv) and DCC (41.3 g, 200 mmol, 1.2 equiv) were added. The reaction mixture was stirred for 12 h, and the precipitate was filtered using a fritted filter funnel (4.5 cm in diameter, Fine). The filtrate was concentrated under reduced pressure by rotary evaporation (50 mmHg, 30 °C water bath), and the residue was purified by flash column chromatography on silica gel (8.0 cm diameter, 20 cm height of silica gel (400 g)) (Note 10). Hexanes/ethyl acetate: 10/1 (100 mL) and hexanes/ethyl acetate: 5/1 (2000 mL) were used as eluents, which were collected in 50 mL fractions (Note 11). The desired product (R_f = 0.40, hexane/ethyl acetate: 5/1) containing minor impurities was obtained (33 g). The material was recrystallized twice from hexanes. Hexanes (100 mL) was added and heated to 60 °C so as to dissolve all the solids. The solution was left to cool slowly to room temperature and subsequently placed in a –23 °C freezer for

approximately 12 h. The crystals were then collected on a Kiriyama filter. The recrystallization was repeated a second time to provide *N*-Boc-glycine *tert*-butyl ester (16.6 g, 71.8 mmol, 41.9%yield)

3. *N*-Bromosuccinimide was purchased from Sigma Aldrich, recrystallized from water, and dried in a vacuum desiccator (1.0 mmHg, 25 °C).
4. Chlorobenzene was purchased from Acros Organics and used without further purification. The submitters and previous checkers used CCl_4. Difficulties in obtaining CCl_4 led the current checkers to make this solvent change.
5. The water bath was coated with aluminum foil.. Foil was wrapped around the entire reaction assembly in order to increase the efficiency of the irradiation. The checkers used a 300W tungsten lamp while the submitters used a 150W tungsten lamp.
6. TLC (ethyl acetate/hexane 1:3; visualized with vanillin/concd. H_2SO_4/heat) (Note 10) reveals complete consumption of the starting material (starting material R_f = 0.47) and only small amounts of impurities are observed by ^{1}H NMR analysis. The crude product is stable for several weeks at −20 °C under argon.
7. Non-stabilized THF was purchased from Fisher Scientific and passed through two packed columns of neutral alumina in a solvent purification system manufactured by SG Water U.S.A., LLC.
8. The equivalents of reagents used in this reaction are based on the amount of starting material used in step A of the reaction sequence. Tetraethyl orthotitanate, (11.5 g, 50 mmol) was purchased from Sigma-Aldrich and distilled at 110–115 °C at 0.1 mmHg. The distilled material is dissolved in dry THF (40 mL) (Note 7). Titanium chloride ($TiCl_4$) (9.5 g, 50 mmol) was purchased from Sigma-Aldrich and distilled at 136 °C under atmospheric pressure. The titanium tetrachloride was added dropwise to the solution of tetraethyl orthotitanate while cooling in an acetone/dry ice bath.
9. Trimethylsiloxybutadiene was purchased from Oakwood Chemical, and employed without further purification.
10. Silica gel SilicaFlash® F60 (40-63 μ / 230-500 mesh) was purchased from Silicycle. Glass-backed extra hard layer TLC plates, 60 Å (250 μm thickness) were also purchased from Silicycle containing F-254 indicator.

11. Ethyl acetate and hexanes (all ACS Grade) used in chromatography were purchased from Fisher Scientific and used as received.
12. The submitters obtained a 33–36% yield over this two-step sequence, and the checkers note that the change in solvent from CCl_4 to chlorobenzene and a change in the lamp from 150 W to 300 W could account for this difference in yield. The physical properties are as follows: ^{1}H NMR ($CDCl_3$) δ: 1.42 (s, 18 H), 3.07 (dd, 1 H, *J* = 4.8, 17.6 Hz), 3.28 (dd, 1 H, *J* = 4.8, 17.6 Hz), 4.44 (m, 1 H), 5.47 (d, 1 H, *J* = 8.0 Hz, N-H), 5.90 (dd, 1 H, *J* = 1.2, 10.4 Hz), 6.24 (dd, 1 H, *J* = 1.2, 17.6 Hz), 6.33 (dd, 1 H, *J* = 10.4, 18.0 Hz); ^{13}C NMR ($CDCl_3$) δ: 27.8, 28.3, 41.5, 50.1, 82.1, 129.3, 136.2, 155.6, 170.2, 198.3; IR (ATR) cm^{-1}: 3370, 2980, 2930, 1710, 1620, 1490. HRMS (+ESI) calcd for $C_{15}H_{25}NO_5$ (M+Na) 322.1630. Found 322.1631.
13. This compound is very hygroscopic.
14. Hydrogen chloride in diethyl ether was prepared by bubbling hydrogen chloride into diethyl ether for 2 h. Hydrogen chloride was generated from treatment of NaCl with H_2SO_4.
15. The physical properties are as follows: ^{1}H NMR (CD_3OD) δ: 3.05 (dt, 2 H, *J* = 1.2, 6.4 Hz), 3.20 (dd, 1 H, *J* = 6.8, 18.8 Hz), 3.29 (dd, 1 H, *J* = 4.0, 19.6 Hz), 3.79 (t, 2 H, *J* = 6.4 Hz), 4.30 (dd, 1 H, *J* = 4.0, 6.8 Hz); ^{1}H NMR (400 MHz, D_2O) δ: 3.11 (t, 2H, *J* = 6.0 Hz), 3.32–3.34 (m, 2H), 3.81 (t, 2H, *J* = 6.0 Hz), 4.27 (dd, 1H, *J* = 4.4, 6.0 Hz); ^{13}C NMR (100 MHz, CD_3OD) δ: 28.0, 38.7, 42.9, 45.7, 48.4, 170.9, 205.6; ^{13}C NMR (100 MHz, D_2O) δ: 37.9, 41.7, 44.1, 48.5, 171.4, 207.8; IR (ATR) cm^{-1}: 3300–2300 broad, 1720, 1580, 1400. HRMS (+ESI) calcd for $C_6H_{11}Cl_2NO_3$ (M-Cl) 180.0422. Found 180.0424. mp = 116 – 118 °C (decomp).
16. In an effort to confirm the structure, the checkers employed the illustrated procedure to convert 1-carboxy-5-chloro-3-oxopentan-1-aminium chloride to 2-*tert*-butyloxycarbonylamino-6-chloro-4-hexanolide, a compound which had been stereoselectively prepared previously by Hesse.[2]

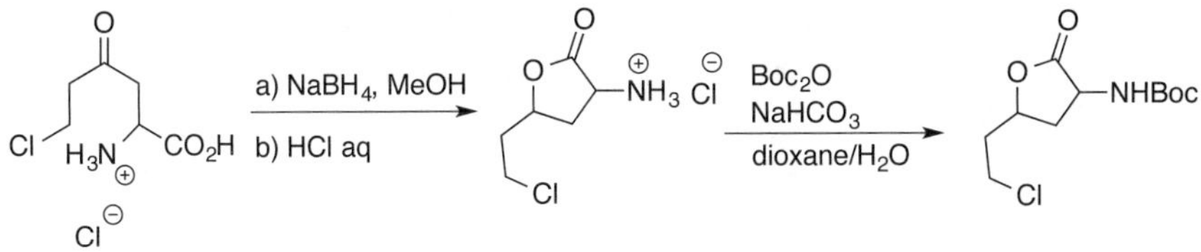

Scheme 1. Conversion to diastereomeric lactones

In a 100 mL one-necked round-bottomed flask equipped with a 2.0 cm x 0.5 cm rod shaped stir bar, 1-carboxy-5-chloro-3-oxopentan-1-aminium chloride (500 mg, 2.3 mmol, 1.0 equiv) is dissolved in MeOH (10 mL, 0.23 M) and cooled in an ice-water bath. $NaBH_4$ (130 mg, 3.5 mmol, 1.5 equiv) is added and the reaction stirred for 2 h. The solvent is removed under reduced pressure by rotary evaporation (50 mmHg, 30 °C water bath). The crude alcohol is then dissolved in 1 M aq. HCl (20 mL) and stirred for 2 h after which the solvent is removed under reduced pressure (1.0 mmHg, 25 °C water bath). The lactonized ammonium salt is subsequently dissolved in dioxane/H_2O (1/1) (50 mL), and treated with $NaHCO_3$ (1.9 g, 23 mmol, 10 equiv) followed by Boc_2O (610 mg, 2.8 mmol, 1.2 equiv). The reaction mixture is stirred for 12 h, after which time the reaction mixture is extracted with ethyl acetate (3 x 30 mL). The combined organic layers are washed with brine (1 x 30 mL), dried over Na_2SO_4, filtered, and concentrated under reduced pressure by rotary evaporation (50 mmHg, 30 °C water bath). The residue is purified by flash column chromatography on silica gel. Hexanes/ethyl acetate:3/1 (500 mL) is used as the eluent and collected in 10 mL fractions. The fractions were monitored by TLC (Rf = 0.18, hexane/ethyl acetate: 3/1). The first fractions were concentrated to give 12.3 mg of the *cis*-isomer (2.0% over 3 steps). The last fractions were concentrated under reduced pressure to give 10.3 mg of the *trans*-isomer (1.7% over 3 steps), and the middle fractions were concentrated under reduced pressure to provide 182 mg of a 47:53 *cis:trans* (30% over 3 steps). In all cases was the 2-*tert*-butyloxycarbonylamino-6-chloro-4-hexanolide obtained as a white solid. *cis*: ^{1}H NMR ($CDCl_3$) δ: 1.45 (s, 9 H), 1.88 (q, 1 H, J = 12.0 Hz), 2.05–2.14 (m, 1 H), 2.17–2.26 (m, 1 H), 2.84–2.93 (m, 1 H), 3.66–3.72 (m, 2 H), 4.35–4.47 (m, 1 H), 4.62–4.69 (m, 1 H), 5.09 (bs, 1 H); ^{13}C NMR (100 MHz, $CDCl_3$) δ: 28.2, 36.4, 38.0, 40.3,

51.4, 74.6, 80.7, 155.3, 174.3; IR (ATR) cm^{-1}: 3420, 2980, 2930, 1770, 1700, 1510. *trans*: ^{1}H NMR ($CDCl_3$) δ: 1.45 (s, 9 H), 1.97–2.07 (m, 1 H), 2.12–2.20 (m, 1 H), 2.34–2.53 (m, 2 H), 3.66 (dd, 2H, *J* = 5.2, 7.6 Hz), 4.32–4.45 (m, 1 H), 4.84–4.93 (m, 1 H), 5.09 (bs, 1 H); ^{13}C NMR (100 MHz, $CDCl_3$) δ: 28.2, 34.3, 38.1, 40.3, 49.2, 74.3, 80.9, 155.4, 174.7; IR (ATR) cm^{-1}: 3380, 2980, 2930, 1780, 1670, 1520, HRMS (+ESI) calcd for $C_{11}H_{18}ClNO_4$ (M+Na) 286.0822. Found 286.0819. Melting Point: 121.4–123.1 °C.

To unambiguously confirm the identity of the lactone product, the checkers prepared single crystals suitable for X-Ray crystallographic analysis using the *trans*-isomer (Figure 3). Single crystals are grown from a DCM solution of the *trans*-isomer by vapor diffusion into hexanes. Accordingly, 10 mg of *trans*-2-*tert*-butyloxycarbonylamino-6-chloro-4-hexanolide are placed in a 1.5 dram vial and dissolved in a minimal amount of DCM (ca. 0.5 mL). This vial is placed into a larger 50 mL jar containing a volume of hexanes such that the meniscus is slightly above that of the DCM in the inner vial. The larger vial is sealed and let stand at room temperature to produce colorless cube crystals (ca. 6 h). Single crystal X-ray diffraction data were

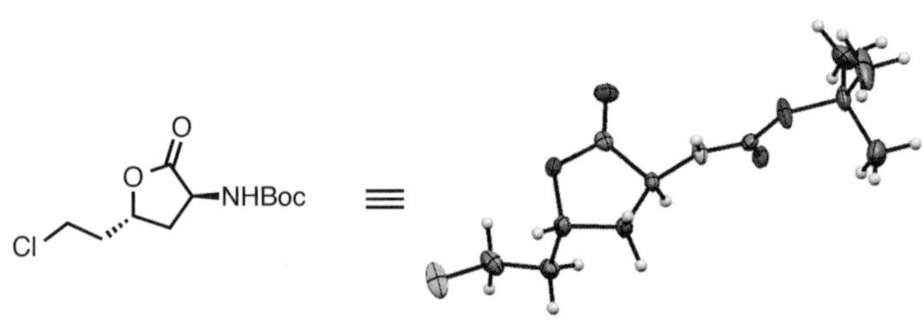

Figure 1. The molecular structure of *trans*-2-*tert*-butyloxy-carbonylamino-6-chloro-4-hexanolide from the crystal structure.

collected on a Bruker Apex II-CCD detector using Mo-K_α radiation (λ= 0.71073 Å). Crystals are removed from the mother liquor and coated in an inert perfluoro-polyether oil, mounted on micromounts then placed in a cold stream of N_2. Structures were solved and refined using SHELXTL (Table 1). These data have been deposited with the Cambridge Crystallographic Data Centre (CCDC) as supplementary publication no. CCDC 1435009 and copies can be obtained in the supporting information or free of charge from the CCDC via www.ccdc.cam.ac.uk/data_request/cif.

Working with Hazardous Chemicals

The procedures in *Organic Syntheses* are intended for use only by persons with proper training in experimental organic chemistry. All hazardous materials should be handled using the standard procedures for work with chemicals described in references such as "Prudent Practices in the Laboratory" (The National Academies Press, Washington, D.C., 2011; the full text can be accessed free of charge at http://www.nap.edu/catalog.php?record_id=12654). All chemical waste should be disposed of in accordance with local regulations. For general guidelines for the management of chemical waste, see Chapter 8 of Prudent Practices.

In some articles in *Organic Syntheses*, chemical-specific hazards are highlighted in red "Caution Notes" within a procedure. It is important to recognize that the absence of a caution note does not imply that no significant hazards are associated with the chemicals involved in that procedure. Prior to performing a reaction, a thorough risk assessment should be carried out that includes a review of the potential hazards associated with each chemical and experimental operation on the scale that is planned for the procedure. Guidelines for carrying out a risk assessment and for analyzing the hazards associated with chemicals can be found in Chapter 4 of Prudent Practices.

The procedures described in *Organic Syntheses* are provided as published and are conducted at one's own risk. *Organic Syntheses, Inc.*, its Editors, and its Board of Directors do not warrant or guarantee the safety of individuals using these procedures and hereby disclaim any liability for any injuries or damages claimed to have resulted from or related in any way to the procedures herein.

Discussion

In February of 2015 Professor RFW Jackson from the University of Sheffield brought to the attention of the *Organic Syntheses* editorial board his suspicions that a structural misassignment had been made for the final product reported in a 1993 *Organic Syntheses* procedure that had been submitted by Jean-Pierre Obrecht and checked by David Coffen.[2] In response to considerable information provided by Professor Jackson and his suggestion that Step C in this procedure converts **1** to **3** rather than the reported cyclized product **2**, the editorial board determined that rechecking of the procedure was in order. The procedure illustrated above reflects this rechecking effort. In the course of these recent investigations every effort was made to reproduce the original report, but due to the difficulty in obtaining carbon tetrachloride a solvent change was necessitated in step A. Although this minor change impacts yield, the physical properties and spectral data obtained for **1** were consistent with the 1993 report. Subjecting this material to conditions identical to those reported in step C of the 1993 *Organic Syntheses* procedure confirmed Professor Jackson's suggestion that this reaction does indeed furnish **3** rather than **2** (Scheme 1).[3]

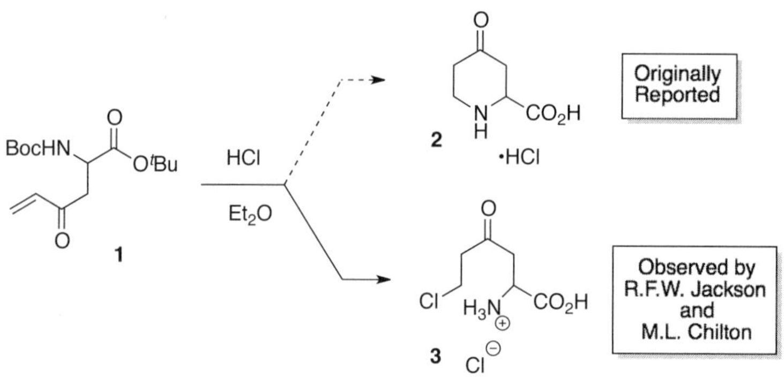

Scheme 2: Discrepancy Discovered by R.F.W. Jackson and M.L. Chilton

Although it is noteworthy that spectral data obtained for the product derived from **1** differs from that reported for **2** in papers published by Burger in 1995 and 1996,[4] positive proof for the structural reassignment was sought via elemental composition data. Mass spectrometry was indicative of structure **3** but efforts to obtain an acceptable elemental analysis furnished results outside the acceptable ± 0.4% window, although the data obtained from multiple EA attempts were all more consistent with structure **3** than for **2**.[5] Given the inability to obtain satisfactory EA data the checkers turned to chemical correlation as a means of providing compelling evidence for structure **3** and thus attempted to convert **3** to lactones (±)-**6** and (±)-**7**, the former of which had been previously prepared in enantioenriched form

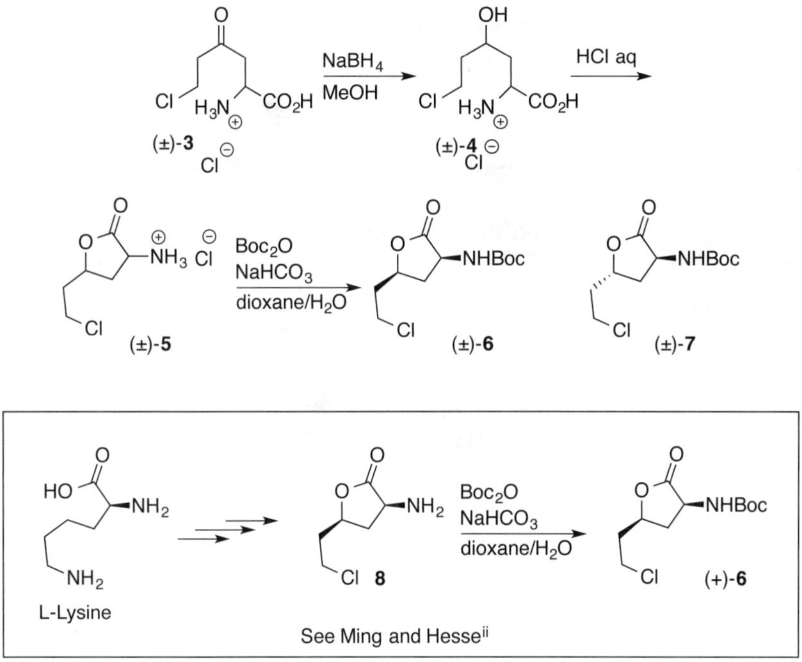

Scheme 3: Chemical correlation to reported synthesis of (+)-6 by Hesse.

via an unrelated route by Hesse (Scheme 2).[2] Although, spectral data obtained by the checkers for (±)-**6** was similar to that reported for (+)-**6** by Hesse, there were clear discrepancies in both the ^{1}H and ^{13}C NMR data. Thus, the checkers turned to X-ray crystallography in order to confirm the structure. To this end *trans*-lactone (±)-**7** was found to furnish crystals suitable for X-ray analysis and the derived structure provided convincing proof that the compound **3** is precursor to **7** and thus does indeed arise upon treatment of **1** with HCl in Et_2O.

References

1. Original article published in *Org. Synth.* **1993**, *71*, 200–206.
2. Meng, Q.; Hesse, M. *Tetrahedron* **1991**, *47*, 6251–6264.
3. Chilton, Martyn L. (2015) *Synthetic Approaches to 2-Substituted 4-Hydroxypiperidines.* Ph.D. Thesis, University of Sheffield.
4. (a) Golubev, A.; Sewald, N.; Burger, K. *Tetrahedron Lett.* **1995**, *36*, 2037–2040. (b) Golubev, A.; Sewald, N.; Burger, K. *Tetrahedron* **1996**, *52*, 14757–14776.
5. R.F.W. Jackson reported to the checkers that they had experienced similar difficulties in obtaining EA data and also observed greater consistency of their data with structure **3**.

Appendix
Chemical Abstracts Nomenclature (Registry Number)

Glycine (513-29-1)
N-Bromosuccinimide (128-08-5)
Titanium chloride (7550-45-0)
Triethylamine (121-44-8)
2-Trimethylsiloxybutadiene (38053-91-7)
Di-*tert*-butyl dicarbonate (24424-99-5)
4-Ketopipecolic acid hydrochloride,
Pipecolic acid, 4-oxo-, hydrochloride (99979-55-2)
2-Bromo-*N*-Boc-glycine *tert*-butyl ester,
N-Boc-2-Bromoglycine *tert*-butyl ester
N-Boc-glycine *tert*-butyl ester,
Boc-Glycine *tert*-butyl ester (111652-20-1)
tert-Butyl [1-(*tert*-butoxycarbonyl)-3-oxo-4-pentenyl]carbamate (117833-62-2)
Dichlorodiethoxytitanium (3582-00-1)
Glycine *tert*-butyl ester hydrochloride (27532-96-3)
Tetraethyl orthotitanate (3087-36-3)
N-Boc-glycine (4530-20-5)

Yasuhiro Shimamoto obtained his undergraduate degree from Kyoto University in 2009 in Engineering. He subsequently obtained his Ph.D. from Professor Masahiro Murakami in Engineering in 2014. He is currently a Post-Doctoral scholar in the laboratory of Professor John L. Wood at Baylor University.

Heemal Dhanjee obtained his B.A. in Mathematics and Molecular and Cell Biology at the University of California, Berkeley in 2007. He subsequently moved to California State University, Northridge and performed research in organic synthesis under the supervision of Professor Thomas Minehan. He is currently pursuing graduate research at Baylor University under the guidance of Professor John L. Wood.

Organic Syntheses

Nickel-Catalyzed Synthesis of Ketones from Alkyl Halides and Acid Chlorides: Preparation of Ethyl 4-Oxododecanoate

Alexander C. Wotal, Donald C. Batesky, and Daniel J. Weix*[1]

University of Rochester, Department of Chemistry, RC Box 270216, Rochester, NY, USA 14627-0216

Checked by Yongda Zhang and Chris H. Senanayake

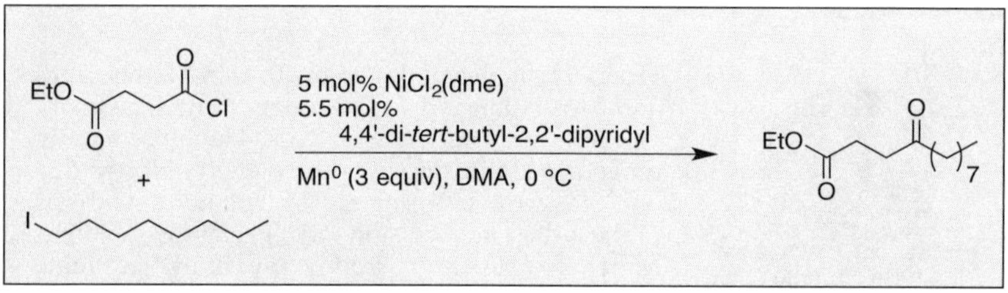

Procedure

A four-necked 500-mL Jacket reactor (Note 1) with 24/40 ground glass joints is equipped with a septum bearing a thermocouple connected to a J kem temperature controller, an addition funnel with a septum, and a nitrogen inlet connected to a Firestone valve. The center neck is equipped for mechanical stirring (**Figure 1**) (Note 2). The reactor is connected to a chiller (Note 3). The reactor is charged with nickel(II) chloride ethylene glycol dimethyl ether complex (1.11 g, 4.90 mmol, 0.050 equiv) (Note 4), 4,4′-di-*tert*-butyl-2,2′-dipyridyl (1.48 g, 5.39 mmol, 0.055 equiv) (Note 5) and manganese powder (16.68 g, 303.6 mmol, 3.1 equiv) (Note 6). The system is evacuated and refilled with nitrogen three times. Anhydrous *N,N*-dimethylacetamide (Note 7) (200 mL) is added to the reactor through the additional funnel by syringe. The reaction mixture is stirred for 1 h under nitrogen flow. During this time the green reaction mixture turns to a dark green/black suspension. The stirred mixture is then cooled to an internal

DOI: 10.15227/orgsyn.093.0050

temperature at 0–2 °C over 0.5 h. 1-Iodooctane (24.0 g, 98.0 mmol, 1.0 equiv) and ethyl succinyl chloride (25.45 g, 146.9 mmol, 1.5 equiv) (Note 8) are successively charged into the reactor via the addition funnel by syringe. The color of the reaction mixture changed to brown in 5 min. The reaction mixture is vigorously stirred at 0 to 2 °C for 14 h and then is allowed to warm to room temperature over 0.5 h and then stirred for another 3–6 h at room temperature (Note 9).

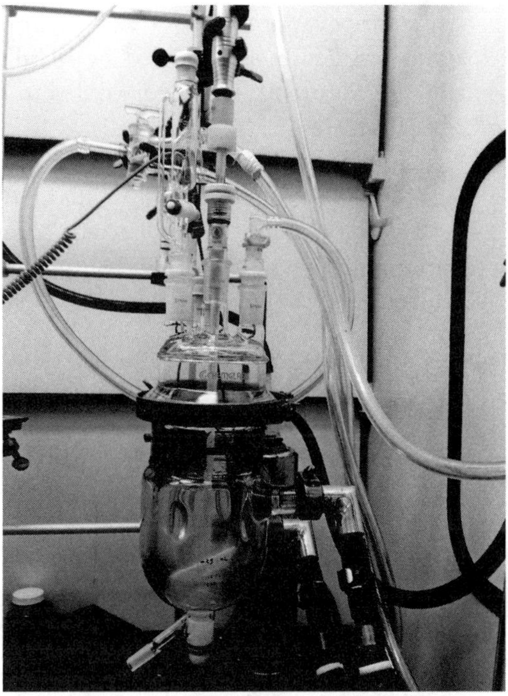

Figure 1. Apparatus Assembly

The stirred reaction mixture is diluted by adding 1:1 hexanes/methyl *tert*-butyl ether (200 mL), resulting in the formation of a dark green/black precipitate. The mixture is drained from the reactor (Notes 10 and 11). The solids are removed by filtration through a celite bed (Note 12). The black solids are scraped off the top of the celite and transferred to a 500-mL Erlenmeyer flask and stirred with 250 mL of deionized water for 1–2 min. The insoluble manganese salts are removed by vacuum filtration through the same celite bed and rinsed with deionized water (20 mL). The water filtrate is combined with the organic filtrate from the reaction mixture. The resulting mixture is transferred to a 1-L separatory funnel. The aqueous layer is extracted with 1:1 hexanes/methyl *tert*-butyl ether (200 mL). The combined organic layers are washed with deionized water (200 mL) followed by a final wash with saturated sodium chloride solution (250 mL). The product solution mixture is transferred to a 1-L Erlenmeyer flask and stirred over 20 g of anhydrous magnesium sulfate for 15–20 min. The magnesium sulfate is removed by vacuum filtration of the dried solution using a 7 cm Büchner funnel. The solvents are distilled from the crude product mixture using a rotary evaporator (30 °C, 25–30 mmHg).

A very small forerun (ca. 3 mL) is removed by distillation at 150 °C under vacuum (5–6 mmHg) (**Figure 2**). The pot is cooled to 23 °C before the vacuum is released, after which the forerun is discarded (Note 11). The brownish crude product is then loaded onto a column pre-prepared with 50 g silica gel in hexanes (Note 13). The product is eluted with 10% MTBE in hexanes (500 mL). All the eluted solution is collected. After the solvent is removed by rotary evaporation (50 °C, 50 mmHg), ethyl 4-oxododecanoate (19.90 g, 84%) is collected as a pale yellow oil (Notes 14, 15, and 16).

Figure 2. Distillation Assembly

Notes

1. The reactor was washed with ~100 mL of concentrated hydrochloric acid followed by 200-300 mL deionized water. Methanol (200 mL) was added to the reactor and then heated to reflux for 2-3 h. After the reactor was cooled to room temperature, MeOH was drained from the reactor and the reactor dried with the jacket temperature of 100 °C under vacuum (150–200 mmHg) for 1–2 h. The reactor was cooled to 23 °C under vacuum and then refilled with nitrogen before use.
2. The stirrer paddle must not be in contact with the bottom or sides of the flask. The stirrer paddle was generally held ~10 mm above the bottom of the flask to prevent mechanical activation of the reducing agent, which could cause lower product yields (see Figure 1). The submitters use a mechanical stirring (glass bearing, glass stirrer rod, 6 cm PTFE paddle, air driven motor). Minimal amounts of Dow Corning high vacuum grease were used to seal all ground glass joints, and heavy mineral oil (mineral oil white, heavy, Mallinckrodt) was used to lubricate the glass stir rod inside the 24/40 standard taper adapter.
3. The model of the chiller is Julabo FS18. The submitters used ice-water in a 10-L container to cool the reaction mixture.
4. Nickel(II) chloride ethylene glycol dimethyl ether complex (dme or glyme) (min. 97%) was obtained from Strem and used as received.
5. 4,4′-*tert*-Butyl-2,2′-dipyridyl (98%) was obtained from Aldrich and used as received.
6. Manganese powder (-325 mesh, ≥99% trace metals basis) was obtained from Aldrich and used as received. Reactions run with a smaller amount of manganese (2 equiv) still proceeded to completion, but took longer to reach completion. Due to the low cost of Mn metal, an excess was used for convenience.
7. Anhydrous *N,N*-dimethylacetamide (99.8%) was obtained from Aldrich and used as received. The submitters used *N,N*-dimethylacetamide (99%) from Alfa, degassed with argon, and dried using cartridge no. VAC 104458 in a VAC Solvent Purifier (Vacuum Atmospheres Co., Hawthorne, CA 90250, USA).
8. Commercial 1-iodooctane (98%, Aldrich) was obtained from Aldrich and used as received. The submitters filtered the commercial 1-iodooctane through a pad of basic alumina to remove any acidity and copper stabilizer. Ethyl succinyl chloride (95%) was obtained from

Aldrich and was used as received. The checker observed mild exotherm during the addition. The internal temperature increased from 1.7 °C to 3.3 °C with the jacket temperature at 0 °C.

9. In two small pilot runs, a sample of the reaction mixture was checked for the presence of 1-iodooctane by gas chromatography, but in these pilot runs, only a trace was seen in the forerun and no monitoring was done on the large-scale runs. The checker monitored the reaction by GC-MS, using an Agilent HP-5MS colum (30 m x 0.25 mm x 0.25 μm). Helium flow rate 1 mL/min; Initial temp: 50 °C (2 min), ramp 25°C/min to 200 °C (hold for 2 min). Retention times are 1-iodooctane (5.47 min) and ethyl 4-oxododecanoate (8.05 min). Only trace amount of 1-iodooctane was observed once the reaction mixture was warmed to room temperature.
10. These solids include remaining manganese, water soluble salts and some product. The filtration is carried out using a 7-cm Büchner funnel with a Celite® bed on a Whatman GF/A glass microfiber filter set on a 1-L filtration flask. The checkers observed slow filtration without the Celite® bed. The Celite® (C212-5W Celite®, 545 filter aid) was 1 cm thick and obtained from Fisher Scientific and used as received.
11. If the reactor is contaminated with residual metal, conc. HCl followed by water rinse can be used to clean the reactor. These rinses are for the purpose of cleaning the glass and are not added to the filtrate.
12. Instead of purification by chromatography, the submitters isolated the product by distillation at reduced pressure. A 50-mL, 2-necked round-bottomed flask is equipped with a thermometer on the side 14/20 joint and a 20-cm vacuum jacketed Vigreux column with a built-in water cooled side arm, a vacuum adapter, and 25-mL receiving flask (**Figure 3**). The apparatus is placed under aspirator vacuum (~18-20 mmHg). A very small forerun of about 2–3 mL is distilled, boiling up to 90 °C. After the last forerun is removed, the pot is allowed to cool to 40 °C before the vacuum released. The forerun is discarded and the 20-cm Vigreux column is exchanged for a shorter 6-cm Vigreux column/head with a vacuum side arm adapter attached to a high-vacuum line and a clean 25-mL receiving flask, cooled in ice water. The hold up using the longer Vigreux column is avoided by switching to the shorter column, hence avoiding overheating and possible degradation of the product (**Figure 4**) The product is distilled at a pot temperature ranging from 120–140 °C, with a head temperature of 101–102 °C (1 mmHg). For this operation, a high-vacuum pump instead of a water aspirator was used.

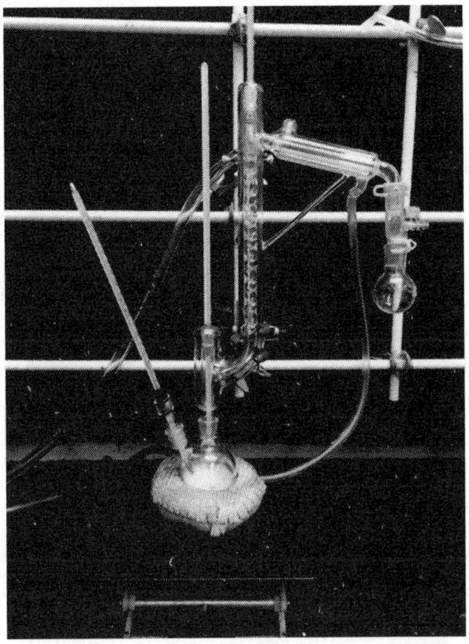

Figure 3. Distillation Assembly

Figure 4. Purification by Distillation

13. Silica gel (technical grade, pores size 60, 230-400 mesh particle size) was received from Aldrich and was used as received. The length of the silica gel is about 10 cm in the column, while the column for flash chromatography has 50 mm ID and 45.7 cm effective length with a sealed coarse porosity fritted disc.
14. The product exhibited the following physical properties: yellow liquid. ^{1}H NMR (400 MHz, $CDCl_3$) δ: 0.88 (t, *J* = 6.5 Hz, 3H), 1.20–1.23 (m, 13H), 1.59 (t, *J* = 7.0 Hz, 2H), 2.44 (t, *J* = 7.5 Hz, 2H), 2.57 (t, *J* = 6.7 Hz, 2H), 2.71 (t, *J* = 6.5 Hz, 2H), 4.14 (q, *J* = 7.2 Hz, 2H). ^{13}C NMR (100 MHz, $CDCl_3$) δ: 14.0, 14.1, 22.6, 23.8, 28.0, 29.1, 29.2, 29.3, 31.8, 37.0, 42.8, 60.5, 172.8, 209.1. FTIR (neat, cm^{-1}): 1734 (C=O of ester), 1715 (C=O of ketone), 1181 (C-O). HRMS: Calcd for $C_{14}H_{27}O_3$ $(M+H^+)$: 243.1960; found: 243.1968.
15. Purity was assessed at 98 wt% by quantitative ^{1}H NMR using dimethyl fumarate as an internal standard.
16. The yield of a second duplicate run was 20.0 g (84%).

Working with Hazardous Chemicals

The procedures in *Organic Syntheses* are intended for use only by persons with proper training in experimental organic chemistry. All hazardous materials should be handled using the standard procedures for work with chemicals described in references such as "Prudent Practices in the Laboratory" (The National Academies Press, Washington, D.C., 2011; the full text can be accessed free of charge at http://www.nap.edu/catalog.php?record_id=12654). All chemical waste should be disposed of in accordance with local regulations. For general guidelines for the management of chemical waste, see Chapter 8 of Prudent Practices.

In some articles in *Organic Syntheses*, chemical-specific hazards are highlighted in red "Caution Notes" within a procedure. It is important to recognize that the absence of a caution note does not imply that no significant hazards are associated with the chemicals involved in that procedure. Prior to performing a reaction, a thorough risk assessment should be carried out that includes a review of the potential hazards associated with each chemical and experimental operation on the scale that is planned for the procedure. Guidelines for carrying out a risk assessment and for analyzing the hazards associated with chemicals can be found in Chapter 4 of Prudent Practices.

The procedures described in *Organic Syntheses* are provided as published and are conducted at one's own risk. *Organic Syntheses, Inc.*, its Editors, and its Board of Directors do not warrant or guarantee the safety of individuals using these procedures and hereby disclaim any liability for any injuries or damages claimed to have resulted from or related in any way to the procedures herein.

Discussion

Because of the prevalence of alkyl alkyl ketones in natural products, their use in biological probes,[2] and their versatility as intermediates in organic synthesis,[3] a wide variety of methods have been developed to access them. The most commonly used methods are the addition of a reactive organometallic reagent to an aldehyde followed by oxidation and the addition of pre-formed organometallic nucleophiles to carboxylic acid derivatives.[2] While both of these protocols are effective, the organometallic reagents can be limiting with complex, functionalized molecules. More recently, cross-coupling methods based upon the less reactive alkyltin,[4] alkylzinc,[5,6] dialkylzinc,[7] or alkylboron[8] reagents have been developed.

Table 1. Scope of Acid Chloride and Alkyl Halide

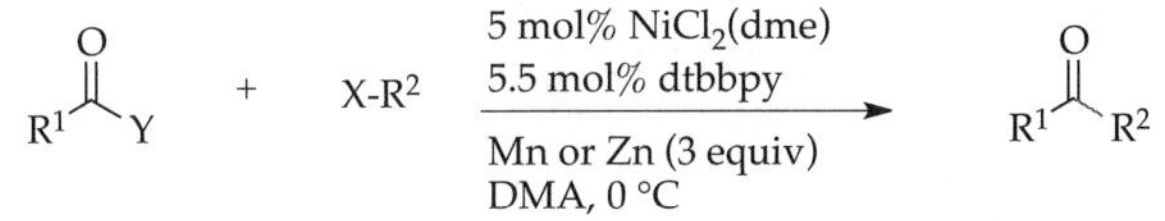

Entry	R^1COY	$X-R^2$	Product	Yield (%)[b]
1	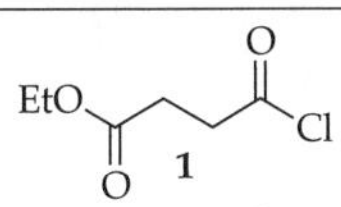 **1**	$I-C_8H_{17}$ **2**	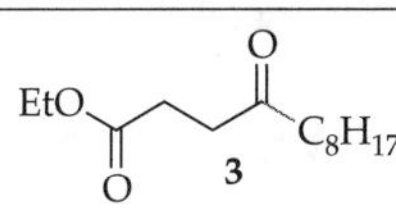 **3**	86
2[c]	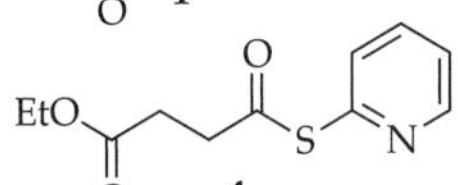 **4**	**2**	**3**	74
3	**1**	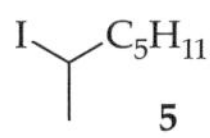 **5**	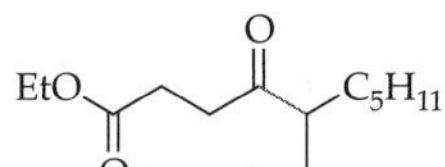 **6**	91
4	**1**	 **7**	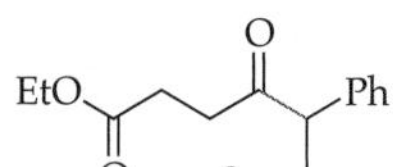 **8**	49[d]
5	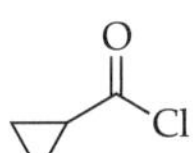 **9**	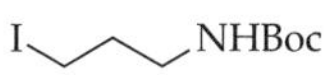 **10**	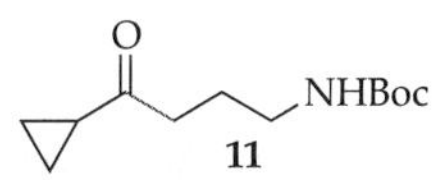 **11**	78
6	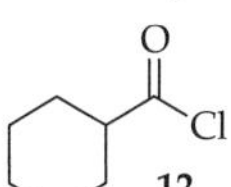 **12**	**5**	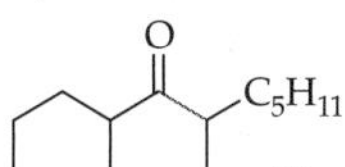 **13**	72
7	**12**	**7**	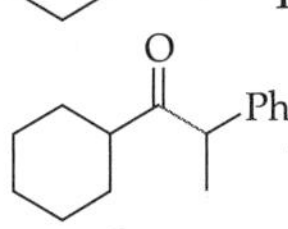 **14**	33
8	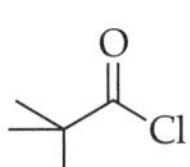 **15**	**10**	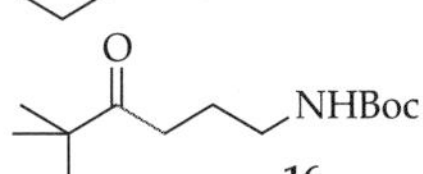 **16**	91
9	**15**	**5**	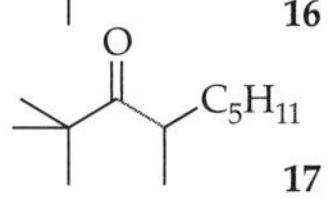 **17**	67

[a] General conditions: 1.5 mmol of R^1COY, 1.0 mmol of R^2-X, 0.05 mmol $NiCl_2$(dme), 0.055 mmol 4,4'-di-*tert*-butyl-2,2'-bipyridine (dtbbpy), 3 mmol Mn, in 2 mL of DMA at 0 °C for 13-30 h. [b] Yield of isolated, purified product. [c] 1 Equiv of R^1COY was used at rt with Zn in place of Mn. [d] Product contained a small amount of diethyl succinate, yield of **8** determined by NMR.

We recently reported a different strategy, the nickel-catalyzed coupling of carboxylic acid chlorides or thioesters with alkyl iodides under reducing conditions (Table 1).[9] This work built upon previous studies of Mukaiyama[10] and electrochemical work from Nédélec, Périchon, Amatore, and Jutand.[11] Relatively sterically hindered ketones could be made by this protocol, a challenge for other coupling methods (Table 1), and the reaction conditions tolerated a variety of functional groups, including protected amines, amide N-H bonds, a vinyl boronic ester, protected alcohols, and protected alkynes (**Tables 1 and 2**).

Table 2. Further Examples of Functionalized Substrates.

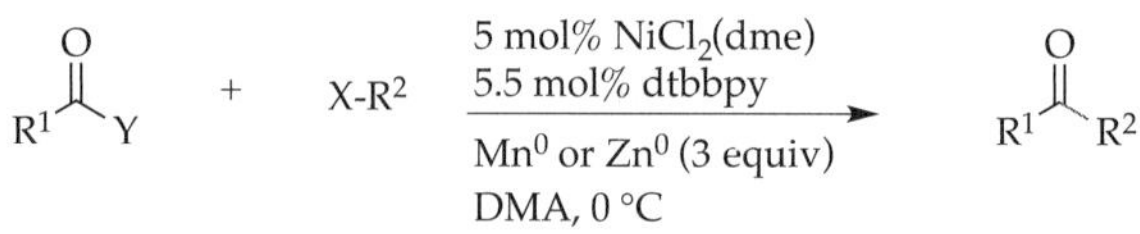

Entry	R¹COY	X-R²	Product	Yield (%)
1	1	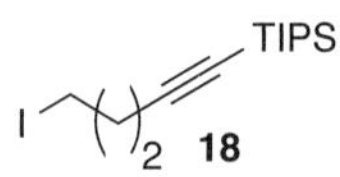 **18**	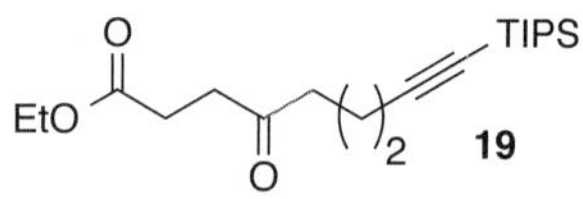 **19**	62
2	1	 **10**	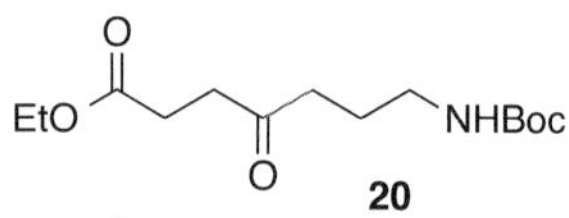 **20**	82
3	1	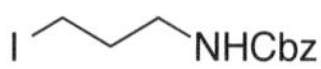 **21**	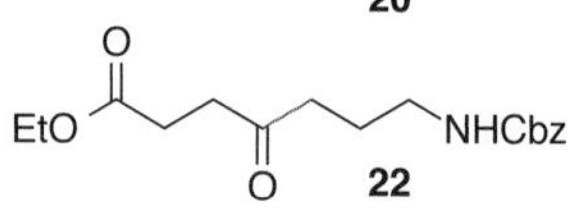 **22**	86
4	1	 **23**	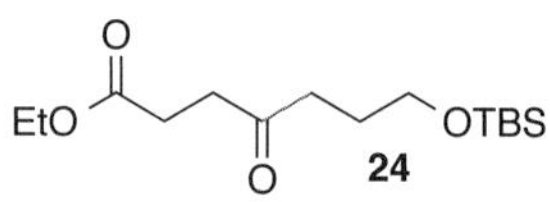 **24**	83
5[c]	4	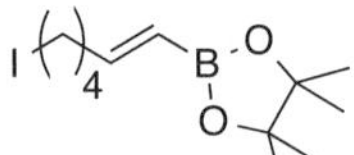 **25**	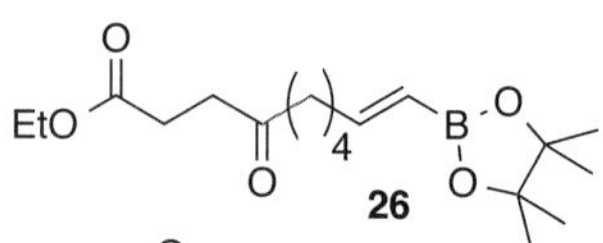 **26**	58
6[c]	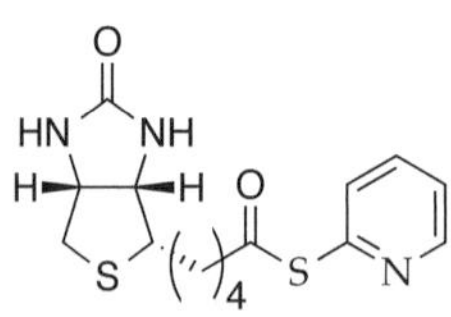 **27**	**18**	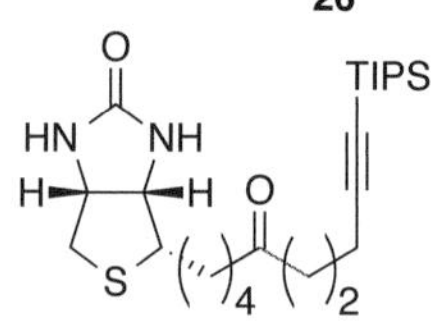 **28**	36

[a] Conditions as in Table 1. [b] Yield of isolated, purified product. [c] 1 Equiv of R¹COY was used at rt with Zn instead of Mn.

These conditions have proven to be general in our hands and in the hands of others. For example, Gong reported modified conditions that enable coupling with anhydrides[12] as well as the use of tertiary alkyl halides,[13] aryl acid chlorides,[14] and methyl *p*-toluenesulfonate ester.[15] It is also important to note that one-pot protocols for the generation and coupling of acid chlorides[14] and anhyrides[13] have been reported. In addition, by substituting a chiral bis(oxazoline) ligand for bipyridine, Reisman was able to realize the enantioconvergent coupling of a secondary benzyl chloride to form enantioenriched α-aryl ketones.[16] Finally, although several mechanisms had been proposed in the literature for these reactions by ourselves and others, recent investigations have led to a revised mechanism. We proposed a connection between the cross-electrophile coupling of aryl halides with alkyl halides based upon mechanistic studies of isolated acylnickel intermediates[17] and subsequent work by Gong[12] suggested that the mechanism is nearly identical to the one we proposed for the coupling of aryl halides with alkyl halides.[18]

Although these studies have demonstrated that cross-electrophile ketone synthesis has promising functional group compatibility and scope, no reports on a preparative scale have appeared. Because these reactions are heterogeneous, we anticipated that stirring method, stirring rate, and activity of the metal powder could all play a role in success. In this case, as in a previous report on a related reaction,[19] avoiding mechanical activation of the metal powder by keeping the stirrer paddle off the bottom of the flask and ensuring turbulent mixing (as with a Morton flask) were the key differences from the smaller-scale procedure. In our previous report,[19] low activity of the zinc powder sometimes led to long induction periods, but no such variability was observed with manganese powder and the reactions proceed even at 0 °C. This difference may be due to *in situ* activation of the metal powder by HCl, formed by reaction of the acid chloride with adventitious water.

References

1. University of Rochester, Department of Chemistry, RC Box 270216, Rochester, NY, USA 14627-0216. email: daniel.weix@rochester.edu. We acknowledge Stephanie Rugg, Nick Gower, and Kierra Huihui (University of Rochester) for experimental and characterization

assistance. This work was supported by the NIH (GM097243) and the University of Rochester (Weissberger Memorial Fellowship to ACW). Analytical data were obtained from the CENTC Elemental Analysis Facility at the University of Rochester, funded by NSF CHE-0650456.

2. (a) Stephanopoulos, N.; Francis, M. B. *Nat. Chem. Biol.* **2011**, *7*, 876–884. (b) Kalia, J.; Raines, R. T. *Curr. Org. Chem.* **2010**, *14*, 138–147. (c) Sletten, E. M.; Bertozzi, C. R., *Angew. Chem., Int. Ed.* **2009**, *48*, 6974–6998.
3. (a) Dieter, R. K. *Tetrahedron* **1999**, *55*, 4177–4236. (b) J. Lawrence, N., *Perkin Trans. 1* **1998**, 1739–1750. (c) Singh, J.; Satyamurthi, N.; Aidhen, I. *J. Prakt. Chem.* **2000**, *342*, 340–347.
4. (a) Milstein, D.; Stille, J. K. *J. Am. Chem. Soc.* **1978**, *100*, 3636–3638. (b) Li, H.; Yang, H.; Liebeskind, L. S. *Org. Lett.* **2008**, *10*, 4375–4378.
5. (a) Sato, T.; Naruse, K.; Enokiya, M.; Fujisawa, T. *Chem. Lett.* **1981**, *10*, 1135–1138. (b) Negishi, E.-i.; Bagheri, V.; Chatterjee, S.; Luo, F.-T.; Miller, J. A.; Stoll, A. T. *Tetrahedron Lett.* **1983**, *24*, 5181–5184. (c) Tamaru, Y.; Ochiai, H.; Sanda, F.; Yoshida, Z.-i. *Tetrahedron Lett.* **1985**, *26*, 5529–5532. (d) Harada, T.; Kotani, Y.; Katsuhira, T.; Oku, A. *Tetrahedron Lett.* **1991**, *32*, 1573–1576. (e) Asaoka, M.; Kosaka, A.; Tanaka, M.; Ueda, T.; Houkawa, T.; Takei, H. *Perkin Trans. 1* **1997**, 2949–2950 (f) Wang, D.; Zhang, Z. *Org. Lett.* **2003**, *5*, 4645–4648. (g) Yus, M.; Ortiz, R. *Eur. J. Org. Chem.* **2004**, *2004*, 3833–3841.
6. (a) Fukuyama, T.; Tokuyama, H., *Aldrichimica Acta* **2005**, *37*, 87–96. For references on industrial uses, see: (b) Mori, Y.; Seki, M. *Adv. Synth. Catal.* **2007**, *349*, 2027–2038.
7. Zhang, Y.; Rovis, T. *J. Am. Chem. Soc.* **2004**, *126*, 15964–15965.
8. (a) Tatamidani, H.; Kakiuchi, F.; Chatani, N. *Org. Lett.* **2004**, *6*, 3597–3599. (b) Yu, Y.; Liebeskind, L. S. *J. Org. Chem.* **2004**, *69*, 3554–3557. (c) Zhang, Z.; Lindale, M. G.; Liebeskind, L. S. *J. Am. Chem. Soc.* **2011**, *133*, 6403–6410.
9. Wotal, A. C.; Weix, D. J. *Org. Lett.* **2012**, *14*, 1476−1479.
10. Onaka, M.; Matsuoka, Y.; Mukaiyama, T. *Chem. Lett.* **1981**, *10*, 531–534.
11. (a) d'Incan, E.; Sibille, S.; Périchon, J.; Moingeon, M.-O.; Chaussard, J. *Tetrahedron Lett.* **1986**, *27*, 4175–4176. (b) Marzouk, H.; Rollin, Y.; Folest, J. C.; Nédélec, J. Y.; Périchon, J. *J. Organomet. Chem.* **1989**, *369*, C47–C50. (c) Amatore, C.; Jutand, A.; Périchon, J.; Rollin, Y. *Monatsh. Chem.* **2000**, *131*, 1293–1304.
12. Yin, H.; Zhao, C.; You, H.; Lin, K.; Gong, H. *Chem. Commun.* **2012**, *48*, 7034–7036.

13. Zhao, C.; Jia, X.; Wang, X.; Gong, H. *J. Am. Chem. Soc.* **2014**, *136*, 17645–17651.
14. Wu, F.; Lu, W.; Qian, Q.; Ren, Q.; Gong, H. *Org. Lett.* **2012**, *14*, 3044–3047.
15. Liang, Z.; Xue, W.; Lin, K.; Gong, H. *Org. Lett.* **2014**, *16*, 5620–5623.
16. Cherney, A. H.; Kadunce, N. T.; Reisman, S. E. *J. Am. Chem. Soc.* **2013**, *135*, 7442–7445.
17. Wotal, A. C.; Ribson, R. D.; Weix, D. J. *Organometallics* **2014**, *33*, 5874–5881.
18. Biswas, S.; Weix, D. J. *J. Am. Chem. Soc.* **2013**, *135*, 16192–16197.
19. Everson, D. A.; George, D. T.; Weix, D. J.; Buergler, J. F.; Wood, J. L. *Org. Synth.* **2013**, *90*, 200–214.

Appendix
Chemical Abstracts Nomenclature (Registry Number)

Nickel (II) chloride ethylene glycol diethylether: Nickel (II) chlorideglyme, Nickel(II) chloride dimethoxyethane complex; (29046-78-4)
4, 4′, Di-*tert*-butyl-2,2′-bipyridine; 4-*tert*-Butyl-2-(4-*tert*-butyl-2-pyridyl)pyridine, (72914-19-3)
N,N-Dimethylacetamide; (127-19-5)
Manganese; (7439-96-5)
1-Iodooctane; (629-27-6)
Ethyl succinyl chloride: Ethyl 4-chloro-4-oxobutyrate; (14794-31-1)

Donald C. Batesky graduated from Purdue University with a B.S. in Chemistry in 1954. Following 38 years with Eastman Kodak Co. as a new product development chemist and in management, Don worked 15 years with Aldrich Chemical Co. developing new products. In 2014, Don began a third career at University of Rochester in the Weix group. In his career spanning over 50 years, he has prepared almost 4000 organic chemicals.

Daniel J. Weix was raised in Oak Creek, WI. After undergraduate training with Tom Katz at Columbia University, graduate work with Jonathan Ellman at Berkeley, and postdoctoral training with John Hartwig at Yale and Illinois, he started his independent career at Rochester in 2008, where he is currently an associate professor. His research program in the area of transition metal catalysis has been focused on developing the concepts of cross-electrophile coupling. Cross-electrophile coupling has proven to be a useful approach to synthesis both in academia and in industry.

Alexander C. Wotal was raised in Palatine, Illinois. In 2006, he began studying chemistry at William Rainey Harper College, Palatine, Illinois under Prof. Daniel J. Stanford. He completed his B.S. in 2009 under Prof. Jeffrey B. Johnson at Hope College, Holland, Michigan. Alexander then joined the group of Daniel J. Weix at University of Rochester, Rochester, New York where he studied Ni-catalyzed ketone-forming cross-electrophile coupling reactions and their mechanisms. In 2014, he participated in a 1-year cooperative position at GlaxoSmithKline pharmaceutical company under Dr. Jared Spletstoser and Dr. Qi Jin developing antibacterial agents.

Yongda Zhang was born in Zhejiang province, China. He received his B.A. degree in chemistry from Hangzhou University in 1994. He earned his Ph.D. in organic chemistry at Fudan University under the guidance of Prof. Dawei Ma (Shanghai Institute of Organic Chemistry, Chinese Academy of Sciences) and Prof. Fenggang Tao (Fudan University) in 2000. After postdoctoral research with Prof. Alan J. Kennan and additional postdoctoral training with Prof. Tomislav Rovis at Colorado State University, he joined Boehringer Ingelheim Pharmaceuticals Inc. in Ridgefield, CT, where currently he is Principal Scientist. His research interests include design and development of innovative and practical processes for large-scale production of active pharmaceutical ingredients and synthetic methodology.

Organic Syntheses

Trichloroboron-promoted Deprotection of Phenolic Benzyl Ether Using Pentamethylbenzene as a Non Lewis-Basic Cation Scavenger

Shun Okaya, Keiichiro Okuyama, Kentaro Okano, and Hidetoshi Tokuyama[1]*

Graduate School of Pharmaceutical Sciences, Tohoku University, Aramaki, Aoba-ku, Sendai 980-8578, Japan

Checked by Brandon Nelson, Nicholas Garcia, and Mohammad Movassaghi

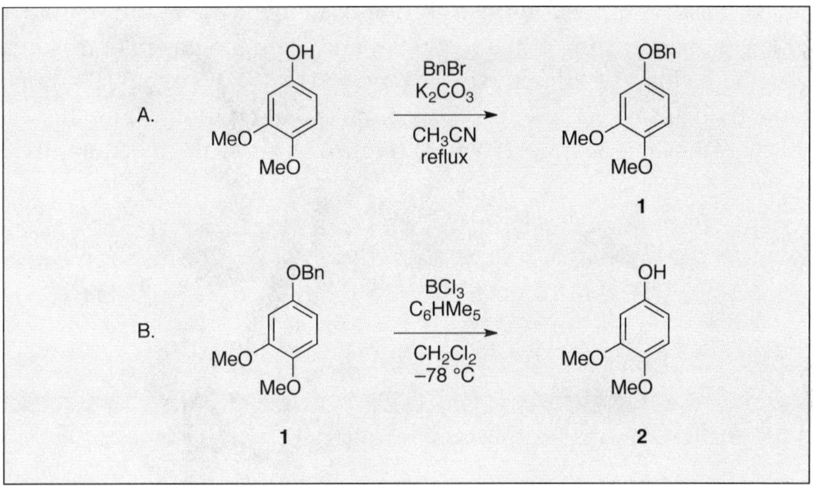

Procedure

A. *4-Benzyloxy-1,2-dimethoxybenzene (**1**).* A 500-mL one-necked round-bottomed flask equipped with a Teflon-coated magnetic stir bar (3.5 x 1.0 cm), with an argon gas inlet is charged with 3,4-dimethoxyphenol (8.32 g, 54.0 mmol) (Note 1), potassium carbonate (8.29 g, 60.0 mmol, 1.11 equiv) (Note 2), and is fitted with a reflux condenser with an argon gas inlet. MeCN (95 mL) (Note 3) is added to the reaction flask. After stirring for

DOI: 10.15227/orgsyn.093.0063
Published on the Web 3/21/2016

10 min at ambient temperature, benzyl bromide (6.54 mL, 55.0 mmol, 1.02 equiv) (Note 4) is added. The reflux condenser is washed with MeCN (5 mL) and the resulting mixture is stirred for 20 min. Then, the reaction mixture is heated to reflux for 2 h (Note 5). The mixture is cooled to ambient temperature, and filtered through a 2 cm pad of Celite; the transfer is completed by washing the reaction flask with EtOAc (2 x 25 mL). The filtrate is concentrated on a rotary evaporator under reduced pressure (30 °C, 8 mmHg) and dried in vacuo (2.5 mmHg) to afford 13.95 g of a crude benzyl ether as an orange solid, which is purified by silica gel column chromatography (elution: 15% EtOAc in hexanes) (Note 6) to provide 4-benzyloxy-1,2-dimethoxybenzene (**1**) as colorless crystals (11.90 g, 90%) (Notes 7, 8, and 9).

B. *3,4-Dimethoxyphenol* (**2**). A 1-L oven-dried three-necked round-bottomed flask equipped with a Teflon-coated magnetic stir bar (3.5 x 1.0 cm), a rubber septum, a glass stopper, and an argon gas inlet is charged with 4-benzyloxy-1,2-dimethoxybenzene (**1**) (7.33 g, 30.0 mmol), pentamethylbenzene (13.34 g, 90.0 mmol, 3.0 equiv) (Note 10), and anhydrous CH_2Cl_2 (150 mL) (Note 11) (**Figure 1**). After the reaction mixture

Figure 1. Apparatus assembly for Step B

is cooled to –78 °C (bath temperature), 1 M boron trichloride in CH_2Cl_2 (60.0 mL, 60.0 mmol, 2.0 equiv) (Note 12) is added to the flask dropwise over 5 min at –78 °C. After stirring for 45 min at –78 °C, the mixture is quenched by syringe addition of chloroform/methanol (10/1, 60.0 mL) (Note 13) at –78 °C and is warmed to ambient temperature. The solution is concentrated on a rotary evaporator under reduced pressure (30 °C, 8 mmHg), and dried *in vacuo* (2.5 mmHg) to afford 24.58 g of a crude phenol as a pale yellow solid, which is purified by silica gel column chromatography (elution: 30% EtOAc, 5% toluene, 65% hexanes) (Notes 14 and 15). The fraction containing the product are collected and concentrated on a rotary evaporator under reduced pressure to yield 3,4-dimethoxyphenol (**2**) as pale yellow crystals (4.22 g, 91.2%) (Notes 16 and 17).

Notes

1. 3,4-Dimethoxyphenol (96%) was purchased from Tokyo Chemical Industry Co., Ltd. and used as received without further purification.
2. Potassium carbonate (99%, powdered, ~325 mesh) was purchased from Sigma-Aldrich Co. and used as received without further purification.
3. Acetonitrile (>99.5%) was purchased from Wako Pure Chemical Industries, Ltd. and used as received without further purification.
4. Benzyl bromide (>98%) was purchased from Wako Pure Chemical Industries, Ltd. and used as received without further purification.
5. The reaction typically requires 2 h to complete and is monitored by TLC analysis on Merck silica gel 60 F_{254} plates eluting with EtOAc/hexanes (1:1). The R_f values of the phenol and benzyl ether are 0.33 and 0.65, respectively (visualized with 254 nm UV lamp and stained with an ethanol solution of *p*-anisaldehyde. After dipping the TLC plate to the solution, the chromatogram is stained by heating.
6. The crude material is dissolved in eluent (20 mL) and CH_2Cl_2 (5 mL) and is charged onto a column (diameter = 10 cm, height = 20 cm) of 400-gram silica gel. The column was eluted with 15% EtOAc/hexanes and 75-mL fractions were collected. Fraction 19-29 were combined and concentrated on a rotary evaporator under reduced pressure.
7. When the reaction was performed on half-scale, the isolated yield was 5.70 g (86%).

8. If crystallization is desired after chromatography, the residue is dissolved in hot EtOAc/hexanes, 1:3 (40 mL) using an oil bath (bath temperature: 95 ℃). The solution is then cooled at 0 ℃ in an ice bath for 30 min. The colorless crystals are collected by filtration (78~86%).
9. The compound (purified by chromatography) exhibits the following physicochemical properties: R_f = 0.65 (hexanes/EtOAc = 1:1); Merck silica gel 60 F_{254} plates (visualized with 254 nm UV lamp and stained with an ethanol solution of *p*-anisaldehyde. After dipping the TLC plate to the solution, the chromatogram is stained by heating); mp 48–48.5 °C (EtOAc/hexanes); IR (CH_2Cl_2, cm^{-1}): 2936, 1596, 1509, 1450, 1280, 1227, 1196, 1159, 1137, 731; ^1H NMR (500 MHz, $CDCl_3$) δ: 3.84 (s, 3H), 3.85 (s, 3H), 5.02 (s, 2H), 6.48 (dd, *J* = 8.7, 2.8 Hz, 1H), 6.61 (d, *J* = 2.8 Hz, 1H), 6.78 (d, *J* = 8.7 Hz, 1H), 7.33 (app-t, *J* = 7.2 Hz, 1H), 7.40 (app-t, *J* = 7.4 Hz, 2H), 7.44 (app-d, *J* = 6.9 Hz, 2H). ^{13}C NMR (125 MHz, $CDCl_3$) δ: 56.5, 57.1, 71.3, 101.9, 104.7, 112.4, 128.3, 128.7, 129.3, 137.8, 144.3, 150.5, 154.1. [M + H] calcd. for $C_{15}H_{16}O_3$: 245.1172, Found: 245.1163. Anal. calcd. for $C_{15}H_{16}O_3$: C, 73.75; H, 6.60. Found: C, 73.51; H, 6.59.
10. Pentamethylbenzene (>99%) was purchased from Tokyo Chemical Industry Co., Ltd. and used as received without further purification.
11. CH_2Cl_2 (>99.5% anhydrous, water content: <0.001%) was purchased from Kanto Chemical Co., Inc. and used as received without further purification.
12. Boron trichloride 1.0 M in CH_2Cl_2 was purchased from Sigma-Aldrich Co. and used as received without further purification.
13. The reaction typically requires 30 min to consume all the 3,4-dimethoxyphenol-1-benzyl ether and is monitored by TLC analysis on Merck silica gel 60 F_{254} plates eluting with EtOAc/hexanes (1:1). The R_f values of the starting material, the undesired 2-benzyl-4,5-dimethoxyphenol, the desired 3,4-dimethoxyphenol are 0.65, 0.40, and 0.33, respectively (visualized with 254 nm UV lamp and stained with an solution of *p*-anisaldehyde. After dipping the TLC plate to the solution, the chromatogram is stained by heating).
14. Integration of the ^1H NMR spectra of the crude material indicated that the molar ratio of desired phenol: benzylpentamethylbenzene: pentamethylbenzene = 1:1.07: 2.20. The crude material is dissolved in eluent (15 mL) and then charged onto a column (diameter = 8 cm, height = 35 cm) of 400-gram silica gel. The column is eluted with 30% EtOAc, 5% toluene, 65% hexanes and 100-mL fractions are collected. Fractions 1-16 contain pentamethylbenzene and

benzylpentamethylbenzene. Fractions 36-42 and 45-60 are combined and concentrated on a rotary evaporator under reduced pressure to provide undesired 2-benzyl-4,5-dimethoxyphenol and desired 3,4-dimethoxyphenol, respectively. Benzylpentamethylbenzene: ^{1}H NMR (400 MHz, $CDCl_3$) δ: 2.17 (s, 3H x 2), 2.25 (s, 3H x 2), 2.27 (s, 3H), 4.11 (s, 2H), 7.04 (d, *J* = 7.2 Hz, 2H), 7.15 (t, *J* = 7.6 Hz, 1H), 7.21-7.26 (m, 2H); ^{13}C NMR (100 MHz, $CDCl_3$) δ: 16.83, 16.85, 16.9, 36.1, 125.6, 127.9, 128.3, 132.5, 132.8, 133.1, 133.8, 140.62.

15. Undesired 2-benzyl-4,5-dimethoxyphenol is obtained as a colorless oil (595 mg, 8%). Its physicochemical properties are as follows: R_f = 0.40 (EtOAc/hexanes = 1:1); Merck silica gel 60 F_{254} plates (visualized with 254-nm UV lamp and stained with an ethanol solution of $Ce_2(SO_4)_3$ and phosphomolybdic acid (Ce-PMA). After dipping the TLC plate to the Ce-PMA solution, the chromatogram is stained by heating); IR (neat, cm^{-1}): 3442, 2935, 1617, 1521, 1451, 1414, 1204, 1109, 998, 702; ^{1}H NMR (400 MHz, $CDCl_3$) δ: 3.80 (s, 3H), 3.83 (s, 3H), 3.93 (s, 2H), 4.34 (s, 1H), 6.45 (s, 1H), 6.66 (s, 1H), 7.17–7.24 (m, 3H), 7.26–7.33 (m, 2H); ^{13}C NMR (100 MHz, $CDCl_3$) δ: 35.8, 55.8, 56.6, 101.3, 114.6, 117.8, 126.2, 128.4, 128.5, 140.1, 142.9, 147.8, 148.3; HRMS (ESI). $[M + Na]^+$ calcd. for $C_{15}H_{16}NaO_3$: 267.0992. Found: 267.0988; Anal. calcd. for $C_{15}H_{16}O_3$: C, 73.75; H, 6.60. Found: C, 72.99; H, 6.65. (The yields and data were provided by the submitters. This byproduct was not characterized by the checkers).
16. When the reaction was performed on half-scale, the isolated yield was 2.09 g (90%).
17. Desired 3,4-dimethoxyphenol is obtained as pale yellow crystals (4.22 g, 91.2%). The compound exhibits the following physicochemical properties: R_f = 0.33 (EtOAc/hexanes = 1:1); Merck silica gel 60 F_{254} plates (visualized with 254-nm UV lamp and stained with an solution of *p*-anisaldehyde. After dipping the TLC plate in the solution, the plate is stained by heating); mp 75.5–76.5 °C (hexanes/EtOAc); IR (neat, cm^{-1}): 3418, 2937, 2835, 1604, 1507, 1455, 1434, 1219, 1194, 1154, 1022, 950, 764; ^{1}H NMR (500 MHz, $CDCl_3$) δ: 3.77 (s, 3H), 3.80 (s, 3H), 6.35 (dd, *J* = 8.6, 2.8 Hz), 6.46 (d, *J* = 2.8 Hz, 1H), 6.71 (d, *J* = 8.6 Hz, 1H); ^{13}C NMR (100 MHz, $CDCl_3$) δ: 56.5, 57.5, 101.4, 106.7, 113.2, 143.6, 150.5, 151.0; [M + H] calcd. for $C_8H_{10}O_3$: 155.0703 Found: 155.0707. Anal. calcd. for $C_8H_{10}O_3$: C, 62.33; H, 6.54. Found: C, 62.42; H, 6.51.

Working with Hazardous Chemicals

The procedures in *Organic Syntheses* are intended for use only by persons with proper training in experimental organic chemistry. All hazardous materials should be handled using the standard procedures for work with chemicals described in references such as "Prudent Practices in the Laboratory" (The National Academies Press, Washington, D.C., 2011; the full text can be accessed free of charge at http://www.nap.edu/catalog.php?record_id=12654). All chemical waste should be disposed of in accordance with local regulations. For general guidelines for the management of chemical waste, see Chapter 8 of Prudent Practices.

In some articles in *Organic Syntheses*, chemical-specific hazards are highlighted in red "Caution Notes" within a procedure. It is important to recognize that the absence of a caution note does not imply that no significant hazards are associated with the chemicals involved in that procedure. Prior to performing a reaction, a thorough risk assessment should be carried out that includes a review of the potential hazards associated with each chemical and experimental operation on the scale that is planned for the procedure. Guidelines for carrying out a risk assessment and for analyzing the hazards associated with chemicals can be found in Chapter 4 of Prudent Practices.

The procedures described in *Organic Syntheses* are provided as published and are conducted at one's own risk. *Organic Syntheses, Inc.*, its Editors, and its Board of Directors do not warrant or guarantee the safety of individuals using these procedures and hereby disclaim any liability for any injuries or damages claimed to have resulted from or related in any way to the procedures herein.

Discussion

The benzyl group is one of the most useful protective groups for a phenolic hydroxy group since phenol benzyl ethers are easily synthesized and tolerate various reaction conditions.[2] Deprotection of a benzyl group is generally performed by palladium-catalyzed hydrogenolysis or acidic treatment. However, these methods are inapplicable for a substrate having an olefinic/acetylenic moiety and acid-labile functional groups. During our synthetic studies on yatakemycin having electron-rich aromatic rings and a thiol ester, we determined that a combination of BCl_3 and pentamethylbenzene (C_6HMe_5) facilitated smooth debenzylation even at −78 °C[3] by modifying Yoshino's conditions using TFA and C_6HMe_5.[4] Pentamethylbenzene is a non-Lewis-basic cation scavenger and does not reduce Lewis acidity of BCl_3 by coordination, unlike general cation scavengers such as PhSMe or Me_2S that have lone electron pairs as a Lewis base.[5] In addition, benzylpentamethylbenzene[4,6] has low polarity, thus easily separated with the desired phenol by column chromatography (see Note 14).

Table 1. Effect of Pentamethylbenzene to Suppress Undesired Migration of Benzyl Group

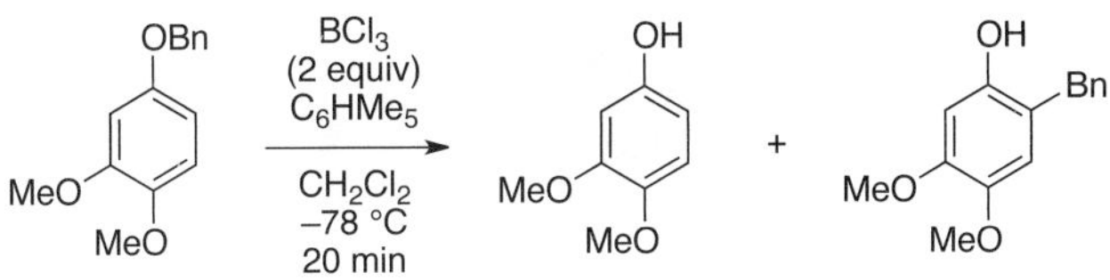

1 2 3

entry	C_6HMe_5 (equiv)	**2** (%)[a]	**3** (%)[a]
1	0	56	43
2	2	81	11
3	3	84	8
4	5	85	8

[a] Isolated yields.

The effectiveness of C_6HMe_5 was examined through a control experiment using 4-benzyloxy-1,2-dimethoxybenzene (**1**) as a substrate (Table 1). Thus, in the absence of pentamethylbenzene, a substantial amount of the undesired product **3** was obtained via electrophilic *C*-benzylation (entry 1). After an extensive optimization of C_6HMe_5 equivalents, we determined that three C_6HMe_5 equivalents were sufficient to suppress the formation of an undesired product.

The deprotection conditions were applicable to a series of phenol benzyl ethers (Table 2). Allyl, iodo, nitro, formyl, acetyl, and thiol ester groups remained intact in this reaction. Selective removal of the benzyl group was also accomplished in the presence of aliphatic benzyl ether, Alloc carbamate, *o*-Ns amide, Cbz carbamate, and Boc carbamate to provide the corresponding phenols in good to excellent yields.[5] In addition, the less polar benzylpentamethylbenzene as a co-product was easily removed by column chromatography.

Table 2. Debenzylation of Various Aryl Benzyl Ethers

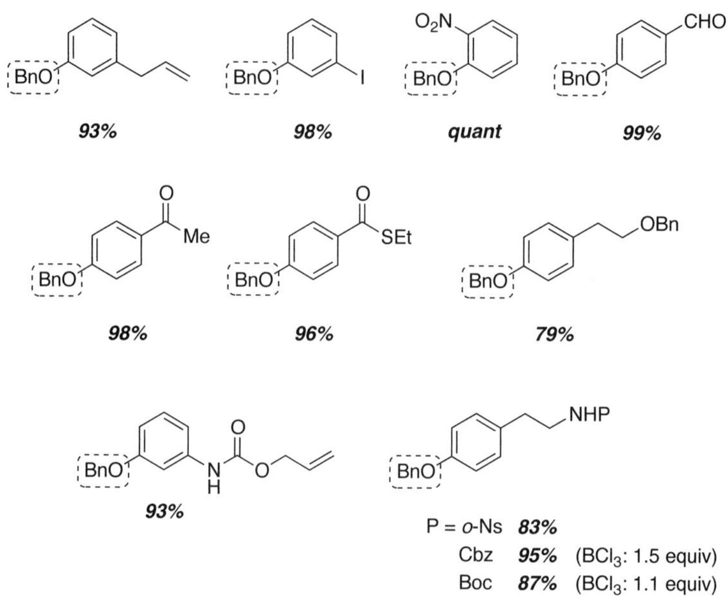

Debenzylation is used in the synthesis of natural products and highly functionalized complex molecules (Table 3). In the total synthesis of yatakemycin, we performed the removal of the two benzyl groups with the labile thiol ester intact.[3] At the later stage in the synthesis of haplophytine, we performed the deprotection of a Cbz carbamate in the presence of α,β-unsaturated imine.[7] Nicolaou and Chen also employed the conditions to a similar substrate and successfully removed both benzyl and Cbz groups.[8] Erhardt performed the deprotection of the two benzyl groups in the final step of the total synthesis of (±)-vestitol and bolusanthin III.[9] They also provided a comment on this step that conventional conditions such as pentamethylbenzene/trifluoroacetic acid[4] and boron tribromide proved to be very harsh, resulting in decomposition of the starting material. Vogel performed a selective deprotection of a benzyl group in the synthesis of polypropionates.[10] Interestingly, they successfully discriminated two benzyl groups according to their electron densities; the more electron-rich benzyl group was removed faster than the other benzyl group bearing the CF_3 group. In addition, *exo*-methylene and enol benzoate remained unreacted after the reaction. Kobayashi and co-workers applied the conditions to remove aliphatic benzyl ethers by replacing BCl_3 with BBr_3.[11] Acid-labile and easily reduced triene remained intact under the conditions to provide the corresponding pentaol in 59% yield.

Table 3. Application of the Mild Deprotection Conditions

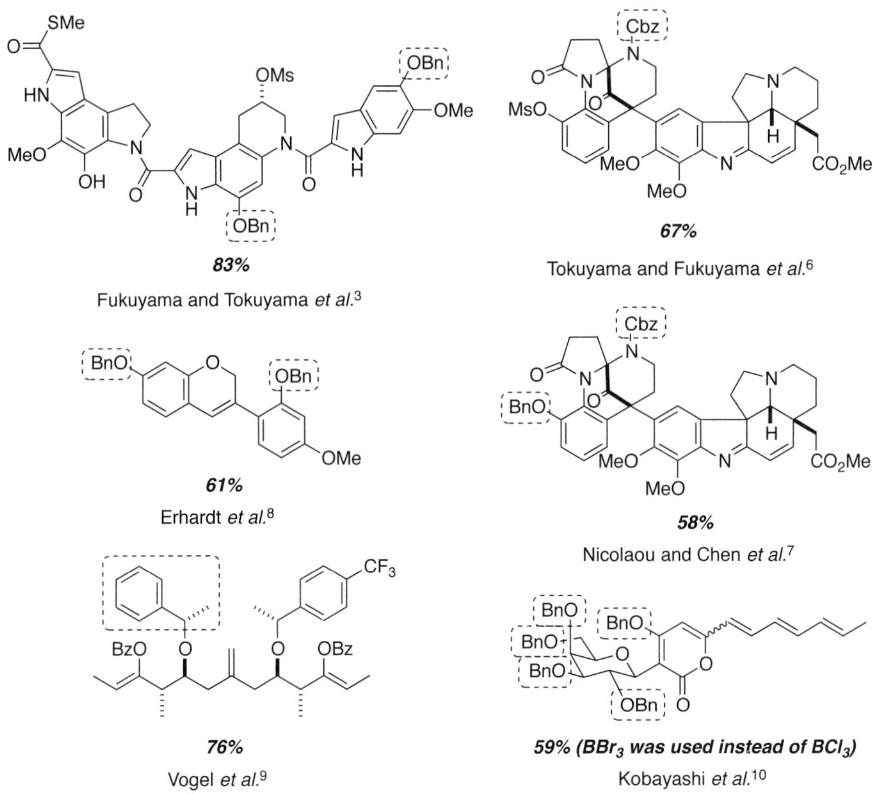

References

1. Graduate School of Pharmaceutical Sciences, Tohoku University, Aramaki, Aoba-ku, Sendai 980-8578, Japan. E-mail: tokuyama@mail.pharm.tohoku.ac.jp
2. (a) Kocienski, P. J. *Protecting Groups*; Thieme: New York, **1994**. (b) Wuts, P. G. M. *Protective Groups in Organic Synthesis*, 5th ed.; Wiley-Interscience: New York, **2014**.
3. (a) Okano, K.; Tokuyama, H.; Fukuyama, T. *J. Am. Chem. Soc.* **2006**, *128*, 7136. (b) Okano, K.; Tokuyama, H.; Fukuyama, T. *Chem. Asian J.* **2008**, *3*, 296.
4. Yoshino, H.; Tsuchiya, Y.; Saito, I.; Tsujii, M. *Chem. Pharm. Bull.* **1987**, *35*, 3438.
5. Okano, K.; Okuyama, K.; Fukuyama, T.; Tokuyama, H. *Synlett* **2008**, 1977.
6. (a) Sun, H.-B.; Li, B.; Chen, S.; Li, J.; Hua, R. *Tetrahedron* **2007**, *63*, 10185. (b) Yamada, K.; Fujita, H.; Kunishima, M. *Org. Lett.* **2012**, *14*, 5026.
7. Ueda, H.; Satoh, H.; Matsumoto, K.; Sugimoto, K.; Fukuyama, T.; Tokuyama, H. *Angew. Chem. Int. Ed.* **2009**, *48*, 7600.
8. Nicolaou, K. C.; Dalby, S. M.; Li, S. L.; Suzuki, T.; Chen, D. Y. K. *Angew. Chem. Int. Ed.* **2009**, *48*, 7616.
9. Luniwal, A.; Erhardt, P. W. *Synlett* **2011**, 1605.
10. Exner, C. J.; Turks, M.; Fonquerne, F.; Vogel, P. *Chem. Eur. J.* **2011**, *17*, 4246.
11. Kanai, A.; Kuramochi, K.; Nakazaki, A.; Kobayashi, S. *Symposium on the Chemistry of Natural Products* **2006**, *48*, 373.

Appendix
Chemical Abstracts Nomenclature (Registry Number)

3,4-Dimethoxyphenol: Phenol, 3,4-dimethoxy-; (2033-89-8)
Benzyl bromide: Benzene, (bromomethyl)-; (100-39-0)
Potassium carbonate: Carbonic acid, potassium salt (1:2); (584-08-7)
Acetonitrile: Acetonitrile; (75-05-8)
Trichloroborane: Borane, trichloro-; (10294-34-5)
Pentamethylbenzene: Benzene, 1,2,3,4,5-pentamethyl-; (700-12-9)
Dichloromethane: Methane, dichloro-; (75-09-2)

Shun Okaya was born in Shizuoka in 1990. He received his B.S. in 2013 from the Faculty of Pharmaceutical Sciences, Tohoku University, where he carried out undergraduate research in the laboratories of Professor Hidetoshi Tokuyama. In the same year, he then began his graduate studies at the Graduate School of Pharmaceutical Sciences, Tohoku University under the supervision of Professor Hidetoshi Tokuyama. His research focuses on the total synthesis of epidithiodiketopiperazine natural products.

.

Keiichiro Okuyama received his B.S. in 2006 from Tohoku Pharmaceutical University, where he carried out undergraduate research under the supervision of Professor Tadashi Kato. He then moved to the laboratories of Professor Hidetoshi Tokuyama, Tohoku University and started his Ph.D. research on synthetic studies toward Haouamine B. In 2011, he received his Ph.D. from Tohoku University under the direction of Professor Hidetoshi Tokuyama. He is currently working for Astellas Pharma Inc. as a drug discovery researcher.

Kentaro Okano was born in Tokyo in 1979. He received his B.S. in 2003 from Kyoto University under the supervision of Professor Tamejiro Hiyama. He then moved to the laboratories of Professor Tohru Fukuyama at the University of Tokyo. In 2007, he joined the faculty at Tohoku University, where he is currently an assistant professor in Professor Hidetoshi Tokuyama's group. In 2014, he visited Professor Amir Hoveyda's laboratories at Boston College as a visiting researcher. His current research interest is natural product synthesis based on the development of new synthetic methodologies.

.

Hidetoshi Tokuyama was born in Yokohama in 1967. He received his Ph.D. in 1994 from Tokyo Institute of Technology under the direction of Professor Ei-ichi Nakamura. He spent one year (1994-1995) at the University of Pennsylvania as a postdoc with Professor Amos B. Smith, III. He joined the group of Professor Tohru Fukuyama at the University of Tokyo in 1995 and was appointed Associate Professor in 2003. In 2006, he was move to Tohoku University, where he is currently Professor of Pharmaceutical Sciences His research interest is on the development of synthetic methodologies and total synthesis of natural products.

Brandon Nelson was born in Streamwood, Illinois in 1991. He received his B.S. in 2013 from Illinois State University where he carried out research in the labs of Professor Hitchcock and Professor McLauchlan. He then moved to the Massachusetts Institute of Technology to pursue his Ph.D. focused on complex natural products total synthesis under the advisement of Professor Movassaghi.

Nicholas Garcia was born in Oklahoma City, Oklahoma in 1994. In 2013, he began working towards his B.S. in Chemistry at the Massachusetts Institute of Technology. In 2015, he joined the laboratories of Professor Mohammad Movassaghi, where he is currently carrying out undergraduate research focused on alkaloid total synthesis.

Enantioselective Synthesis of α,α-Disubstituted Lactones via a Chiral Brønsted Acid Catalyzed Intramolecular Cyclization

Jennifer E. Wilent, Ghassan Qabaja, and Kimberly S. Petersen[1*]

Department of Chemistry and Biochemistry, University of North Carolina at Greensboro, Greensboro, NC 27412

Checked by Bo Qu and Chris H. Senanayake

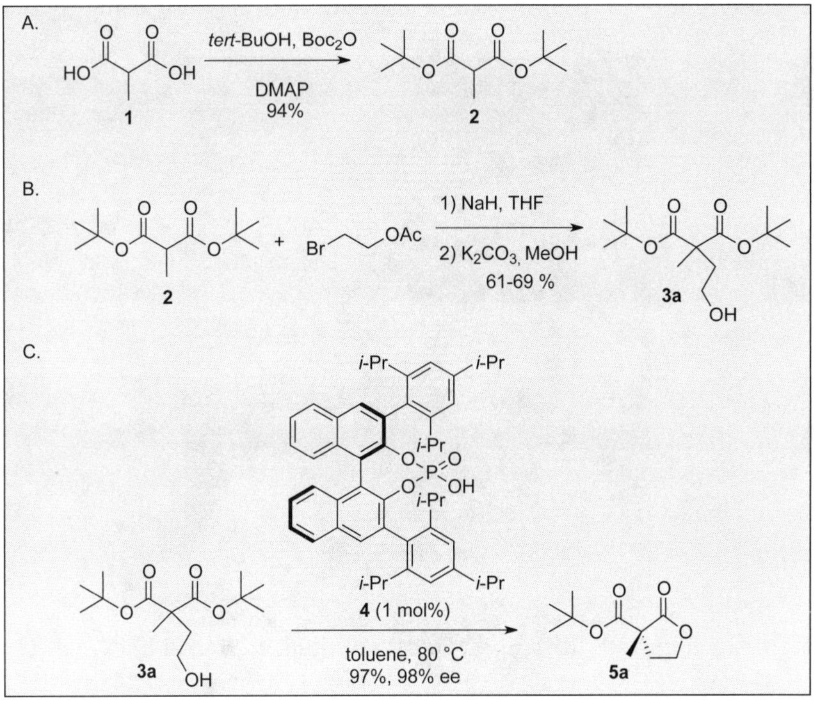

Procedure

A. *Di-tert-butyl-2-methylmalonate* (**2**). An oven-dried 500-mL three-necked round-bottomed flask equipped with an egg-shaped stirring bar

(30 mm x 15 mm) is fitted with two rubber septa and placed under an Ar atmosphere through a gas inlet. After cooling to 23 °C, methyl malonic acid (**1**) (12.0 g, 97.6 mmol, 1.0 equiv) (Note 1) is added followed by addition of diethyl ether (50 mL) (Note 2), 4-(dimethylamino)pyridine (1.2 g, 9.8 mmol, 0.1 equiv) (Note 3), *tert*-butyl alcohol (150 mL) (Note 4), and solid di-*tert*-butyl dicarbonate (48.3 g, 214.6 mmol, 2.2 equiv) (Note 5). The heterogeneous slurry is stirred (500 rpm) at room temperature (23 °C) for 48 h, after which time the reaction mixture became a clear pale yellow mixture (Figure 1). The reaction mixture is then diluted by addition of diethyl ether (150 mL), transferred to a 1000-mL separatory funnel and washed with water (2 x 50 mL) and HCl (1.0 M, 2 x 50 mL). The organic layer is dried over $MgSO_4$, gravity filtered using coarse filter paper, rinsed with diethyl ether (2 x 20 mL), and concentrated *in vacuo* using a rotary

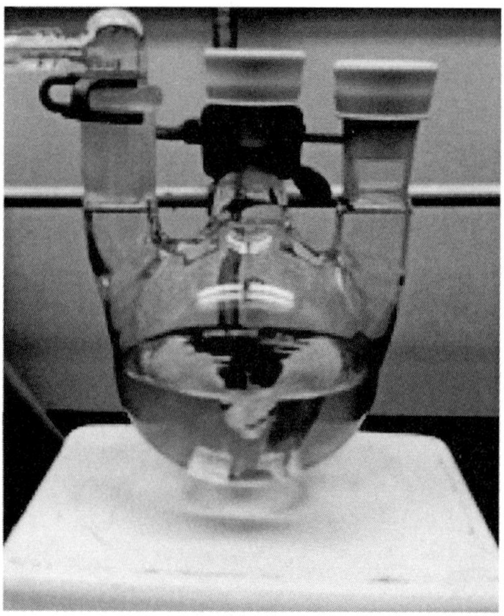

Figure 1. Reaction Mixture in Step A after 48 h

evaporator (27 °C, 7.5 mmHg) to give an oil, which is dissolved in 200 mL of hexanes and ethyl acetate (10:1) and passed through a plug of silica gel (3 x 5.2 cm) (Note 6) using a 60-mL medium fritted filter funnel. The colorless filtrate is received into a 500-mL round-bottomed flask and concentrated *in vacuo* using a rotary evaporator (27 °C, 7.5 mmHg), the colorless oil is put under high vacuum (23 °C, 0.4 mmHg) for 30 min to yield 21.1 g of the product **2** as a colorless oil at >99% purity (94% yield) (Notes 7 and 8).

B. *Di-tert-butyl-2-(2-hydroxyethyl)-2-methylmalonate (**3a**).* An oven-dried 500-mL three-necked round-bottomed flask equipped with an egg-shaped stirring bar (30 mm x 15 mm) is fitted with two rubber septa and placed under an Ar atmosphere through a gas inlet. After cooling to 23 °C, tetrahydrofuran (100 mL) (Note 9) is added via a syringe and the reaction flask is cooled with an ice-water bath. After stirring for 15 min, NaH (3.13 g, 78.2 mmol, 1.2 equiv) (Note 10) is added in one portion and the slurry is stirred for an additional 15 min. Di-*tert*-butyl-2-methylmalonate (**2**) (15.0 g, 65.1 mmol, 1.0 equiv) is dissolved in tetrahydrofuran (30 mL) in a 100-mL oven dried pear-shaped flask and this solution is transferred by cannula into the sodium hydride slurry in portions over 15 min. After 30 min 2-bromoethyl acetate (8.9 mL, 78.2 mmol, 1.2 equiv) (Note 11) is added drop wise via a syringe over a period of 5 min. The reaction mixture is allowed to slowly warm to room temperature (23 °C). The progress of the reaction is monitored by TLC analysis (Note 12). TLC analysis indicated consumption of the di-*tert*-butyl-2-methylmalonate (**2**) after 16 h. The reaction flask is cooled in an ice-water bath for 15 min, and a solution of saturated NH_4Cl solution (10 mL) is added drop wise via syringe. The mixture is diluted with diethyl ether (300 mL) and transferred to a 1000-mL separatory funnel. The ether layer is washed with water (2 x 50 mL) and saturated NaCl solution (50 mL), dried over $MgSO_4$, gravity filtered using coarse filter paper, rinsed with diethyl ether (2 x 20 mL), and concentrated *in vacuo* using a rotary evaporator (27 °C, 7.5 mmHg) to give a pale yellow oil which is dissolved in methanol (50 mL) (Note 13). The solution is transferred to an oven dried 500-mL three-necked round-bottomed flask equipped with an egg-shaped stir bar (30 mm x 15 mm) and two septa. A temperature probe is placed through the septum in the side neck to monitor the internal temperature of the reaction (Figure 2). The reaction flask is filled with argon and held under an atmosphere of Ar. The reaction flask is cooled to −20 to −10 °C using acetonitrile/dry ice bath (Note 14) and K_2CO_3 (9.0 g, 65.1 mmol) (Note 15) is added in five portions such that the internal temperature stays below −5 °C. After 2 h, the reaction mixture is diluted with diethyl ether (200 mL) and deionized water (50 mL) is added drop wise such that the internal temperature stays below −5 °C. The contents of the flask are transferred to a 1000-mL separatory funnel and hexanes (100 mL) are added. The organic layer is washed with water (3 x 50 mL) and saturated NaCl solution (50 mL), dried over $MgSO_4$, gravity filtered using coarse filter paper, rinsed with diethyl ether (2 x 20 mL), and concentrated *in vacuo* using a rotary evaporator (27 °C, 7.5 mmHg) to give a

Figure 2. Step B Assembly with Temperature Probe Inserted

light yellow oil. The oil is dissolved into hexanes (100 mL) and the flask is cooled with an ice-water bath for 1 h to yield a layer of white solids. The solids are broken via a spatula and collected via vacuum filtration into a 100 mL ceramic Buchner funnel equipped with filter paper of moderate porosity and washed with ice-cold hexanes (25 mL). A second crop of solids are obtained after the filtrate is concentrated under vacuum and the residue is dissolved in hexanes (20 mL), cooled with an ice water bath for 2 h and filtered as before. The two batches are combined to yield 10.9 g of product **3a** as a white crystalline solid (61% yield) (Notes 16 and 17).

C. *(S)-tert-Butyl-3-methyl-2-oxotetrahydrofuran-3-carboxylate (**5a**).* A 500-mL two-necked, round-bottomed flask equipped with an egg-shaped magnetic stir bar (38 x 16 mm) is flame-dried under vacuum. After cooling to 23 °C, di-*tert*-butyl 2-(2-hydroxyethyl)-2-methylmalonate (**3a**) (8.5 g, 31 mmol) is added to the round-bottomed flask, a rubber septum is fitted, the reaction flask is put under an atmosphere of Ar, and toluene (310 mL) (Note 18) is added via syringe. A thermometer probe is placed through the septum in the side neck to monitor internal temperature of the reaction. To this clear colorless solution, (*R*)-3, 3′-bis(2,4,6-triisopropylphenyl)-1,1′-binapthyl-2,2′-diylhydrogenphosphate (**4**, TRIP) (0.23 g, 0.31 mmol, 0.01 equiv) (Note 19) is added as a solid and an oven-dried reflux condenser is attached. Using an oil bath, the reaction mixture is heated to 80 °C (internal reaction temperature 76–80 °C) with stirring. The reaction is stirred at 500 rpm. The progress of the reaction is monitored by TLC analysis (Note 20). TLC analysis indicated consumption of the di-*tert*-butyl 2-(2-hydroxyethyl)-2-methylmalonate after 48 h. The round-bottomed flask is allowed to cool slowly in the oil bath to 23 °C. The reflux condenser is removed and EtOAc (50 mL) is added. After stirring for 5 min, the reaction mixture is transferred to a 500-mL separatory funnel and washed with deionized water (250 mL). After the layers separate, the lower, clear water layer is drained and the upper, cloudy EtOAc layer is collected in a 500 mL Erlenmeyer flask. The water layer is returned to the separatory funnel and reextracted with EtOAc (200 mL). The combined EtOAc layers are dried over $MgSO_4$, gravity filtered through coarse filter paper into a 1000 mL round bottomed flask and the $MgSO_4$ is rinsed with EtOAc (50 mL). The 1000 mL flask is concentrated *in vacuo* using a rotary evaporator (28 °C, 7.5 mmHg) and the crude reaction mixture is purified via flash chromatography (Note 21) resulting in the isolation of 6.02 g (97% yield) of product **5a** as a white crystalline solid (98% ee) (Notes 22, 23, and 24).

Notes

1. Methyl malonic acid (96%) was purchased from Acros and used as received.
2. Diethyl ether was purchased from Pharmaco-AAper (Reagent Grade ACS Anhydrous) and used without purification.
3. DMAP (4-(Dimethylamino)pyridine, ReagentPlus®, ≥99%) was purchased from Aldrich and used as received.
4. *tert*-Butyl alcohol (99.5%, extra pure) was purchased from Acros and used as received.
5. Di-*tert*-butyl dicarbonate (97%) was purchased from Acros and used as received.

6. Silica gel was purchased from Silicylce, Inc. (SiliaFlash P60 (230-400 mesh)) and used as received.
7. Physical characteristics of di-*tert*-butyl-2-methylmalonate (**2**): colorless liquid with >99% purity as determined by quantitative ^{1}H NMR using dimethyl fumarate as the internal standard; ^{1}H NMR (500 MHz, $CDCl_3$) δ: 1.32 (d, J = 7.4 Hz, 3H), 1.46 (s, 18H), 3.24 (q, J = 7.5 Hz, 1H); ^{13}C NMR (126 MHz, $CDCl_3$) δ: 13.4, 27.9, 48.1, 81.1, 169.6; IR (neat) cm^{-1} 2978, 1725, 1136, 848; HRMS ($C_{12}H_{22}O_4$, ESI): calculated 253.1415 $[M+Na]^{+1}$, found 253.1423;
8. A second run produced 21.3 g of the product **2** as colorless oil, 95% yield, >99% purity.
9. THF (tetrahydrofuran, minimum 99%, inhibited with <0.025% butlylated hydroxytoluene) was purchased from Acros and distilled under Ar over benzophenone and sodium.
10. Sodium hydride (60% dispersion in mineral oil in soluble bags, in resealable cans) was purchased from Acros and used as received.
11. 2-Bromoethyl acetate (97%) was purchased from Acros and used without further purification.
12. TLC analysis was performed on silica gel with 10:1 hexanes-EtOAc as eluent and visualization with *p*-anisaldehyde. The starting material has R_f = 0.38 (blue) and the product has R_f = 0.22 (blue). TLC plates were purchased from Silicycle, Inc. (Glass Backed TLC Extra Hard Layer, 60Å).
13. Anhydrous methanol (anhydrous, 99.8%, AcroSeal (TM)) was purchased from Acros and used received.
14. Saturated NaCl solution/ice bath was used by the submitting authors. The checkers observed that temperature is difficult to be kept under –5 °C using this cooling bath at this scale after addition of base K_2CO_3.
15. Anhydrous potassium carbonate (Certified ACS Granular Powder) was purchased from Fisher and used as received.
16. NMR spectra and purity of the first batch and second batch of di-*tert*-butyl 2-(2-hydroxyethyl)-2-methylmalonate (**3a**) are the same: white crystalline solid with 98% purity as determined by quantitative ^{1}H NMR using dimethyl fumarate as an internal standard; mp = 38–39 °C; ^{1}H NMR (500 MHz $CDCl_3$) δ: 1.40 (s, 3H), 1.48 (s, 18H), 2.09 (t, J = 6.3 Hz, 2H), 2.13 (br s, 1H), 3.75 (t, J = 6.3 Hz, 2H); ^{13}C NMR (126 MHz, $CDCl_3$) δ: 20.2, 27.8, 38.2, 53.7, 59.0, 81.5, 172.0; IR (neat) cm^{-1} 3446, 2977, 1710,

1367, 1154, 1112, 1019, 844 ; HRMS ($C_{14}H_{26}O_5$, ESI): calculated 297.1672 $[M+Na]^{+1}$, found 297.1665;

17. A second run produced 12.4 g of product **3a**, 69% yield, 98% purity.
18. Toluene was purchased from Innovative Technology and passed through a Pure Solv MD 5 Solvent Purification System. The checkers used toluene purchased from Aldrich and distilled over CaH_2.
19. (*R*)-3,3′-Bis(2,4,6-triisopropylphenyl)-1,1′-binapthyl-2,2′-diyl hydrogen phosphate (**4**) was purchased from Corvinus Chemicals (currently Ark Chemicals) and used as received.
20. TLC analysis was performed on silica gel with 30% EtOAc-hexanes as eluent and visualization with *p*-anisaldehyde. The starting material has R_f = 0.37 (blue) and the product has R_f = 0.58 (blue). TLC plates were purchased from Silicycle, Inc. (Glass Backed TLC Extra Hard Layer, 60Å).
21. Column diameter: 3.8 cm, silica: 84.0 g (Silicycle Inc., silica 60 (230-400 mesh), eluant: 500 mL 4:1 EtOAc:hexane), fraction size: 20 mL (16 x 150 mm test tubes), product typically found in factions 6-20.
22. Physical characteristics of (*S*)-*tert*-butyl 3-methyl-2-oxotetrahydrofuran-3-carboxylate (**5a**): white solid with 99% purity, as determined by quantitative ^{1}H NMR using dimethyl fumarate as an internal standard; mp = 49–50 °C; $[\alpha]_D^{24.5}$ = +7.08° (c = 1.08, $CHCl_3$); ^{1}H NMR (400 MHz, $CDCl_3$) δ: 1.49 (s, 12 H), 2.18 (m, 1H), 2.71 (m, 1H), 4.35 (m, 2H); ^{13}C NMR (100 MHz, $CDCl_3$) : 20.1, 27.8, 35.1, 50.5, 65.8, 82.8, 169.4, 176.3; IR (neat) cm^{-1} 2980, 1763, 1730, 1426, 1370, 1227, 1132, 1025, 845; HRMS ($C_{10}H_{20}NO_4$, ESI): calculated 218.13868 $[M+NH_4]^{+1}$, found 218.1385;
23. Enantiomeric purity was assessed at 98% ee by chiral HPLC: Chiralcel OJ-3, 4.6 x 150 mm, 3 micro; isocratic: 1% of IPA in heptane; eluent Rate: 1.3 mL/ min; Temperature 5 °C; UV detector 220 nm; RT_{major} = 10.2 min, RT_{minor} = 11.2 min.
24. A second run produced 5.97 g of product **5a**, 96% yield, 99% purity, 98% ee.

Working with Hazardous Chemicals

The procedures in *Organic Syntheses* are intended for use only by persons with proper training in experimental organic chemistry. All hazardous materials should be handled using the standard procedures for work with chemicals described in references such as "Prudent Practices in the Laboratory" (The National Academies Press, Washington, D.C., 2011; the full text can be accessed free of charge at http://www.nap.edu/catalog.php?record_id=12654). All chemical waste should be disposed of in accordance with local regulations. For general guidelines for the management of chemical waste, see Chapter 8 of Prudent Practices.

In some articles in *Organic Syntheses*, chemical-specific hazards are highlighted in red "Caution Notes" within a procedure. It is important to recognize that the absence of a caution note does not imply that no significant hazards are associated with the chemicals involved in that procedure. Prior to performing a reaction, a thorough risk assessment should be carried out that includes a review of the potential hazards associated with each chemical and experimental operation on the scale that is planned for the procedure. Guidelines for carrying out a risk assessment and for analyzing the hazards associated with chemicals can be found in Chapter 4 of Prudent Practices.

The procedures described in *Organic Syntheses* are provided as published and are conducted at one's own risk. *Organic Syntheses, Inc.*, its Editors, and its Board of Directors do not warrant or guarantee the safety of individuals using these procedures and hereby disclaim any liability for any injuries or damages claimed to have resulted from or related in any way to the procedures herein.

Discussion

The development of single-enantiomer chiral compounds has been a trend in the drug industry due to the potentially diverse pharmacology and toxicology of different stereoisomers of a molecule. The field of asymmetric methodology development has thus gained importance over the past several decades.[2] Often the key to an enantioselective synthesis is the availability of small chiral building blocks. Enantioenriched α-substituted lactones are common structural motifs and/or key intermediates in the synthesis of many biologically active compounds.[3] Organocatalysis and specifically chiral Brønsted acid and hydrogen bonding catalysts as a strategy to generate single enantiomer compounds through asymmetric synthesis has become a widely explored field.[4]

Recently, we have developed an enantioselective desymmetrization of prochiral diesters through a Brønsted acid catalyzed intramolecular cyclization to yield lactones containing an α-quaternary center.[5] Disubstituted hydroxy *tert*-butyl malonates **3a–3g** were readily prepared in 3 steps and in the presence of Binol based chiral Brønsted acid, (*R*)- or (*S*)-TRIP (**4**), enantioenriched lactones **5a–5g** were formed (Table 1). Coordination of the chiral phosphoric acid through one or more hydrogen bonds to the substrate likely leads to a rigid transition state which allows for discrimination of the enantiotopic ester groups.

Table 1. Substrate Scope

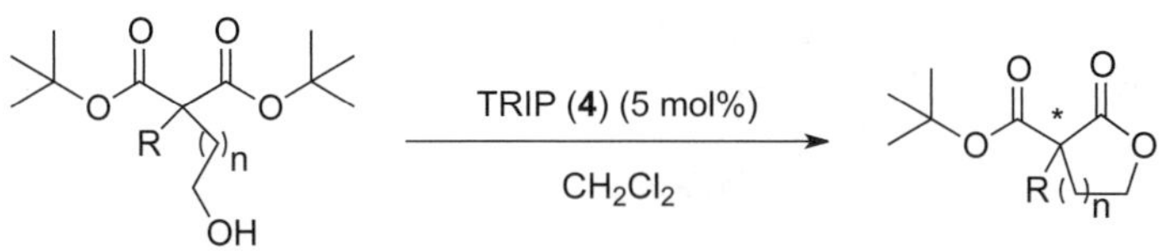

3a–3g **5a–5g**

entry	R	n	catalyst	time (h)	temp (°C)	yield (%)[a]	ee (%)[b]
1	Me (**5a**)	1	*S*	120	25	95	98
2	**Me (5a)**	**1**	***R* (1 mol%)**	**48**	**80**[c]	**96**	**98**
3	H (**5b**)	1	*S*	72	5	93	91
4	Et (**5c**)	1	*S*	120	25	93	95
5	*i*-Pr (**5d**)	1	*R*	192	35	89	90
6	allyl (**5e**)	1	*R*	168	35	96	97
7	benzyl (**5f**)	1	*S*	155	35	67	94
8	Me (**5g**)	2	*S*	144	25	84	86

[a] isolated yield; [b] determined by GC or HPLC analysis with a chiral support column; [c] toluene used as solvent.

A bulky ester is necessary for the process, however the R group can vary in length and branching. Small alkyl groups such as methyl (**5a**, entry 1), ethyl (**5c**, entry 4), and allyl (**5e**, entry 6) as well as larger groups such as isopropyl (**5d**, entry 5) and benzyl (**5f**, entry 7) all generate γ-lactones in good to excellent yields (67–96%) and enantioselectivities (90–98%). A one carbon homologation of the hydroxy chain led to the formation of δ-lactone with minimal loss in enantioselectivity (**5ge**, entry 8).

The cyclization was originally performed at room temperature using 5 mol% of catalyst (entry 1). The process has now been optimized such that catalyst loading can be lowered to 1 mol% with heating at 80 °C in toluene with no loss in ee or yield (entry 2). Additionally, the improved reaction conditions vastly lowered the reaction time from 120 to 48 hours.

Absolute configuration of lactones **5a–5g** was determined by comparison of the reduced diol of lactone **5f** to known values.[6]

Lactones such as **5a** where shown to be versatile intermediates through transformation into highly functionalized small building blocks (Scheme 1). Each transformation exhibited good yields while maintaining excellent enantioselectivity. Selective reduction of lactone (–)-**5a** with lithium tri-*tert*-butoxyaluminum hydride yielded diol (–)-**6** in 92% yield and without loss

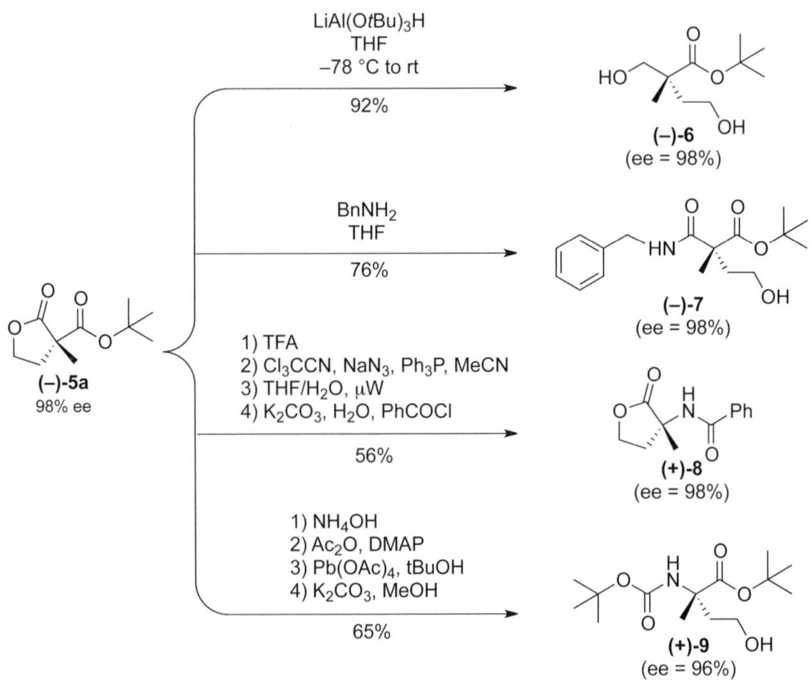

Scheme 1. Transformations of the γ-Lactones

of enantiopurity (ee = 98%).[7] Treatment of lactone **5a** with the benzyl amine gave amide ester (–)-**7** in 76% yield and 98% ee.[8] Curtius rearrangement of intermediate acyl azide yielded amido lactone (+)-**8** in 56% overall yield and no loss of enantiopurity.[9] Treatment of lactone **5a** with aqueous ammonium hydroxide followed by acetylation of the resulting alcohol yielded an amide ester that then underwent a Hofmann rearrangement with lead (IV) acetate and hydrolysis with potassium carbonate to give α-amino ester (+)-**9** in 65% overall yield and 96% ee.[10]

In conclusion, we have developed a highly efficient and generalized procedure for the synthesis of enantioenriched lactones containing a quaternary center through a Brønsted acid catalyzed desymmetrization of hydroxy diesters. The process has been improved to utilize only 1 mol% of Binol based phosphoric acid and proceed in under 48 hours.

References

1. Department of Chemistry and Biochemistry, University of North Carolina, Greensboro, NC 27412, kspeters@uncg.edu. Financial support is gratefully acknowledged from the National Institutes of Health (GM116041) and the American Chemical Society Petroleum Research Fund (53916-DNI).
2. (a) Jacobsen, E. N.; Pfaltz, A.; Yamamoto, H. *Comprehensive Asymmetric Catalysis*; Springer: Berlin, 1999; Vol. 1–3; (b) Ojima, I. *Catalytic Asymmetric Synthesis*; Wiley: New York, 2000; Vol. 2; (c) von Kieseritzky, F.; Wang, Y.; Axelson, M. *Org. Process Res. Dev.* **2014**, *18*, 643–645; (d) Serra, S.; Fuganti, C.; Brenna, E. *Trends Biotechnol.* **2005**, *23*, 193–198.
3. Akiyama, T.; Itoh, J.; Fuchibe, K. *Adv. Synth. Catal.* **2006**, *348*, 999–1010.
4. Janecki, T. Natural Lactones and Lactams: Synthesis, Occurrence and Biological Activity; Wiley-VCH Verlag GmbH & Co. KGaA: Weinheim, 2014.
5. Wilent, J.; Petersen, K. S. *J. Org. Chem.* **2014**, *79*, 2303–2307.
6. Ha, M. W.; Lee, H.; Yi, H. Y.; Park, Y.; Kim, S.; Hong, S.; Lee, M.; Kim, M.; Kim, T.; Park, H. *Adv. Synth. Catal.* **2013**, *355*, 637–642.
7. Reddy, D. S.; Shibata, N.; Nagai, J.; Nakamura, S.; Toru, T.; Kanemasa, S. *Angew. Chem. Int. Ed.* **2008**, *47*, 164–168.
8. Suzuki, T.; Goto, T.; Hamashima, Y.; Sodeoka, M. *J. Org. Chem.* **2007**, *72*, 246–250.

9. Volp, K. A.; Johnson, D. M.; Harned, A. M. *Org. Lett.* **2011**, *13*, 4486–4489.
10. Burgess, K.; Ho, K. *J. Org. Chem.* **1992**, *57*, 5931–5936.

Appendix
Chemical Abstracts Nomenclature (Registry Number)

Methyl malonic acid: Propanedioic acid, methyl-; (516-05-2)
4-(Dimethylamino)pyridine: 4-Pyridinamine, *N,N*-dimethyl-; (1122-58-3)
tert-butanol: 2-Propanol, 2-methyl-; (75-65-0)
Di-*tert*-butyl dicarbonate: Dicarbonic acid, bis(1,1-dimethylethyl) ester (24424-99-5)
Sodium hydride; (7646-69-7)
2-Bromoethyl acetate; (927-68-4)
(*R*)-3, 3′-Bis(2,4,6-triisopropylphenyl)-1,1′-binapthyl-2,2′-diylhydrogenphosphate; (791616-63-2)

Kimberly S. Petersen received her B.S. in chemistry from the University of Wisconsin, Madison. She then earned her Ph.D. at Johns Hopkins University under the supervision of Professor Gary Posner followed by postdoctoral research at the California Institute of Technology with Professor Brian Stoltz. In 2011, she began her independent career at the University of North Carolina at Greensboro.

Jennifer E. Wilent received her B.S. and M.S. in chemistry from the University of North Carolina at Wilmington. In 2011, she began her doctoral studies under the supervision of Dr. Kimberly S. Petersen. Her Ph.D. studies focus on asymmetric enantioselective syntheses of novel heterocycles.

Dr. Ghassan Qabaja was born and raised in West Bank, Palestine. He received his undergraduate degree from Yarmouk University in Jordan. In 2000, he received his Ph.D. from Northeastern University under the supervision of Graham Jones. His graduate studies focused on multistep synthesis of complex natural products and designed compounds. He worked as a medicinal chemistry in the biotechnology sector prior to his work at the University of North Carolina at Greensboro. He began as a research scientist for Dr. Kimberly S. Petersen in 2011.

Dr. Bo Qu received her Ph.D. with Professor Richard Adams at the University of South Carolina in 2002 followed by postdoctoral fellow with Professor David Collum at Cornell University. In 2006 she joined Chemical Development at Boehringer Ingelheim Pharmaceuticals in Ridgefield, CT, where she is currently a Principal Scientist. Dr. Qu's research interests focus on development of new catalytic transformations for efficient chemical processes. She has authored more than 80 papers and patents.

Organic Syntheses

Enantioselective Synthesis of α-Bromonitroalkanes for Umpolung Amide Synthesis: Preparation of *tert*-Butyl ((1*R*)-1-(4-(benzyloxy)phenyl)-2-bromo-2-nitroethyl)carbamate

Victoria T. Lim, Sergey V. Tsukanov, Amanda B. Stephens, and Jeffrey N. Johnston[1*]

Department of Chemistry and Vanderbilt Institute of Chemical Biology, Vanderbilt University, Nashville, Tennessee 37235

Checked by Fabian Konrad, Estíbaliz Merino and Cristina Nevado

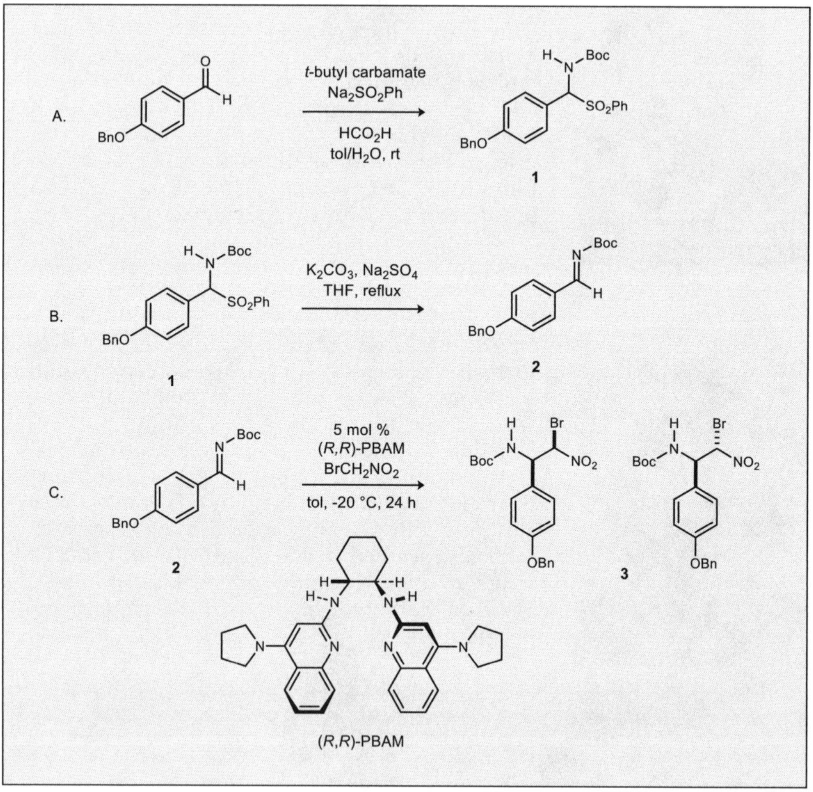

DOI: 10.15227/orgsyn.093.0088
Published on the Web 3/31/2016

Procedure

> *Caution! Bromonitromethane is a strong oxidizing agent with properties similar to nitromethane. Keep away from heat and avoid mixing with combustible materials.*

A. *tert-Butyl ((4-(benzyloxy)phenyl)(phenylsulfonyl)methyl)carbamate* (**1**). A 500-mL round-bottomed flask (29/32 neck) is equipped with a teflon-coated, egg-shaped stir bar (30 x 16 mm). The flask is charged with 4-benzyloxybenzaldehyde (10.0 g, 47.1 mmol, 1.2 equiv) and *tert*-butyl carbamate (4.60 g, 39.3 mmol, 1.0 equiv) (Note 1), then toluene (157 mL) (Note 2) is added, and the mixture is stirred until homogeneity is achieved. Benzenesulfinic acid sodium salt dihydrate (12.9 g, 64.4 mmol, 1.6 equiv) (Note 3) is introduced and then water (33 mL) is slowly added until the salt is completely dissolved. The flask is fitted with a rubber septum and an argon inlet needle. Formic acid (2.84 mL, 75.3 mmol, 2.0 equiv) (Note 4) is added, and the reaction is stirred for one week at room temperature (Note 5) Figure 1). The suspension is filtered through a Büchner funnel (7 cm diameter), fitted with filter paper. The reaction flask is rinsed with ether (2 x 100 mL and 1 x 50 mL) (Note 6) and the solid is soaked with the rinsates (2 x 100 mL and 1 x 50 mL) at room temperature for 30 min (Note 7). The suspension is stirred with a spatula every 5 min. Reduced pressure is used to facilitate the filtering process. The resulting crude product is

Figure 1. Initial appearance of Step A, appearance after 44 h, and appearance after 1 week.

transferred to a 500 mL Erlenmeyer flask. Residual solid is washed from the filter with dichloromethane (50 mL) (Note 8). Additional dichloromethane (200 mL) is added, and the flask is stirred until the crude product is fully dissolved. The resulting solution is dried over $MgSO_4$ (17.0 g) (Note 9), then filtered through a 350-mL fritted filter funnel into a pre-weighed 500-mL round-bottomed flask. The flask and the filter were rinsed with additional dichloromethane (2 x 50 mL). The dichloromethane solution is concentrated by rotary evaporation (40 °C, 23 mmHg) and further dried under vacuum (1-2 mmHg) for at least 10 h to leave 9.45–9.98 g of a white solid (Notes 7 and 10). The solid is used in step B without further purification.

B. *(E)-tert-Butyl 4-(benzyloxy)benzylidenecarbamate* (**2**). A 500-mL round-bottomed flask (29/32 neck) is equipped with a teflon-coated, egg-shaped stir bar (30 x 16 mm) (Note 11), charged with K_2CO_3 (18.2 g, 131.7 mmol, 6.0 equiv) and Na_2SO_4 (22.5 g, 158.3 mmol, 7.2 equiv) (Note 12). Sulfone (**1**) (9.97 g, 22.0 mmol) (Note 7) and THF (235 mL) (Note 13) were added. A condenser is attached to the reaction vessel, and the system is placed under an argon atmosphere. The reaction vessel is placed in an oil bath (85 °C), and heated to reflux for 3 h. The resulting mixture is cooled to room temperature (25 °C) and filtered through Celite® (1 inch layer) (Note 14) using a 350-mL fritted-filter funnel, washed with diethyl ether (2 x 100 mL) (Note 6) and additional diethyl ether (2 x 50 mL) is used to rinse residual material from the reaction vessel. The clear, colorless solution is collected in a pre-weighed 500-mL round-bottomed flask and concentrated by rotary evaporation (40 °C, 23 mmHg). The resulting product is further dried under vacuum (1-2 mmHg) to afford a white solid (6.80–6.94 g) (Note 15), which is used in step C without further purification (Note 16).

C. *tert-Butyl ((1R)-1-(4-(benzyloxy)phenyl)-2-bromo-2-nitroethyl)carbamate* (**3**). A 250-mL round-bottomed flask (29/32 neck) equipped with a teflon-coated, egg-shaped stir bar (30 x 16 mm) (Note 17), a rubber septum, and an argon inlet needle is charged with imine (**2**) (6.80 g, 21.9 mmol), toluene (120 mL) (Note 18), and (1*R*,2*R*)-N^1,N^2-bis[4-(1-pyrrolidinyl)-2-quinolinyl]-1,2-cyclohexanediamine, ((*R*,*R*)-PBAM) (0.54 g, 1.10 mmol, 0.05 equiv) (Note 19). The reaction vessel is placed under argon atmosphere and the solution is cooled to –20 °C (Note 20) followed by the addition of bromonitromethane (1.81 mL, 26.1 mmol, 1.2 equiv) (Note 21). The reaction mixture is stirred at –20 °C for 24 h and the resulting yellow-orange solution (Figure 2) is concentrated using rotary evaporation (40 °C, 23 mmHg).

The product is redissolved in 50 mL of hot ethyl acetate (Note 22), filtered through a silica plug (Note 23), and washed with additional hot ethyl acetate (3 x 100 mL) into a 1000-mL receiving flask. The resulting clear, light yellow solution is concentrated by rotary evaporation (40 °C, 23 mmHg) and is dried under vacuum (1-2 mmHg) for 1 h. The resulting white solid is then transferred to a 50-mL Erlenmeyer flask and redissolved

Figure 2. Appearance of Step C after 24 h.

in hot ethyl acetate (ca. 30 mL). The solution is allowed to slowly cool to room temperature before it is placed in an ice-water bath. After 1 h the resulting crystals were vacuum filtered through a Büchner funnel (5 cm diameter). The crystallized product is washed with ice-cold ethyl acetate on the filter. The mother liquor and crystal washings were concentrated and subjected to the same recrystallization procedure twice to obtain additional product. The solid is transferred to a 100-mL round-bottomed flask and dried under vacuum (1-2 mmHg) to leave the product **3** as a 1.2:1 mixture of diastereoisomers (6.26–7.52 g, 34–40% overall yield) (Notes 25).

Notes

1. 4-Benzyloxyaldehyde (98%) was purchased from Alfa Aesar and *tert*-butyl carbamate (98%) was purchased from Sigma-Aldrich.
2. Toluene (99.9%) was purchased from Fisher Scientific Company and used as received.

3. Benzenesulfinic acid sodium salt dihydrate (97%) was purchased from Alfa Aesar. Benzene acid sodium salt anhydrous (Sigma Aldrich, 123579) also can be used with similar results.
4. Formic acid (≥95%) was purchased from Sigma-Aldrich.
5. A clear, pale yellow mixture was observed after addition of all reagents, and the product started to precipitate as a white solid after 1-2 h. By the second day, a thick white slurry of the product was formed.
6. Ethyl ether (99.9%) was purchased from Fisher Scientific Company and used as received.
7. The checkers observed more residual aldehyde (11-22%) than the submitters (<3%) by ^{1}H-NMR of the crude reaction mixture. The submitters note that the suflone is insoluble in ether, allowing fresh ether to be used to thoroughly soak and wash the solid during the filtration process.
8. Dichloromethane (99.9%) was purchased from Fisher Scientific Company and used as received.
9. Magnesium sulfate was purchased from Fisher Scientific Company and used as received.
10. Data of product **1**: R_f = 0.46 (30% EtOAc/hexanes); ^{1}H NMR (400 MHz, $CDCl_3$) δ: 1.25 (s, 9H), 5.09 (s, 2H), 5.74 (d, *J* = 10.6 Hz, 1H), 5.88 (d, *J* = 10.0 Hz, 1H), 7.01 (d, *J* = 8.8 Hz, 2H), 7.31–7.43 (m, 7H), 7.52 (m, 2H), 7.63 (t, *J* = 7.4 Hz, 1H), 7.91 (d, *J* = 7.6 Hz, 2H); ^{13}C NMR (101 MHz, $CDCl_3$) δ: 28.0, 70.1, 73.5, 81.2, 115.2, 122.0, 124.9, 127.4, 128.1, 128.6, 129.0, 129.4, 130.2, 133.8, 136.5, 137.0, 153.5, 160.0.
11. The apparatus with the salts was flame dried under reduced pressure (1-2 mmHg) and was allowed to cool to room temperature.
12. Potassium carbonate and sodium sulfate were purchased from Sigma-Aldrich and used as received.
13. Tetrahydrofuran (THF) was dried by passage through a column of activated alumina as described by Grubbs.[2]
14. Celite was purchased from Sigma-Aldrich and used as received.
15. The imine product was stored in a –78 °C freezer. ^{1}H NMR (400 MHz, $CDCl_3$) δ: 1.59 (s, 9H), 5.13 (s, 2H), 7.04 (d, *J* = 9.0 Hz, 2H), 7.33–7.37 (m, 1H), 7.40–7.42 (m, 4H), 7.89 (d, *J* = 8.8 Hz, 2H), 8.89 (s, 1H); ^{13}C NMR (101 MHz, $CDCl_3$) δ: 27.9, 70.2, 81.9, 115.2, 127.1, 127.5, 128.2, 128.7, 132.5, 136.0, 162.8, 163.3, 169.6; IR (CH_2Cl_2): 1698, 1507, 1496, 1455, 1366, 1236, 1162, 1016, 881, 830, 735, 696 cm^{-1}. HRMS (ESI): Exact mass calcd for $C_{19}H_{22}NO_3$ $[M+H]^+$ 312.1594, found 312.1598.

16. The material was used as a reactant to prepare **3** and contained 3-16% of 4-benzyloxybenzaldehyde.
17. The flask was flame dried under reduced pressure (1-2 mmHg) and was allowed to cool to room temperature.
18. Toluene was dried by passage through a column of activated alumina as described by Grubbs.[2]
19. (*R*,*R*)-PBAM was prepared according to a published procedure.[4]
20. The reaction was cooled using a cryostat (at –20 °C) with an isopropanol bath.
21. Bromonitromethane (90%, tech.) was purchased from Sigma-Aldrich and used as received.
22. Ethyl acetate (99.9%) was purchased from Fisher Scientific Company and used as received.
23. A 350-mL fritted-glass funnel was packed with Celite® (1 inch layer) and silica gel (50 g).
24. A sample of racemic **3** can be prepared by stirring the imine and bromonitromethane in toluene with DMAP (10 mol%) for 10 min at room temperature, followed by its filtration through silica gel.
25. The enantiomeric excess of the major and minor diastereomers was determined to be 97% and 95% ee, respectively, by chiral HPLC analysis (Chiralcel AD-H, 20% ^{i}PrOH/hexanes, 1 mL/min, t_r(d_1e_1, major) = 15.6 min, t_r(d_1e_2, minor) = 19.33 min, t_r(d_2e_2, minor) = 20.4 min), t_r(d_2e_1, major) = 26.1 min. R_f = 0.38 (20% EtOAc/hexanes); mp = 149–151 °C; R_f = 0.38 (20% EtOAc/hexanes); IR (film) 3359, 1687, 1561, 1509, 1352, 1251, 1162, 1039, 1027, 834, 736 cm^{-1}; ^{1}H NMR (400 MHz, $CDCl_3$, 1.2:1 mixture of diastereomers) δ: 1.45 (s, 9H), 1.46 (s, 9H), 5.05 (s, 2H), 5.06 (s, 2H), 5.38 (d, *J* = 9.2 Hz, 1H), 5.42 (br s, 1H), 5.57 (dd, *J* = 8.9 and 4.3 Hz, 1H), 5.68 (br s, 1H), 6.27 (d, *J* = 2.5 Hz, 1H), 6.31 (d, *J* = 4.1 Hz, 1H), 6.96 (m, 2H), 6.99 (m, 2H), 7.21 (d, *J* = 8.6 Hz, 2H), 7.24 (d, *J* = 8.6 Hz, 2H), 7.31–7.45 (m, 10H); ^{13}C NMR (101 MHz, $CDCl_3$) δ: 28.1, 28.2, 57.6, 57.8, 70.1 (2C), 81.0, 81.1, 82.0 (2C), 85.1, 115.2, 115.3, 126.8 (2C), 127.4 (2C), 128.1 (2C), 128.2 (2C), 128.3 (2C), 128.7 (2C), 136.5 (2C), 154.4, 154.7, 159.2, 159.3; HRMS (ESI): Exact mass calcd for $C_{20}H_{23}BrN_2NaO_5$ $[M+Na]^+$ 473.0683, found 473.0685; Anal. calcd. for $C_{20}H_{23}BrN_2O_5$: C, 53.23; H, 5.14; N, 6.21; found: C, 53.04; H, 5.17; N, 6.13.

Working with Hazardous Chemicals

The procedures in *Organic Syntheses* are intended for use only by persons with proper training in experimental organic chemistry. All hazardous materials should be handled using the standard procedures for work with chemicals described in references such as "Prudent Practices in the Laboratory" (The National Academies Press, Washington, D.C., 2011; the full text can be accessed free of charge at http://www.nap.edu/catalog.php?record_id=12654). All chemical waste should be disposed of in accordance with local regulations. For general guidelines for the management of chemical waste, see Chapter 8 of Prudent Practices.

In some articles in *Organic Syntheses*, chemical-specific hazards are highlighted in red "Caution Notes" within a procedure. It is important to recognize that the absence of a caution note does not imply that no significant hazards are associated with the chemicals involved in that procedure. Prior to performing a reaction, a thorough risk assessment should be carried out that includes a review of the potential hazards associated with each chemical and experimental operation on the scale that is planned for the procedure. Guidelines for carrying out a risk assessment and for analyzing the hazards associated with chemicals can be found in Chapter 4 of Prudent Practices.

The procedures described in *Organic Syntheses* are provided as published and are conducted at one's own risk. *Organic Syntheses, Inc.*, its Editors, and its Board of Directors do not warrant or guarantee the safety of individuals using these procedures and hereby disclaim any liability for any injuries or damages claimed to have resulted from or related in any way to the procedures herein.

Discussion

The amide bond might be considered a treasured functional group within organic synthesis, owing to its natural abundance among small molecules and peptidic biomolecules, its utility within the process of drug development, and the existence of an array of methods and reagents to prepare amides. The role of synthetic peptides and peptidomimetics among marketed pharmaceuticals continues to increase, with over 50 approved as medicines.[4] This has resulted in substantial pressure to improve amide bond-forming methods, particularly those reliant on 'coupling reagents' used for carboxylic acid-amine condensations.[5] This has accelerated innovation in amide bond synthesis, leading to a range of new methods.[6]

We discovered that α-bromo nitroalkanes could engage amines in amide bond formation with the assistance of *N*-iodosuccinimide.[7] Evidence collected during the initial study, and conistent with subsequent mechanistic studies, suggested that the key carbon-nitrogen bond-forming step involved a nucleophilic nitronate carbon and an electrophilic iodamine nitrogen.[8] This inversion of polarity relative to all other known amide bond-forming reactions led to our description of this reaction as Umpolung Amide Synthesis (UmAS).[9]

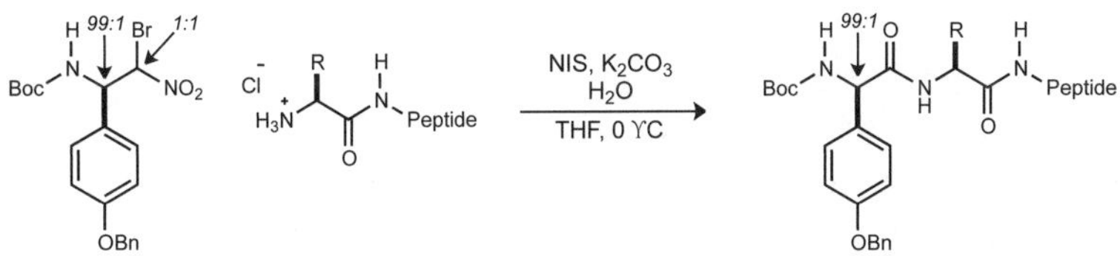

Scheme 1. Use of an Umpolung Amide Synthesis reaction (UmAS) to homologate with a *para*-hydroxyl phenyl glycine residue (Hpg) using 3.

This preparation highlights the utility of bis(amidine)-based organocatalysis in enantioselective α-amino amide synthesis. Oxygenated aryl glycines are observed in a range of interesting peptidic natural products, including the vancomycin class of antibiotics. *para*-Hydroxy phenyl glycine (Hpg) is a key component of cephadroxil and feglymycin,[10] and in vancomycin where it is crosslinked and/or halogenated.[11] Use of the Sharpless aminohydroxylation reaction remains a prevailing approach to Hpg based on enantioselective catalysis, targeting the protected α-amino acid for the homologation step.[12] The preparation described above provides gram-scale access to α-bromo nitroalkanes **3** which converge during UmAS to achieve homologation with (protected) D-Hpg. The coupling of aryl glycines using conventional dehydrating agents and a carboxylic acid donor is subject to competitive epimerization of the α-stereocenter.[13] Preservation of this stereochemical integrity is a key advantage of UmAS due to its unique mechanism which avoids an electrophilic active ester and acidic α-C–H bond.

This practical three-step protocol is straightforward, is characterized by a good overall yield, and does not require any chromatographic purification steps. In the first two steps the corresponding aldehyde was converted into an imine via an α-amido sulfone intermediate by a similar procedure previously described.[3] The final enantioselective organocatalytic aza-Henry step with bromonitromethane was conducted in the presence of 5 mol % of PBAM catalyst.[4] Filtration of the reaction through a plug of silica gel alleviates the need for chromatography and simplifies the isolation of the product. The α-bromonitroalkanes **3** are obtained with high ee (95-97%) and good overall yield for use directly in an UmAS coupling.

References

1. Department of Chemistry and Vanderbilt Institute of Chemical Biology, Vanderbilt University, Nashville, Tennessee 37235, USA; Fax: 615-343-6361, E-mail: jeffrey.n.johnston@vanderbilt.edu. This work was financially supported by the National Institutes of Health, General Medical Sciences (GM 063557 & GM 084333).
2. Pangborn, A. B.; Giardello, M. A.; Grubbs, R. H.; Rosen, R. K.; Timmers, F. J. *Organometallics* **1996**, *15*, 1518–1520.
3. Yang, J. W.; Pan, S. C.; List, B. *Org. Synth.* **2009**, *86*, 11.
4. Davis, T. A.; Dobish, M. C.; Schwieter, K. E.; Chun, A. C.; Johnston, J. N. *Org. Synth.* **2012**, *89*, 380–393.
5. Bray, B. L. *Nat. Rev. Drug Discov.* **2003**, *2*, 587–593.
6. a) Constable, D. J. C.; Dunn, P. J.; Hayler, J. D.; Humphrey, G. R.; Leazer, J. J. L.; Linderman, R. J.; Lorenz, K.; Manley, J.; Pearlman, B. A.; Wells, A.; Zaks, A.; Zhang, T. Y. *Green Chem.* **2007**, *9*, 411–420; b) Valeur, E.; Bradley, M. *Chem. Soc. Rev.* **2009**, *38*, 606–631.
7. a) Pattabiraman, V. R.; Bode, J. W. *Nature* **2011**, *480*, 471–479; b) Albericio, F. *Curr. Opin. Chem. Biol.* **2004**, *8*, 211–221.
8. Shen, B.; Makley, D. M.; Johnston, J. N. *Nature* **2010**, *465*, 1027-1032.
9. Shackleford, J. P.; Shen, B.; Johnston, J. N. *Proc. Natl. Acad. Sci. U. S. A.* **2012**, *109*, 44–46.

10. For an example of umpolung hydroxamic acid synthesis, see: Wong, F. T.; Patra, P. K.; Seayad, J.; Zhang, Y.; Ying, J. Y. *Org. Lett.* **2008**, *10*, 2333–2336.
11. a) Bunkóczi, G.; Vértesy, L.; Sheldrick, G. M. *Angew. Chem. Int. Ed.* **2005**, *44*, 1340–1342. b) Vertesy, L.; Aretz, W.; Knauf, M.; Markus, A.; Vogel, M.; Wink, J. *J. Antibiot.* **1999**, *52*, 374–382.
12. Williams, R. M.; Hendrix, J. A. *Chem. Rev. (Washington, DC, U. S.)* **1992**, *92*, 889–917.
13. Dettner, F.; Hänchen, A.; Schols, D.; Toti, L.; Nußer, A.; Süssmuth, Roderich D. *Angew. Chem. Int. Ed.* **2009**, *48*, 1856–1861.
14. Smith, G. G.; Sivakua, T. *J. Org. Chem.* **1983**, *48*, 627–634.

Appendix
Chemical Abstracts Nomenclature (Registry Number)

4-(Phenylmethoxy)-benzaldehyde; (4397-53-9)
Benzenesulfinic acid sodium salt dihydrate; (25932-11-0)
Carbamic acid, 1,1-dimethylethyl ester, *tert*-butyl carbamate; (4248-19-5)
Formic acid; (64-18-6)
Sulfuric acid sodium salt (1:2), Sodium sulfate; (7757-82-6)
Carbonic acid, potassium salt (1:2), Potassium carbonate; (584-08-7)
Bromonitromethane; (563-70-2)
((*R*,*R*)-PBAM): (1*R*,2*R*)-N^1,N^2-bis[4-(1-pyrrolidinyl)-2-quinolinyl]- 1,2-cyclohexanediamine; (1214287-91-8)

Victoria Lim, a native of Spring Hill, TN, received her B.S. degree in both chemistry and mathematics from Belmont University in 2015. She was an undergraduate researcher in the Johnston laboratory, and matriculated at the University of California, Irvine, in 2015 to pursue research in computational chemistry.

Sergey Tsukanov completed his Master's Degree from Moscow State Academy of Fine Chemical Technology in 2007. Then he received his Ph.D. in Organic Chemistry under the guidance of Prof. Daniel L. Comins at North Carolina State University where he studied the total synthesis of complex alkaloids. In 2012 he started his postdoctoral training at Vanderbilt University focusing on synthesis of the peptidic natural product feglymycin using Umpolung Amide Synthesis and enantioselective aza-Henry chemistry. He received a Lilly Innovation Fellowship Award in 2013 and led a collaborative project to develop a continuous flow paradigm appropriate for nitroalkane synthesis and enantioselective organocatalysis.

Amanda Stephens (née Doody) completed her B.S. Chemistry degree in 2008 at Wofford College. She was an NSF-REU fellow at Columbia University in Summer 2007, where she worked with Prof. Gerard Parkin. Amanda completed her Ph.D. in Organic Chemistry at Vanderbilt Universtiy in 2014 under the mentorship of Prof. Jeffrey N. Johnston, where she focused on the applications of Brønsted acid-catalyzed additions of azides and diazo-ylides to electrophiles, and Umpolung Amide Synthesis to the synthesis of alkaloid and peptidic natural products. Amanda is currently an Academic Professional at the Georgia Institute of Technology, where she coordinates and teaches the organic and inorganic synthesis laboratories for the School of Chemistry & Biochemistry.

Jeffrey N. Johnston completed his B.S. Chemistry degree at Xavier University in 1992, and a Ph.D. in organic chemistry at The Ohio State University in 1997 with Prof. Leo A. Paquette. He then worked as an NIH postdoctoral fellow with Prof. David A. Evans at Harvard University. He began his independent career at Indiana University, where he was promoted to Professor of Chemistry in 2005. In 2006, his research program moved to Vanderbilt University. His group has developed a range of new reactions and reagents that are used to streamline the chemical synthesis of complex small molecules. The integration of new enantioselective Brønsted acid-catalyzed reactions with target-oriented synthesis is an ongoing investigational theme.

Fabian Konrad completed his education as a laboratory technician at the Organic Chemistry Institute of the University of Zürich in 2009. He worked in Virometix AG and Merck before joining Prof. Cristina Nevado´s group at the University of Zürich (2013-2015).

Estíbaliz Merino obtained her PhD degree from the Autónoma University (Madrid-Spain). After a postdoctoral stay with Prof. Magnus Rueping at Goethe University Frankfurt and RWTH-Aachen University in Germany, she worked with Prof. Avelino Corma at Instituto de Tecnología Química-CSIC (Valencia) and Prof. Félix Sánchez at Instituto de Química Orgánica General-CSIC (Madrid) in Spain. At present, she works as research associate in Prof. Cristina Nevado´s group at the University of Zürich. She is interested in the synthesis of natural products using catalytic tools and in the development of new materials with application in heterogeneous catalysis.

Organic Syntheses

Synthesis of 1-Bromopyrene and 1-Pyrenecarbaldehyde

Matthias Schulze,[¶] Alexander Scherer,[¶] Colin Diner,[§] and Rik R. Tykwinski[1¶*]

Department of Chemistry and Pharmacy & Interdisciplinary Center for Molecular Materials (ICMM), Friedrich-Alexander-Universität Erlangen-Nürnberg (FAU), Henkestraße 42, 91054 Erlangen, Germany; [§]Department of Chemistry, University of Alberta, Edmonton, AB, T6G 2G2 Canada

Checked by Jeffrey D. Mighion and Huw M. L. Davies

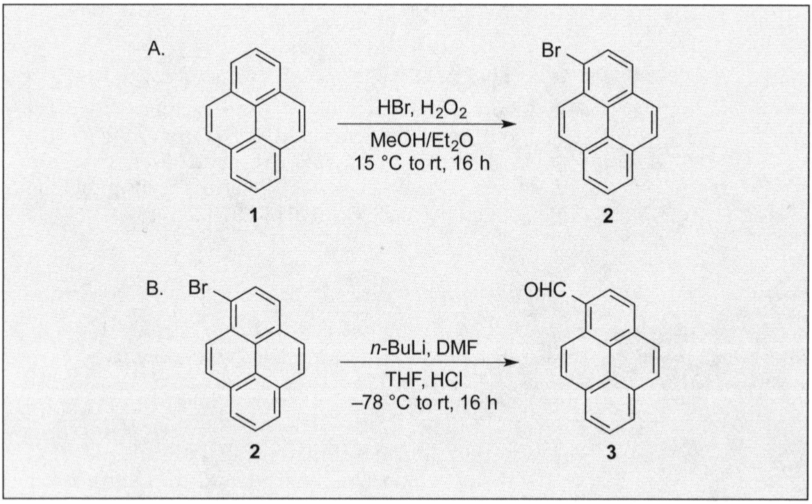

Procedure

A. *1-Bromopyrene* (**2**). A 500-mL, one-necked, round-bottomed flask was equipped with an octagonal Teflon-coated magnetic stirbar (25 mm x 10 mm), and a mixture of MeOH (Note 1) and Et_2O (250 mL, 1:1 v/v) (Note 2) is added. Pyrene (**1**) (20.0 g, 98.9 mmol, 1.00 equiv) (Note 3) is added with vigorous stirring, followed by addition of HBr (48% w/w aq, 8.80 g, 12.3 mL, 109 mmol, 1.10 equiv) (Note 4) *via* syringe. The round-

DOI: 10.15227/orgsyn.093.0100
Published on the Web 4/4/2016

bottomed flask is equipped with a 25-mL pressure-equalizing addition funnel, which is then charged with H_2O_2 (30% w/w aq, 3.54 g, 10.4 mL, 104 mmol, 1.05 equiv) (Note 5). The reaction mixture is cooled to 15 °C using a water bath (Note 6), and the H_2O_2 slowly added over a period of 0.5 h (ca. 7 drops/min) (Note 7). The addition funnel is then replaced by a gas bubbler, the cooling bath removed, and the mixture stirred for 16 h at ambient temperature (Figure 1). Water (150 mL) is added, and the mixture is extracted with CH_2Cl_2 (2 x 250 mL) (Note 8) in a 1-L separatory funnel. The combined organic extracts are washed with NaOH (1 M aq, 150 mL) (Note 9) and saturated aq NaCl (2 x 150 mL) (Note 10) in a 1-L separatory

Figure 1. Reaction assembly for Step A before addition of H_2O_2, after addition of H_2O_2, and after mixture has stirred for 16 h

funnel, dried over $MgSO_4$ (20 g, 5 min) (Note 11), and filtered through a plug of silica gel (Note 12), which is rinsed with CH_2Cl_2 (2 x 100 mL) (Note 8). The solvent is removed by rotary evaporation (40 °C, 675–20 mmHg) and the resulting solid is dried *in vacuo* (0.01 mmHg) for 2 h. The crude material is placed in a Soxhlet extractor (Note 13) and extracted with *n*-pentane (750 mL) (Note 14) for 48 h (Figure 2). After cooling to room temperature, the *n*-pentane fraction, already containing crude precipitated pale yellow product, is concentrated by rotary evaporation (40 °C, 675 mmHg) to 500 ± 25 mL and placed in a refrigerator (ca. 2 °C) overnight. The formed precipitate is collected by gravity filtration, dried *in vacuo* (0.01 mmHg) for 2 h, dissolved in boiling hexanes (1.0 L) (Note 15) in a 2-L, round-bottomed flask, and placed in a freezer (ca. –20 °C) overnight. The formed precipitate is collected by filtration on a Büchner funnel and dried *in vacuo* (0.01 mmHg) for 8 h to give 18.9 g of the title compound **2** as a pale yellow, air-stable powder. The mother liquor is concentrated to 400 ± 20 mL by rotary evaporation (40 °C, 315–245 mmHg) and cooled in a freezer (ca. –20 °C) overnight. The mixture is filtered using a Büchner funnel and the product dried *in vacuo* (0.01 mmHg) for 8 h to obtain 2.5 g of additional material. The combined yield from both recrystallizations is 21.4 g (77%) (Notes 16 and 17).

Figure 2. Soxhlet extraction assembly before extraction and after 48 h extraction.

B. *1-Pyrenecarbaldehyde* (**3**). A 500-mL, one-necked, round-bottomed Schlenk flask is equipped with an octagonal Teflon-coated magnetic stirbar (50 mm x 10 mm). After connecting to the vacuum/nitrogen line, 1-bromopyrene (**2**) (21.4 g, 0.076 mol, 1.0 equiv) is added, and the flask equipped with a rubber septum. The flask is evacuated under vacuum and purged with nitrogen (three times). Dry deoxygenated THF (300 mL) (Note 18) is added under a nitrogen atmosphere. After cooling to –78 °C (Note 19), *n*-BuLi (1.6 M in hexane, 62 mL, 0.99 mol, 1.3 equiv) (Note 20) is added *via* syringe under a nitrogen atmosphere over a period of 10 min to the light brown solution (Note 21). After stirring for 1 h, anhydrous DMF (7.6 g, 8.0 mL, 0.99 mol, 1.3 equiv) (Note 22) is added *via* syringe within 5 s (Note 23), and the mixture is allowed to reach room temperature while stirring for 16 h overnight under a nitrogen atmosphere. The mixture is carefully poured over 10 min at room temperature into a rapidly stirred mixture of conc HCl and H_2O (500 mL, 1:1 v/v) (Note 24) in a 2-L one-necked, round-bottomed flask equipped with an octagonal Teflon-coated magnetic stirbar (50 mm x 10 mm) (Figure 3). Diethyl ether (500 mL) (Note 2) is added. The organic phase is separated in a 2-L separatory funnel and washed with H_2O (3 × 300 mL). After drying over $MgSO_4$ (10 g, 5 min) (Note 11), the mixture

Figure 3. Reaction assembly for Step B after the addition of *n*-BuLi, after addition of DMF, and after mixture is poured into aqueous HCl

is filtered by gravity filtration, the solvent removed from the filtrate by rotary evaporation (40 °C, 650–10 mmHg) giving an orange solid, which is dried *in vacuo* (0.01 mmHg) for 2 h. The orange solid is dissolved in boiling EtOH (450 mL) (Note 25) in a 500-mL, one-necked, round-bottomed flask and after addition of H_2O (5 mL) *via* syringe and cooling to room temperature, the mixture is placed in a freezer (ca. –20 °C) overnight. The formed precipitate is collected by suction filtration using a medium porosity sintered glass funnel and dried *in vacuo* (0.01 mmHg) for 8 h to give 7.3 g of the title compound **3** as a dark yellow, crystalline, air-stable solid. The mother liquor is concentrated to 250 ± 15 mL by rotary evaporation (40 °C, 130 mmHg) and cooled in a freezer (ca. –20 °C) overnight. The formed precipitate is collected by suction filtration using a medium porosity sintered glass funnel and dried *in vacuo* (0.01 mmHg) for 8 h to obtain 3.3 g of additional material. The mother liquor is concentrated to 100 ± 10 mL and placed in a freezer (ca. –20 °C) overnight. The precipitate is collected and dried as described above to obtain 2.4 g of additional material. The combined yield from the three recrystallizations is 13.0 g (74%) (Notes 26 and 27) (Figure 4).

Figure 4. Product of Step B

Notes

1. Methanol was purchased from Fisher (HPLC grade, 99.9% pure) and was used as received.
2. Diethyl ether was purchased from Fisher (99.9% BHT stabilized) and was used as received.
3. Pyrene (**1**) was purchased from Sigma-Aldrich (98%) and was used as received.
4. Hydrobromic acid (48% w/w aq) was purchased from Fluka (purum p.a.; ≥48%) and was used as received.
5. Hydrogen peroxide (30% w/w aq) was purchased from Sigma-Aldrich and was used as received.
6. To cool to 15 °C, the solution was stirred for 10 min with the round-bottomed flask immersed in a water/ice bath. Not all starting material is dissolved.
7. During addition of hydrogen peroxide the color of the reaction mixture darkens slightly to deep-yellow/orange and a colorless/grayish precipitate slowly starts to appear; the amount of precipitate increases while stirring for 16 h overnight.
8. Dichloromethane was purchased from Fisher (99.9% pure) and was used as received.
9. Sodium hydroxide was purchased from Macron (95% pure) and was used as received.

10. Sodium chloride was purchased from Sigma-Aldrich (>99.0% pure) and was used as received.
11. Magnesium sulfate (anhydrous) was purchased from Fisher (powder/certified) and was used as received.
12. The size of the silica plug was 5 cm (diameter) x 1 cm. Silica gel 60 (0.048–0.063 mm) was purchased from Silicycle.
13. Soxhlet apparatus: 19 cm length x 4.5 cm diameter on a 1-L, one-necked, round-bottomed flask filled with *n*-pentane (Note 14). The crude material was placed as a solid in the Soxhlet-cartridge and the *n*-pentane was heated to 50–55 °C *via* an oil bath.
14. *n*-Pentane was purchased from Fisher (99.6% pure) and was used as received.
15. Hexanes (isomeric mixture, 99.6%) was purchased from Fisher and was used as received.
16. When the reaction was performed on half scale, the isolated yield was 10.6 g (76%) with purity of >99%.
17. Physical properties and spectroscopic analysis of **2**: mp 93–94 °C (for both crops of material). IR (ATR) 3047, 1588, 1481, 1429, 178, 1015, 966, 835, 751, 706 cm^{-1}; ^{1}H NMR (400 MHz, $CDCl_3$) δ: 7.99–8.06 (m, 3H), 8.09 (d, *J* = 8.9 Hz, 1H), 8.17 (d, *J* = 9.2 Hz, 1H), 8.20–8.24 (m, 3H), 8.43 (d, *J* = 9.2 Hz, 1H); ^{13}C NMR (100 MHz, $CDCl_3$) δ: 120.0, 124.1, 125.6, 125.7, 125.9, 126.0, 126.1, 126.6, 127.2, 127.8, 129.1, 129.7, 130.1, 130.7, 131.1, 131.3. APCI HRMS *m*/*z* calcd. for $C_{16}H_9Br$ $[M+H]^+$ 279.9882, found 279.9885. Purity analysis of **2**: HPLC > 99 area % purity at 254 nm detection for both crops of material, using an Agilent 1260 Infinity instrument package (pump/autosampler/detector P4000/AS 3000/UV6000LP); HPLC conditions, Macherey-Nagel Nucleosil 100-5 C18 column (4.6 x 250 mm), 5 μM particle size, 1.00 mL/min flow; eluent (acetonitrile/water = 80/20); product elutes at 17.0 min. Residual pyrene (HPLC = <1 area % at 254 nm detection; product elutes at 9.0 min) was identified.
18. Tetrahydrofuran was purchased from Fisher (HPLC grade, 99.9% pure) and passed through an activated alumina column under an argon atmosphere prior to use.
19. The solution was cooled *via* a dry ice-acetone bath.
20. *n*-Butyllithium (1.6 M in hexane) was obtained from Acros Organics and stored at ca. 2 °C in a refrigerator. The reagent solution was used as received.
21. During addition of *n*-butyllithium, a yellow precipitate appears.

22. *N,N*-Dimethylformamide anhydrous, Drisolv® (water <50 ppm) was obtained from EMD Millipore and was used as received.
23. With the addition of *N,N*-dimethylformamide, the yellow precipitate disappears and transforms into a grey mixture, which changes color to brown by stirring overnight.
24. Hydrochloric acid (37.0%), obtained from Fisher, was used as received. Quenching the *N,N*-dimethylformamide complex with HCl/H_2O gives a brown organic phase and a yellow aqueous phase. The solution becomes warm during quenching and is cooled to room temperature prior to addition of diethyl ether.
25. Ethanol was purchased from Decon Labs (200 proof).
26. When the reaction was performed on half-scale, the isolated yield was 5.82 g (67%) with purity of 95%.
27. Physical properties and spectroscopic analysis of **3**: mp 125–126 °C (1st crop), 124–125 °C (2nd crop), 123–124 °C (3rd crop). IR (ATR) 3041, 2857, 2716, 1678, 1596, 1508, 1227, 1201, 845 cm^{-1}; ^1H NMR (500 MHz, $CDCl_3$, at 60 mg/mL, ^1H NMR is concentration dependent) δ: 7.89 (d, *J* = 8.9 Hz, 1H), 7.93–8.08 (m, 3H), 8.11 (d, *J* = 7.6 Hz, 1H), 8.16 (d, *J* = 7.6 Hz, 2H), 8.23 (d, *J* = 7.9 Hz, 1H), 9.20 (d, *J* = 9.3 Hz, 1H), 10.63 (s, 1H); ^{13}C NMR (125 MHz, $CDCl_3$) δ: 123.2, 124.2, 124.7, 124.8, 126.7, 127.0, 127.2, 127.3, 127.5, 130.6, 130.9, 131.0, 131.1, 131.2, 131.5, 135.7,193.2. APCI HRMS *m/z* calcd. for $C_{17}H_{11}O$ $[M+H]^+$ 231.0804, found 231.0806. Purity analysis of **3**: HPLC 95 area % purity (1st crop), 97% area % purity (2nd crop), and 97% area % purity (3rd crop) at 254 nm detection, using an Agilent 1260 Infinity instrument package (pump/autosampler/detector P4000/AS 3000/UV6000LP); HPLC conditions, Macherey-Nagel Nucleosil 100-5 C18 column (4.6 x 250 mm), 5 μM particle size, 1.00 mL/min flow; eluent (acetonitrile/water = 80/20); product elutes at 7.6 min.

Working with Hazardous Chemicals

The procedures in *Organic Syntheses* are intended for use only by persons with proper training in experimental organic chemistry. All hazardous materials should be handled using the standard procedures for work with chemicals described in references such as "Prudent Practices in the Laboratory" (The National Academies Press, Washington, D.C., 2011; the full text can be accessed free of charge at http://www.nap.edu/catalog.php?record_id=12654). All chemical waste should be disposed of in accordance with local regulations. For general guidelines for the management of chemical waste, see Chapter 8 of Prudent Practices.

In some articles in *Organic Syntheses*, chemical-specific hazards are highlighted in red "Caution Notes" within a procedure. It is important to recognize that the absence of a caution note does not imply that no significant hazards are associated with the chemicals involved in that procedure. Prior to performing a reaction, a thorough risk assessment should be carried out that includes a review of the potential hazards associated with each chemical and experimental operation on the scale that is planned for the procedure. Guidelines for carrying out a risk assessment and for analyzing the hazards associated with chemicals can be found in Chapter 4 of Prudent Practices.

The procedures described in *Organic Syntheses* are provided as published and are conducted at one's own risk. *Organic Syntheses, Inc.*, its Editors, and its Board of Directors do not warrant or guarantee the safety of individuals using these procedures and hereby disclaim any liability for any injuries or damages claimed to have resulted from or related in any way to the procedures herein.

Discussion

Due to the useful photophysical attributes of pyrene (**1**),[2] it is a desirable component of many functional materials.[3] Thus, 1-bromopyrene (**2**) is a key building block for a large portion of the chemical community.[4] The first synthesis of **2** was described in 1937 by Lock, *via* the bromination of **1**.[5] Similar approaches have since been described using reagents such as NBS,[6] $CuBr_2$,[7] and HBr with H_2O_2.[8] The use of column chromatography in these protocols, however, limits the reaction scale.[8] Fulfilling the demand for 1-bromopyrene in larger amounts, a simple and chromatography free procedure has been reported in 1968 by Gumprecht.[9] Unfortunately, this procedure (and many others) require the use of CCl_4 as solvent,[5–7,9] which has been prohibited in many countries[10] due to toxicological[11] and environmental concerns.[12] Thus, there is considerable demand for a procedure with similar synthetic utility that does not require CCl_4 or column chromatography. Herein, we report such a protocol for the synthesis of **2** that has been reliably reproduced on a 10–20 g scale by undergraduate students and that tolerates the use of inexpensive commercially available technical grade pyrene as precursor material.

For the design of functional materials, **2** can be easily transferred in analogy to other haloarenes into reactive species e.g., by lithiation,[13] magnesiation,[14] or borylation.[15] Consequently, **2** has been used in a large variety of reactions (Scheme 1).[13–15]

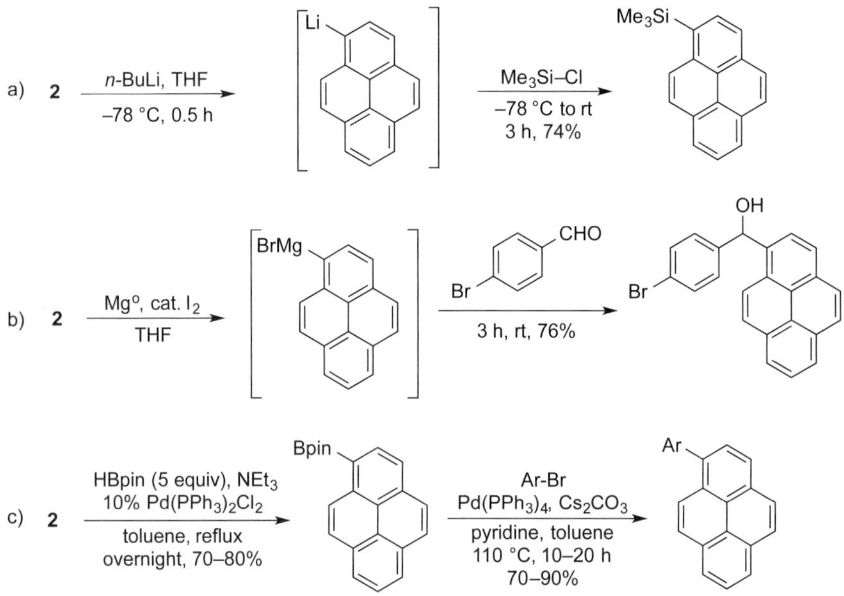

Scheme 1. Functionalization of 2 via reactive species using routes: a) lithiation,[13] b) Grignard reaction,[14] and c) borylation.[15]

Bromide **2** can also be used directly in transition metal-catalyzed cross-coupling reactions. For example, heteroatomic or organic nucleophiles can be directly linked to **2** *via* Buchwald–Hartwig amination[16] or Sonogashira cross-coupling[17] (Scheme 2).

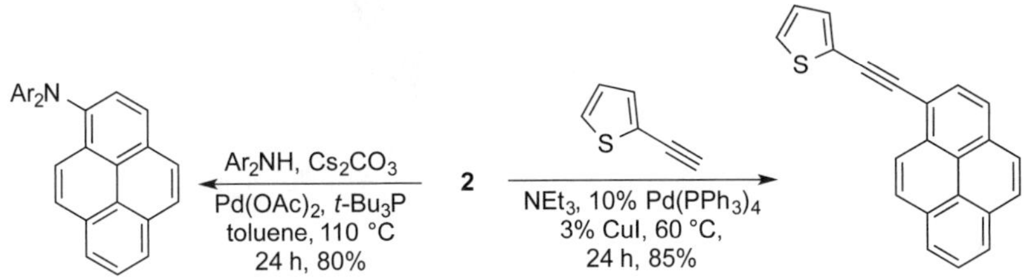

Scheme 2. Cross-coupling reactions of 2: Buchwald-Hartwig amination (Ar = *p*-OMePh)[16] (left) and Sonogashira reaction[17] (right).

Beside **2**, 1-pyrenecarbaldehyde (**3**) is also of considerable interest as building block for the synthesis of functional materials.[18,19] Several synthetic pathways toward **3** are already known. Vollmann reported the synthesis of **3** in 1937 starting from pyrene in a Vilsmeier-Haack[20] reaction with *N*-methylformanilide and $POCl_3$.[21] Other more recently reported methods use slightly modified Vilsmeier-Haack procedures,[18b,19c] or metal-catalyzed reactions with $AlCl_3$,[22] $TiCl_4$,[23] or a Ru-Co-Ce catalyst.[24] Most of these synthetic procedures, however, suffer from disadvantages, such as poor to moderate yields,[18b,19c,21,22] the necessity of toxic reagents (e.g., haloethers),[22,23,25] or the use of expensive metals.[24] With the interest of the chemical community in **3**,[18,19] we developed an inexpensive and facile synthesis, based on the lithiation of **2**,[13] and subsequent quenching with DMF. After acidic workup, aldehyde **3** is isolated in good yields by recrystallization. In combination with the synthesis of **2** described above, the preparation of **3** is efficiently conducted on a 10–15 g scale and does not require the use of expensive or toxic reagents.

Aldehyde **3** has been used as a building block for the synthesis of molecular sensors based on the unique UV-vis and emission characteristics of the pyrene chromophore.[18,19] These syntheses often rely on the use of an amine condensation reactions with **3**, as in Scheme 3a.[19a] Condensation reaction of aldehyde **3** with 2-naphthalenamine, followed by reaction with 5-α-cholestanone was used to form a benzoquinoline derivative (Scheme 3b).[26a] In this case, the pyrene moiety was designed to serve as an aromatic constituent of a petroleum model compound for thermal cracking experiments.[26b] Finally, the aldehyde moiety of **3** can serve as an electrophile in addition reactions toward the formation of carbon-rich molecules, for example, as in Scheme 3c.[27–29]

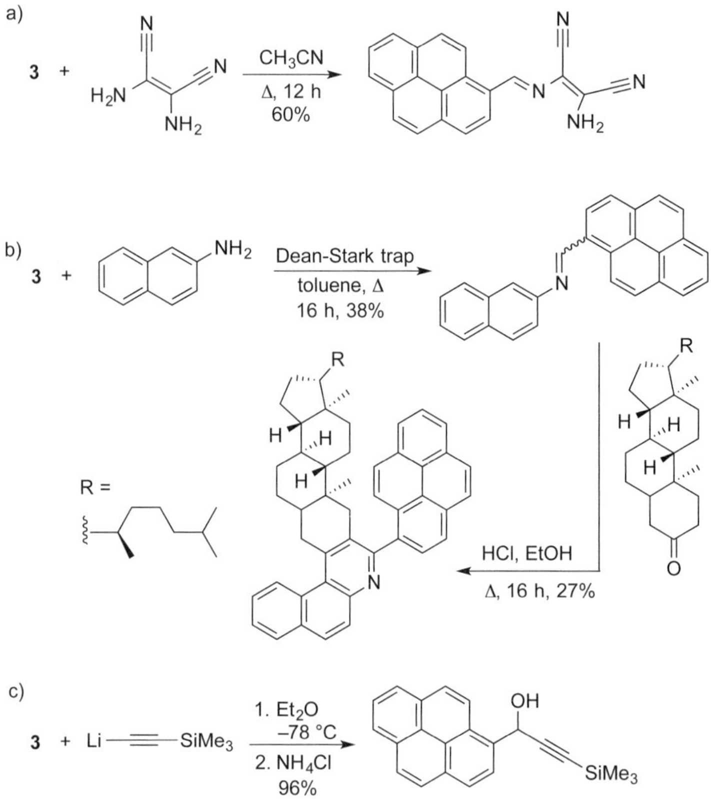

Scheme 3. Functionalization of 3 *via* a condensation reaction leading to a) a fluorescence sensor[19a] or b) a petroleum model compound.[26] c) Reaction of 3 with a Li-acetylide, toward formation of carbon-rich compounds.[27]

References

1. Matthias Schulze, Alexander Scherer, and Rik R. Tykwinski, Department of Chemistry and Pharmacy & Interdisciplinary Center of Molecular Materials (ICMM), Friedrich-Alexander-Universität Erlangen-Nürnberg (FAU), Henkestraße 42, 91054 Erlangen, Germany; Colin Diner, Department of Chemistry, University of Alberta, Edmonton, AB, T6G 2G2 Canada. Email: rik.tykwinski@fau.de. We are

grateful to FAU and the Institute for Oil Sands Innovation at the University of Alberta (IOSI) for funding this work, and we thank Ms. Ann-Kristin Steiner for helping to optimize the reactions.

2. Kalyanasundaram, K.; Thomas, J. K. *J. Am. Chem. Soc.* **1977**, *99*, 2039.
3. Figueira-Duarte, T. M.; Müllen, K. *Chem. Rev.* **2011**, *111*, 7260.
4. Winnik, F. M. *Chem. Rev.* **1993**, *93*, 587.
5. Lock, G. *Ber. Dtsch. Chem. Ges.* **1937**, *70*, 926.
6. Djerassi, C. *Chem. Rev.* **1948**, *43*, 271.
7. Nonhebel, D. C. *J. Chem. Soc.* **1963**, 1216.
8. Yukawa, S.; Omayu, A.; Matsumoto, A. *Macromol. Chem. Phys.* **2009**, *210*, 1776.
9. Gumprecht, W. H. *Org. Synth.* **1968**, *48*, 30.
10. Rae, I. D. *J. Chem. Educ.* **2009**, *86*, 689.
11. U.S. Environmental Protection Agency, *Toxicological Review of Carbon Tetrachloride*, Washington DC, **2010**.
12. Health Canada, *Guidelines for Canadian Drinking Water Quality: Guideline Technical Document – Carbon Tetrachloride*. Water, Air and Climate Change Bureau, Health Canada, Ottawa, Ontario, **2010**. (Catalogue No. H128-1/11-661E).
13. Niko, Y.; Kawauchi, S.; Otsu, S.; Tokumaru, K.; Konishi, G.-i. *J. Org. Chem.* **2013**, *78*, 3196.
14. Oseki, Y.; Fujitsuka, M.; Sakamoto, M.; Majima, T. *J. Phys. Chem. A* **2007**, *111*, 9781.
15. Diev, V. V.; Schlenker, C. W.; Hanson, K.; Zhong, Q.; Zimmerman, J. D.; Forrest, S. R.; Thompson, M. E. *J. Org. Chem.* **2012**, *77*, 143.
16. Jeon, N. J.; Lee, J.; Noh, J. H.; Nazeeruddin, M. K.; Grätzel, M.; Seok, S. I. *J. Am. Chem. Soc.* **2013**, *135*, 19087.
17. Maeda, H.; Maeda, T.; Mizuno, K.; Fujimoto, K.; Shimizu, H.; Inouye, M. *Chem. Eur. J.* **2006**, *12*, 824.
18. a) Chansuvarn, W.; Panich, S.; Imyim, A. *Spectrochim. Acta A Mol. Biomol. Spectrosc.* **2013**, *113*, 154; b) Yang, Y.; Ji, S.; Zhou, F.; Zhao, J. *Biosens. Bioelectron.* **2009**, *24*, 3442; c) Pinheiro, D.; de Castro, C. S.; Seixas de Melo, J. S.; Oliveira, E.; Nuñez, C.; Fernández-Lodeiro, A.; Capelo, J. L.; Lodeiro, C. *Dyes Pigm.* **2014**, *110*, 152.
19. a) Wu, S.-P.; Wang, T.-H.; Liu, S.-R. *Tetrahedron* **2010**, *66*, 9655; erratum: Wu, S.-P.; Wang, T.-H.; Liu, S.-R. *Tetrahedron* **2011**, *67*, 2474; b) Das, S.; Sahana, A.; Banerjee, A.; Lohar, S.; Safin, D. A.; Babashkina, M. G.; Bolte, M.; Garcia, Y.; Hauli, I.; Mukhopadhyay, S. K.; Das, D. *Dalton Trans.* **2013**, *42*, 4757; c) Zeng, Z.; Spiccia, L. *Chem. Eur. J.* **2009**, *15*, 12941.

20. Rajput, A. P.; Girase, P. D. *IJPCBS* **2012**, *3*, 25.
21. Vollmann, H.; Becker, H.; Corell, M.; Streeck, H. *Liebigs Ann. Chem.* **1937**, *531*, 1.
22. Chen, J. J.; Wang, I. J. *Dyes Pigm.* **1995**, *29*, 305.
23. Yamato, T.; Fujimoto, M.; Nagano, Y.; Miyazawa, A.; Tashiro, M. *Org. Prep. Proced. Int.* **1997**, *29*, 321.
24. Ji, H.; Mizugaki, T.; Ebitani, K.; Kaneda, K. *Tetrahedron Lett.* **2002**, *43*, 7179.
25. van Duuren, B. L.; Katz, C.; Goldschmidt, B. M.; Frenkel, K.; Sivak, A. *J. Natl. Cancer Inst.* **1972**, *48*, 1431.
26. a) Scherer, A.; Hampel, F.; Gray, M. R.; Stryker, J. M.; Tykwinski, R. R. *J. Phys. Org. Chem.* **2012**, *25*, 597; b) Alshareef, A. H.; Scherer, A.; Stryker, J. M.; Tykwinski, R. R.; Gray, M. R. *Energy Fuels* **2012**, *26*, 3592.
27. Shi Shun, A. L. K.; Chernick, E. T.; Eisler, S.; Tykwinski, R. R. *J. Org. Chem.* **2003**, *68*, 1339.
28. Telitel, S.; Dumur, F.; Gigmes, D.; Graff, B.; Fouassier, J. P.; Lalevée, J. *Polymer* **2013**, *54*, 2857.
29. Tyson, D. S.; Castellano, F. N. *J. Phys. Chem. A* **1999**, *103*, 10955.

Appendix
Chemical Abstracts Nomenclature (Registry Number)

Pyrene: Benzo[*def*]phenanthrene; (129-00-0)
HBr: Hydrobromic acid; (10035-10-6)
H_2O_2: Hydrogen peroxide solution; (7722-84-1)
1-Bromopyrene; (1714-29-0)
n-BuLi: *n*-Butyllithium solution; (109-72-8)
DMF: *N*,*N*-Dimethylformamide; (68-12-2)
HCl: Hydrochloric acid; (7647-01-0)
1-Pyrenecarbaldehyde; (3029-19-4)

Matthias Schulze was born in Lauf a.d. Pegnitz (Germany) in 1986. He received his B. Sc. degree in chemistry in 2009 and his M. Sc. degree in 2011 from the Friedrich-Alexander-Universität Erlangen-Nürnberg (FAU). He is currently a doctoral student at FAU under the supervision of Prof. Rik R. Tykwinski working on the synthesis of asphaltene model compounds and their supramolecular interactions.

Alexander Scherer was born in Bamberg (Germany) in 1980. He received his diploma in chemistry in 2005 and his doctoral degree in chemistry in 2009 from the Friedrich-Alexander-Universität Erlangen-Nürnberg (FAU). He was a post-doctoral researcher at the University of Alberta in Edmonton (Canada) and then at FAU (2009–2012), focusing on the aggregation of heavy oil, especially model compounds for the asphaltenes. In 2012 he became research associate (Akademischer Rat) at FAU.

Colin Diner was born in Boulder, Colorado in 1985. He received his B. S. in chemistry in 2009 from UW Madison and his doctorate in 2015 from the Stryker Group at the University of Alberta, Canada. He is currently a postdoctoral researcher at Stockholm University, Sweden under the supervision of Prof. Kalman Szábo. His research interests broadly include functional materials, asymmetric catalysis, and practical chemical synthesis. When not in the lab Colin can be found rock climbing, road cycling, and cooking.

Rik R. Tykwinski was born in Marshall, MN and completed his Ph. D. in 1994 at the University of Utah working with Prof. Peter Stang. He did post-doctoral research at ETH-Zürich with Prof. François Diederich (1994–1997), and in 1997 he joined the faculty at the University of Alberta. In 2009, he moved to the Friedrich-Alexander-Universität Erlangen-Nürnberg (FAU) as Chair of Organic Chemistry. His interests focus on the development of synthetic methods for carbon-rich molecules and allotropes, characterization of their electronic properties, asphaltene model compounds, as well as mountain biking and entertaining his two sons.

Jeffrey Mighion was born in Concord, North Carolina in 1986. He received his B. S. in chemistry in 2008 from UNC Chapel Hill. In 2014, he completed his Ph. D. at Princeton University with Prof. Erik Sorensen in the area of natural products synthesis. He is currently a post-doctoral researcher at Emory University with Huw Davies where his research focuses on development and application of C-H functionalization.

Organic Syntheses

Intermolecular [2+2] Cycloaddition of Alkynes with Alkenes Catalyzed by Gold(I)

M. Elena de Orbe, and Antonio M. Echavarren*[1]

Institute of Chemical Research of Catalonia (ICIQ),
Av. Països Catalans 16, 43007 Tarragona, Spain.

Checked by Giulia Rusconi, Estíbaliz Merino and Cristina Nevado

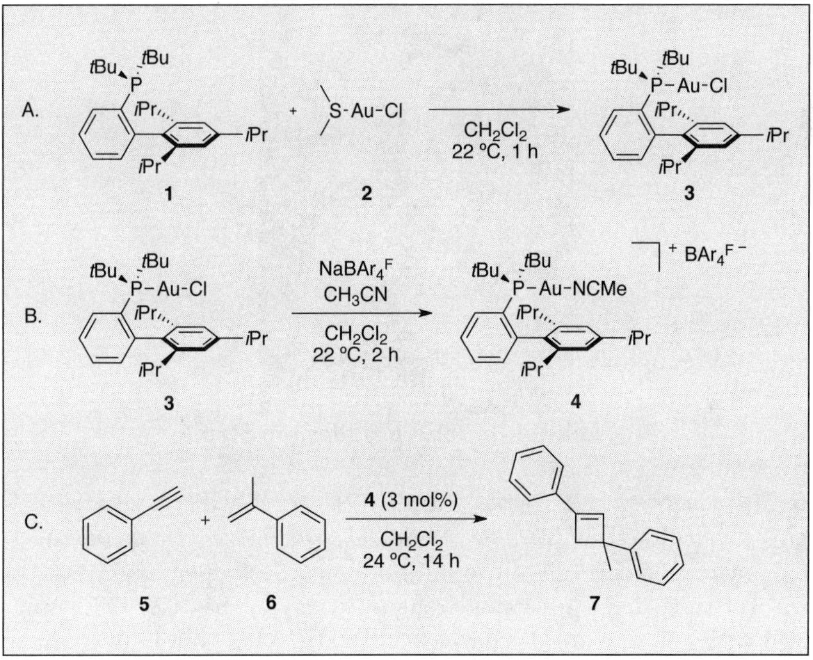

Procedure

A. *Chloro[(2′,4′,6′-triisopropyl-1,1′-biphenyl-2-yl)di-tert-butylphosphine]-gold(I) (3)*. A one-necked (B14, diameter: 2.0 cm) 10-mL round-bottomed

Published on the Web 5/25/2016

flask open to air, equipped with a 1-cm Teflon-coated cylindrical magnetic stirring bar is charged with di-*tert*-butyl(2',4',6'-triisopropyl-[1,1'-biphenyl]-2-yl)phosphine (**1**) (0.500 g, 1.14 mmol, 1.00 equiv) and chloro(dimethylsulfide)gold(I) (**2**) (0.345 g, 1.14 mmol, 1.00 equiv) (Note 1). Dichloromethane (6 mL) is added via syringe (Note 2) and the flask is fitted with a glass stopper. The resulting colorless solution is stirred (600 rpm) at 22 °C for 1 h. The volatiles are removed by rotatory evaporation (300 mmHg, 40 °C bath temperature) and then, under vacuum for 19 h (0.2 mmHg) to afford gold(I) chloride complex **3** (735 mg, 98% yield) as a white solid (Note 3) (Figure 1).

Figure 1. White solid produced in Step A

B. *(Acetonitrile)[(2′,4′,6′-triisopropyl-1,1′-biphenyl-2-yl)di-tert-butylphosphine]gold(I) tetrakis[3,5-bis(trifluoromethyl)phenyl] borate (**4**)*. A one-necked (B29, diameter: 3.5 cm) 100-mL round-bottomed flask open to air, equipped with a Teflon-coated egg-shaped magnetic stirring bar (19 × 10 mm) is charged with gold(I) chloride complex **3** (0.600 g, 0.913 mmol, 1.0 equiv) and dichloromethane (45 mL) (Note 2). Acetonitrile (119 μL, 2.28 mmol, 2.5 equiv) is added via Hamilton syringe (Note 4), the flask is fitted with a septum and the resulting colorless solution (Figure 2A) is stirred (600 rpm) at 22 °C for 5 min. Then, sodium tetrakis(3,5-bis(trifluoromethyl)phenyl)borate (876 mg, 0.959 mmol, 1.05 equiv) is added in one portion (Note 5) and the flask is again fitted with the septum. The resulting white suspension is stirred (600 rpm) at 22 °C for 2 h (in this time, the solution becomes purple) (Figure 2B) (Note 6). The reaction mixture is

Figure 2A. Colorless Solution

Figure 2B. Purple Solution in Step B

filtered under vacuum through a plug of Celite (20 mm diameter × 20 mm height) (Figure 3A) (Note 7), which is washed with dichloromethane (1 x 25 mL) (Note 2). The filtrate is concentrated by rotatory evaporation (300 mmHg, 40 °C bath temperature) to a volume of approximately 2 mL and filtered through two 0.2 µm Teflon filters (13 mm diameter), which are washed with dichloromethane (1 x 4 mL) (Note 2). The solvent of the filtrate is removed by rotatory evaporation (300 mmHg, 40 °C bath temperature), and then the residue is dried under vacuum (0.3 mmHg) for 19 h to afford cationic gold(I) complex **4** (1.37 g, >99% yield) as a white solid (Figure 3B) (Note 8).

C. *1,3-Diphenyl-3-methylcyclobut-1-ene* (**7**). A one-necked (B29, diameter: 3.5 cm) 100-mL round-bottomed flask open to air, equipped with a Teflon-coated egg-shaped magnetic stirring bar (19 × 10 mm) is charged with dichloromethane (50 mL) (Note 2). α-Methylstyrene (**6**) (7.00 mL, 6.39 g, 53.5 mmol, 2.00 equiv) and ethynylbenzene (**5**) (3.00 mL, 2.79 g, 26.8 mmol, 1.00 equiv) are successively added via syringe (Note 9) while stirring (800 rpm) at 24 °C. Afterwards, cationic gold(I) complex **4** (1.225 g, 0.803 mmol, 0.03 equiv) is added in one portion, the flask is fitted with a septum and the solution is stirred (800 rpm) at 24 °C for 14 h (Note 10). During the course of the reaction, the solution turns progressively from

 DOI: 10.15227/orgsyn.093.0115

Figure 3A. Filtration of Step B

Figure 3B. White product of Step B

colorless to yellow and finally to orange (Figure 4). Triethylamine (1.6 mL) is added via syringe (Note 11) to quench the reaction while stirring. At this point, the solution becomes yellow. Volatiles are removed using a rotary evaporator under reduced pressure (375 mmHg and then 20 mmHg, 40 °C bath temperature). The resulting residue is dissolved in dichloromethane (20 mL) and poured into a 250-mL round-bottomed flask containing florisil (22 g) (Note 12). After removing the solvent by rotatory evaporation (375 mbar and then 20 mbar, 40 °C bath temperature), the crude material adsorbed on florisil is charged on a 390 g silica gel column (7 cm diameter × 25 cm height) (Note 13), packed with cyclohexane (600 mL) and eluted with cyclohexane (about 6 L) under compressed air (1.5 atm) (Note 14 and 15). Fractions 26-40 containing pure product are concentrated by rotary evaporation (110 mmHg and then 20 mmHg, 40 °C bath temperature) and the resultant residue is dried under vacuum (0.2 mmHg) for 20 h to afford cyclobutene **7** (5.8 g, 98% yield) as a dense yellow oil (Note 16).

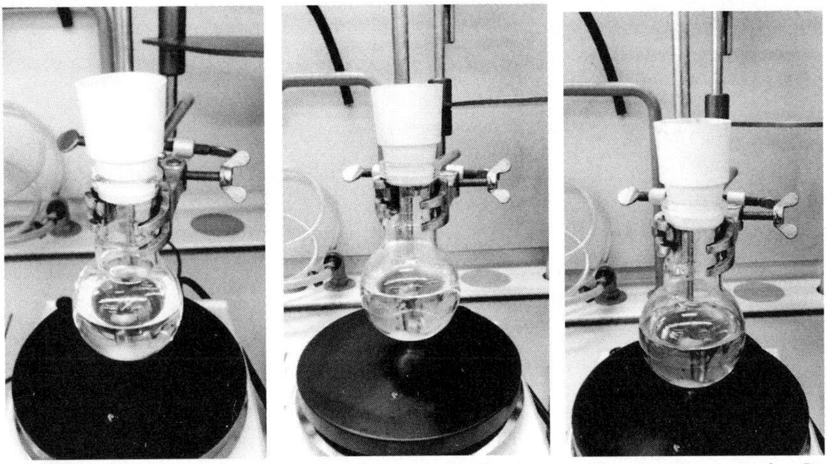

Figure 4. Progression of color from colorless to yellow to orange in Step C

Notes

1. Di-*tert*-butyl(2',4',6'-triisopropyl-[1,1'-biphenyl]-2-yl)phosphine (97%) and chloro(dimethylsulfide)gold(I) (97.5%) were purchased from Sigma-Aldrich Chemical Co. (Ref. 638080 and 420727, respectively) and used as received.
2. Dichloromethane (CHROMASOLV®, for HPLC, ≥99.8%) was purchased from Sigma-Aldrich Chemical Co. (Ref. 34856) and used as received.
3. The physical and spectroscopic properties of gold(I) chloride complex **3** are as follows: mp = 255–257 °C. ^{1}H NMR (400 MHz, CD_2Cl_2) δ: 0.91 (d, J = 6.6 Hz, 6H), 1.28 (d, J = 6.8 Hz, 6H), 1.35 (d, J = 6.9 Hz, 6H), 1.39 (s, 9H), 1.44 (s, 9H), 2.36 (hept, J = 6.7 Hz, 2H), 2.95 (hept, J = 6.9 Hz, 1H), 7.06 (s, 2H), 7.27–7.32 (m, 1H), 7.47–7.57 (m, 2H), 7.88–7.94 (m, 1H). ^{13}C {^{1}H} NMR (101 MHz, CD_2Cl_2) δ: 23.0 (s), 24.5 (s), 26.3 (s), 31.1 (s), 31.4 (d, J (^{13}C-^{31}P) = 6.4 Hz), 34.7 (s), 38.6 (d, J (^{13}C-^{31}P) = 26.6 Hz), 122.0 (s), 126.9 (d, J (^{13}C-^{31}P) = 6.9 Hz), 128.3 (s), 128.8 (s), 130.7 (d, J (^{13}C-^{31}P) = 2.4 Hz), 135.0 (d, J (^{13}C-^{31}P) = 3.0 Hz), 135.2 (d, J (^{13}C-^{31}P) = 8.1 Hz), 136.3 (d, J (^{13}C-^{31}P) = 5.5 Hz), 146.3 (s), 148.5 (d, J (^{13}C-^{31}P) = 14.0 Hz), 150.3 (s). ^{31}P {^{1}H} NMR (162 MHz, CD_2Cl_2) δ: 58.69. IR (neat) : 2970, 1739, 1444, 1365, 1228, 1217, 1206, 775, 528 cm^{-1}. HRMS-ESI^{+} *m/z* calcd for $C_{29}H_{45}AuClNaP^{+}$ $[M+Na]^{+}$ 679.2505, found 679.2516. Anal. calcd for $C_{29}H_{45}AuClP$: C, 53.01; H, 6.90; found: C, 52.65; H, 6.77.

4. Acetonitrile (CHROMASOLV® gradient grade, for HPLC, ≥99.9%) was purchased from Sigma-Aldrich Chemical Co. (Ref. 34851) and used as received.
5. Sodium tetrakis(3,5-bis(trifluoromethyl)phenyl)borate (97%, may contain 1-5% water) was purchased from Alfa Aesar (Ref. H30014) and used as received.
6. **Important:** It is necessary to use an excess of acetonitrile (2.5 equiv) and sodium tetrakis(3,5-bis(trifluoromethyl)phenyl)borate (1.05 equiv) in order to obtain pure cationic gold(I) complex **4**. If not, the reaction outcome can be a mixture of desired product **4**, unreacted gold(I) chloride complex **3** and other different species.
7. Merck Celite 545 (particle size 0.02-0.1 mm) was used.
8. The physical and spectroscopic properties of cationic gold(I) complex **4** are as follows: mp = 148–150 °C. ^{1}H NMR (400 MHz, CD_2Cl_2) δ: 0.93 (d, J = 6.6 Hz, 6H), 1.26 (d, J = 6.8 Hz, 6H), 1.32 (d, J = 6.9 Hz, 6H), 1.40 (d, J = 16.3 Hz, 18H), 2.22 (broad s, 3H), 2.34 (hept, J = 6.7 Hz, 2H), 2.95 (hept, J = 6.9 Hz, 1H), 7.17 (s, 2H), 7.31–7.35 (m, 1H), 7.58 (s, 4H), 7.56–7.62 (m, 2H), 7.74–7.75 (m, 8H), 7.87–7.91 (m, 1H). ^{13}C NMR (101 MHz, CD_2Cl_2) δ: 3.0 (s), 23.2 (s), 24.3 (s), 26.1 (s), 31.3 (d, J (^{13}C-^{31}P) = 5.5 Hz), 31.4 (s), 34.5 (s), 39.1 (d, J (^{13}C-^{31}P) = 28.2 Hz), 117.9 (p, J (^{13}C-^{19}F) = 3.9 Hz), 118.9 (s), 122.3 (s), 125.1 (q, J (^{13}C-^{19}F) = 272.4 Hz), 125.9 (d, J (^{13}C-^{31}P) = 49.8 Hz), 128.0 (d, J (^{13}C-^{31}P) = 7.9 Hz), 129.3 (qq, J (^{13}C-^{19}F) = 31.6 Hz and J (^{13}C-^{11}B) = 2.7 Hz), 130.3 (s), 132.0 (d, J (^{13}C-^{31}P) = 2.4 Hz), 134.6 (d, J (^{13}C-^{31}P) = 4.3 Hz), 135.2 (s), 135.3 (s), 136.6 (d, J (^{13}C-^{31}P) = 6.3 Hz), 147.4 (d, J (^{13}C-^{31}P) = 11.7 Hz), 147.9 (s), 150.3 (s), 162.2 (q, J (^{13}C-^{11}B) = 49.8 Hz). ^{31}P {^{1}H} NMR (162 MHz, CD_2Cl_2) δ: 55.35. ^{19}F {^{1}H} NMR (376 MHz, CD_2Cl_2) δ: –62.83. ^{11}B {^{1}H} NMR (128 MHz, CD_2Cl_2) δ: –6.56. ESI$^+$ m/z calcd for $C_{31}H_{48}AuNP^+$ $[M-C_{32}H_{12}BF_{24}]^+$ 662.3184, found 662.3186. Anal. calcd for $C_{63}H_{60}AuBF_{24}NP$: C, 49.59; H, 3.96; N, 0.92; found: C, 49.52; H, 3.95; N, 0.58.
9. α-Methylstyrene (99%, stab. with 10-20 ppm 4-*tert*-butylcatechol) was purchased from Alfa Aesar (Ref. L03609). Ethynylbenzene (>98.0%) was purchased from Tokyo Chemical Industry Co. (Ref. E0196). Both were used as received.
10. The gold(I)-catalyst is neither sensitive to oxygen nor moisture, so the reaction was performed under air (with *ca.* 80% relative humidity).
11. Triethylamine (puriss. p.a., ≥99.5%) was purchased from Sigma-Aldrich Chemical Co. (Ref. 90340) and used as received.
12. Florisil (60-100 mesh) was purchased from Acros Organics.

13. Silica gel (technical grade, pore size 60 Å, 230-400 mesh particle size, 40-63 μm particle size) was purchased from Sigma-Aldrich Chemical Co.
14. Column fractions were checked by TLC analysis on Merck Silica Gel 60 F_{254} aluminium sheets, using cyclohexane as eluent. Visualization is accomplished with 254 nm UV light and subsequently with vanillin staining (solution of 12 g of vanillin and 2 mL of concentrated sulphuric acid in 200 mL of absolute ethanol) by immersion and heating with a heat gun. The starting α-methylstyrene has Rf = 0.71 (UV active; blue after staining) whereas the cyclobutene product **7** has Rf = 0.50 (UV active; purple after staining).
15. Fractions of 60 mL (about 1.5 L, 25 fractions in total) are collected until all α-methylstyrene is eluted and then, fractions of 60 mL (about 0.9 L, 15 fractions) and 500 mL (about 4 L, 8 fractions in total) are collected until no more product elutes.
16. The physical and spectroscopic properties of cyclobutene **7** are as follows: ^{1}H NMR (400 MHz, $CDCl_3$) δ: 1.64 (s, 3H), 2.96 (q, *J* = 12.5 Hz, 2H), 6.74 (s, 1H), 7.18–7.23 (m, 1H), 7.25–7.29 (m, 1H), 7.31–7.37 (m, 4H), 7.39–7.42 (m, 4H). ^{13}C {^{1}H} NMR (101 MHz, $CDCl_3$) δ: 27.6, 44.3, 46.0, 124.6, 125.7, 125.8, 127.8, 128.1, 128.3, 133.7, 134.7, 143.8, 147.7. IR (neat): 3025, 2953, 1741, 1736, 1727, 1490, 1445, 1368, 1229, 1216, 1206, 748, 690 cm^{-1}. HRMS-APCI *m/z* calculated for $C_{17}H_{17}$ $[M+H]^+$ 221.1325, found 221.1324. Anal. calcd for $C_{17}H_{16}$: C, 92.68; H, 7.32; found: C, 92.69; H, 7.30.

Working with Hazardous Chemicals

The procedures in *Organic Syntheses* are intended for use only by persons with proper training in experimental organic chemistry. All hazardous materials should be handled using the standard procedures for work with chemicals described in references such as "Prudent Practices in the Laboratory" (The National Academies Press, Washington, D.C., 2011; the full text can be accessed free of charge at http://www.nap.edu/catalog.php?record_id=12654). All chemical waste should be disposed of in accordance with local regulations. For general guidelines for the management of chemical waste, see Chapter 8 of Prudent Practices.

In some articles in *Organic Syntheses*, chemical-specific hazards are highlighted in red "Caution Notes" within a procedure. It is important to recognize that the absence of a caution note does not imply that no significant hazards are associated with the chemicals involved in that procedure. Prior to performing a reaction, a thorough risk assessment should be carried out that includes a review of the potential hazards associated with each chemical and experimental operation on the scale that is planned for the procedure. Guidelines for carrying out a risk assessment and for analyzing the hazards associated with chemicals can be found in Chapter 4 of Prudent Practices.

The procedures described in *Organic Syntheses* are provided as published and are conducted at one's own risk. *Organic Syntheses, Inc.*, its Editors, and its Board of Directors do not warrant or guarantee the safety of individuals using these procedures and hereby disclaim any liability for any injuries or damages claimed to have resulted from or related in any way to the procedures herein.

Discussion

The synthesis of cyclobutenes by metal-catalyzed intermolecular [2+2] cycloaddition of alkynes with alkenes is normally restricted to strained bicyclic olefins.[2] Otherwise, internal or terminal alkynes react with electron-deficient alkenes to give cyclobutenes using rhodium (I)[3] or (II)[4] catalysts, respectively. Cyclobutenes are also obtained in the intermolecular reaction of 1,3-enynes with alkenes catalyzed by nickel[5] or cobalt[6] complexes. Apart from that, in the presence of cobalt catalysts, only cyclopentene and cycloheptene were found to undergo cycloaddition with internal alkynes to generate cyclobutene derivatives.[7] In this context, the gold(I)-catalyzed intermolecular [2+2] cycloaddition of terminal alkynes with differently-substituted alkenes, whose procedure is reported herein, is an important method for the formation of cyclobutenes.[8]

The key point for the development of this transformation was the use of a gold(I)-catalyst bearing a sterically hindered ligand such as *t*BuXPhos, which allows the selective activation of the alkyne towards nucleophilic attack of the alkene and reduces competitive pathways leading to oligomers.[8a] Further mechanistic studies revealed the impact of the more bulky, non-coordinating and less basic BAr_4^{F-} as counterion in the gold catalyst in order to decrease the deprotonation of terminal alkynes forming unreactive α,π-(acetylide)digold complexes.[8b]

The optimized [*t*BuXPhosAu(MeCN)]BAr_4^F catalyst can be easily synthesized in two steps, starting from the phosphine *t*BuXPhos and chloro(dimethylsulfide)gold(I) to form the gold(I) chloride complex[9] and then, treating with slight excess of $NaBAr_4^F$ and acetonitrile to obtain the corresponding cationic gold(I) complex.

Table 1. Other cyclobutenes synthesized following this procedure.

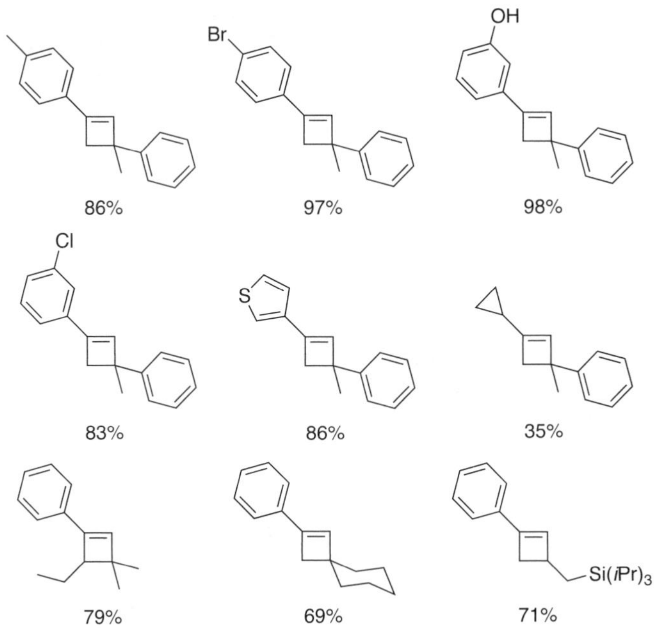

In the presence of this efficient gold(I)-catalyst, terminal arylalkynes react with mono-, di- and trisubstituted alkenes to build regioselectively cyclobutenes containing a wide variety of functional groups in moderate to excellent yields (Table 1). Ethynylcyclopropane also undergoes the cycloaddition in moderate yields, but internal alkynes such as 1-phenyl-1-propyne are unreactive under these conditions.

It is worth mentioning that this gold(I)-catalyzed transformation proceeds at room temperature and under air. Thus, our procedure represents a facile method for the preparation of substituted cyclobutenes under mild conditions.

References

1. Institute of Chemical Research of Catalonia (ICIQ), Av. Països Catalans 16, 43007 Tarragona, Spain. E-mail: aechavarren@iciq.es. We thank MINECO (project CTQ2013-42106-P, Severo Ochoa Excellence Accreditation SEV-2013-0319 and Severo Ochoa predoctoral fellowship), European Research Council (Advanced Grant No. 321066), AGAUR (2014 SGR 818, and the ICIQ Foundation for financial support.
2. Using nickel: (a) Huang, D.-J.; Rayabarapu, D. K.; Li, L.-P.; Sambaiah, T.; Cheng, C.-H. *Chem. Eur. J.* **2000**, *6*, 3706–3713. Employing ruthenium: (b) Jordan, R. W.; Tam, W. *Org. Lett.* **2000**, *2*, 3031–3034. Utilizing cobalt: (c) Chao, K. C.; Rayabarapu, D. K.; Wang, C.-C.; Cheng, C-H. *J. Org. Chem.* **2001**, *66*, 8804–8810. (d) Treutwein, J.; Hilt, G. *Angew. Chem. Int. Ed.* **2008**, *47*, 6811–6813. Using rhodium: (e) Shibata, T.; Takami, K.; Kawachi, A. *Org. Lett.* **2006**, *8*, 1343–1345. Employing iridium: (f) Fan, B.-M.; Li, X. J.; Peng, F.-Z.; Zhang, H.-B.; Chan, A. S. C.; Shao, Z. H. *Org. Lett.* **2010**, *12*, 304–306.
3. Motokura, K.; Nakayama, K.; Miyaji, A.; Baba, T. *ChemCatChem* **2011**, *3*, 1419–1421.
4. Sakai, K.; Kochi, T.; Kakiuchi, F. *Org. Lett.* **2013**, *15*, 1024–1027.
5. Nishimura, A.; Ohashi, M.; Ogoshi, S. *J. Am. Chem. Soc.* **2012**, *134*, 15692–15695.
6. Nishimura, A.; Tamai, E.; Ohashi, M.; Ogoshi, S. *Chem. Eur. J.* **2014** , *20*, 6613–6617.
7. Hilt, G.; Paul, A.; Treutwein, J. *Org. Lett.* **2010**, *12*, 1536–1539.
8. (a) López-Carrillo, V.; Echavarren, A. M. *J. Am. Chem. Soc.* **2010**, *132*, 9292–9294. (b) Homs, A.; Obradors, C; Leboeuf, D.; Echavarren, A. M. *Adv. Synth. Catal.* **2014**, *356*, 221–228.
9. Other methods to prepare phosphine gold(I) chlorides: (a) Nieto-Oberhuber, C.; Muñoz, M. P.; López, S.; Jiménez-Núñez, E.; Nevado, C.; Herrero-Gómez, E.; Raducan, M.; Echavarren, A. M. *Chem. Eur. J.* **2006**, *12*, 1677–1693. (b) Hashmi, A. S. K.; Bechem, B.; Loos, A.; Hamzic, M.; Rominger, F.; Rabaa, H. *Aust. J. Chem.* **2014**, *67*, 481–499.

Appendix
Chemical Abstracts Nomenclature (Registry Number)

Di-*tert*-butyl(2',4',6'-triisopropyl-[1,1'-biphenyl]-2-yl)phosphine: Phosphine, bis(1,1-dimethylethyl)[2',4',6'-tris(1-methylethyl)[1,1'-biphenyl]-2-yl]-; (564483-19-8)

Chloro(dimethylsulfide)gold(I): Gold, chloro[thiobis[methane]]-; (29892-37-3)

Chloro[(2′,4′,6′-triisopropyl-1,1′-biphenyl-2-yl)di-tert-butylphosphine]gold(I): Gold, [bis(1,1-dimethylethyl)[2',4',6'-tris(1-methylethyl)[1,1'-biphenyl]-2-yl]phosphine]chloro-; (1312108-97-6)

Sodium tetrakis(3,5-bis(trifluoromethyl)phenyl)borate: Borate(1-), tetrakis[3,5-bis(trifluoromethyl)phenyl]-, sodium (1:1) (79060-88-1)

(Acetonitrile)[(2′,4′,6′-triisopropyl-1,1′-biphenyl-2-yl)di-tert-butylphosphine]gold(I) tetrakis[3,5-bis(trifluoromethyl)phenyl] borate; (1610790-72-1)

Ethynylbenzene: Benzene, ethynyl-; (536-74-3)

α-Methylstyrene: Benzene, (1-methylethenyl)-; (98-83-9)

(1,3-Diphenyl-3-methylcyclobut-1-ene: Benzene, 1,1'-(3-methyl-1-cyclobutene-1,3-diyl)bis-; (1235453-87-8)

M. Elena de Orbe was born in Madrid (Spain) in 1991. She graduated with a degree in Chemistry from the Universidad Complutense de Madrid with the Extraordinary Award in 2013. In 2014, she completed the Master in Synthesis, Catalysis and Molecular Design at the Universitat Rovira i Virgili (Tarragona, Spain) and joined the group of Prof. Antonio M. Echavarren at the Institute of Chemical Research of Catalonia (ICIQ) to perform her Ph.D. studies.

Antonio M. Echavarren was born in Bilbao (Spain) in 1955 and obtained his Ph.D. at the Universidad Autónoma de Madrid in 1982 with Francisco Fariña. After postdoctoral stays at Boston College with T. Ross Kelly and at Colorado State University with John K. Stille he joined the Institute of Organic Chemistry (CSIC) in Madrid, where he stayed until 1992. That year he returned to the UAM as a Professor of Organic Chemistry. He moved in 2004 to Tarragona as a Group Leader at the Institute of Chemical Research of Catalonia (ICIQ).

Giulia Rusconi graduated with a degree in Chemistry from the Università degli Studi di Milano in 2010. In 2012, she completed the Master in Organic Chemistry at the same university and joined the group of Prof. Cristina Nevado at the University of Zurich to perform her Ph.D. studies.

Estíbaliz Merino obtained her Ph.D. at the University Autónoma de Madrid. After a postdoctoral stay with Prof. Magnus Rueping at Goethe University Frankfurt and RWTH-Aachen University in Germany, she worked with Prof. Avelino Corma in Instituto de Tecnología Química-CSIC (Valencia) and Prof. Félix Sánchez in Instituto de Química Orgánica General-CSIC (Madrid) Spain. At present, she is a research associate in Prof. Cristina Nevado´s group in University of Zürich.

Photochemical Benzannulation of *N*-Phosphoryl Ynamides and α-Diazo Ketones in Continuous Flow

John M. Read, Yu-Pu Wang, and Rick L. Danheiser*[1]

Department of Chemistry, Massachusetts Institute of Technology, Cambridge, MA 02139

Checked by Christian Ebner and Erick M. Carreira

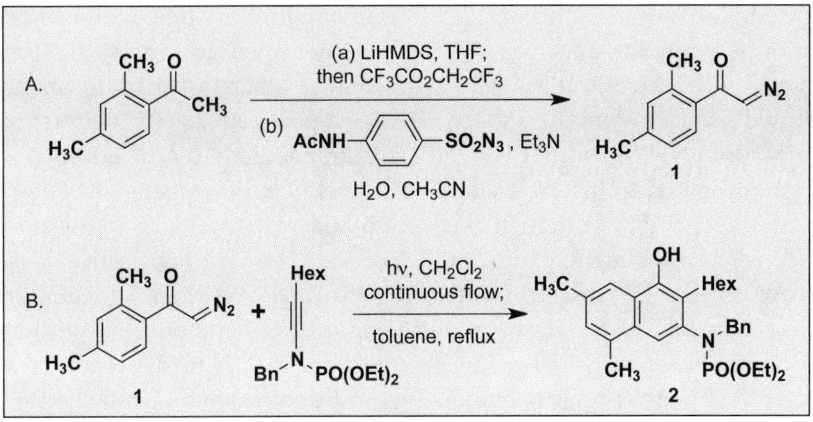

Procedure

A. *2-Diazo-1-(2,4-dimethylphenyl)ethan-1-one* (**1**). A 250-mL, three-necked, round-bottomed flask (Note 1) equipped with a 25 x 10 mm, Teflon-coated, octagonal magnetic stir bar is placed in a glove box. Solid LiHMDS (1.81 g, 13.5 mmol, 1.1 equiv) (Note 2) was weighed out in the glove box, added to the flask, and three septa are attached. The flask is removed from the glove box and equipped with an argon inlet adapter and a 50-mL pressure-equalizing addition funnel fitted with a rubber septum. The third septum is fitted with a thermocouple temperature probe. Tetrahydrofuran (35 mL) (Note 3) is added by syringe and the solution cooled to –78 °C in a

DOI: 10.15227/orgsyn.93.0127
Published on the Web 6/1/2016

dry ice-acetone bath (Note 4). A solution of 2′,4′-dimethylacetophenone (1.82 mL, 1.81 g, 12.2 mmol, 1.0 equiv) (Note 5) in 27 mL of THF is added via the addition funnel over 15 min while the internal temperature of the reaction mixture is kept below -70 °C. The addition funnel is rinsed with two 5-mL portions of THF. The reaction mixture is stirred at -78 °C for 1 h and then 2,2,2-trifluoroethyl trifluoroacetate (1.97 mL, 2.88 g, 14.7 mmol, 1.2 equiv) (Note 6) is added rapidly in one portion via syringe. After 30 min, the yellow solution is poured into a 500-mL separatory funnel containing 120 mL of diethyl ether and 120 mL of 5% aqueous HCl solution. The aqueous layer is separated and extracted with two 60-mL portions of diethyl ether. The combined organic layers are washed with 120 mL of saturated NaCl solution, dried over 10 g of anhydrous $MgSO_4$, and filtered through a 30-mL sintered glass Büchner funnel (medium porosity, 30 mm diameter). The $MgSO_4$ is washed with diethyl ether (3 x 10 mL) and the combined filtrate is concentrated by rotary evaporation (20 °C, 20 mmHg) to afford 4 g of a yellow oil. This material is immediately dissolved in 15 mL of acetonitrile and transferred via glass funnel (9 cm diameter) to a 250-mL single-necked, round-bottomed flask equipped with a 25 x 10 mm, Teflon-coated, octagonal magnetic stir bar (Note 7). The original flask is rinsed with acetonitrile (2 x 5 mL), which is transferred via glass funnel (9 cm diameter) to the 250-mL flask. Water (0.22 mL, 12.2 mmol, 1.0 equiv) and triethylamine (2.56 mL, 1.86 g, 18.4 mmol, 1.5 equiv) (Note 8) are added via syringe. A 50-mL pressure-equalizing addition funnel is attached. A solution of 4-acetamidobenzenesulfonyl azide (4.41 g, 18.4 mmol, 1.5 equiv) (Note 9) in 25 mL of acetonitrile is then added via the addition funnel over 15 min. The addition funnel is rinsed with 5 mL of acetonitrile and the reaction flask is wrapped with aluminum foil. The resulting dark yellow solution is stirred at room temperature for 2 h, during which time a white precipitate appears (Figure 1).

The solution is then concentrated by rotary evaporation (40 °C, 20 mmHg) to provide 10.7-16.6 g of a thick yellow suspension. This material is diluted with 1:1 diethyl ether-hexanes (100 mL) and filtered through 5 g of Celite in a 30-mL sintered glass funnel (medium porosity, 30 mm diameter) into a 250-mL, round-bottomed flask. The solid material is washed with 1:1 diethyl ether-hexanes (3 x 15 mL), and the filtrate is then concentrated by rotary evaporation (20 °C, 20 mmHg) to yield 2.8-4.7 g of a dark yellow oil.

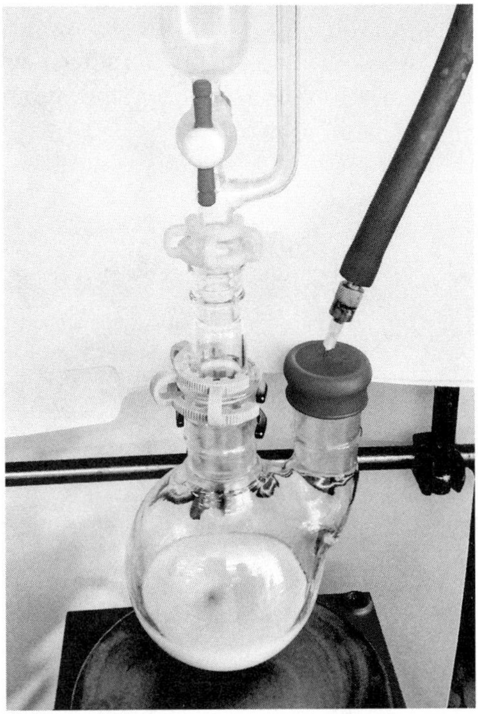

Figure 1. White precipitate formed in Step A

This material is dissolved in a minimum amount of 1:7 ethyl acetate-hexanes (ca. 6 mL) and loaded onto a column (64 mm diameter) of 200 g of silica gel (Note 10) prepared as a slurry in 1:7 ethyl acetate-hexanes. Elution with 1:7 ethyl acetate-hexanes (35 mL fractions collected in 20 x 150 mm test tubes) affords the product in fractions 21-57. These fractions are combined and the solvent is removed by rotary evaporation (20 °C, 20 mmHg). Further concentration at room temperature, 0.05 mmHg for 1 h provides 1.77 g (83%) of diazo ketone **1** as a yellow crystalline solid (Notes 11 and 12).

B. *Diethyl benzyl(3-hexyl-4-hydroxy-6,8-dimethylnaphthalen-2-yl)phosphoramidate* (**2**). Figure 2 shows the continuous flow photochemical reactor employed for this reaction. Fluorinated ethylene propylene (FEP) tubing, o.d. = 1.59 mm, i.d. = 0.76 mm, length = 1520 cm (Note 13), is wrapped around a quartz immersion well in tightly packed coils leaving 90 cm of tubing free at each end. The length of tubing wrapped around the well is 1340 cm (internal volume = 6.1 mL). The ends of the tubing are secured to the well with Teflon tape. The top end of the tubing is fitted with a nut, ferrule, and a thread to a female Luer adapter for attachment to a syringe as shown in Figure 2. The bottom end of the tubing is connected

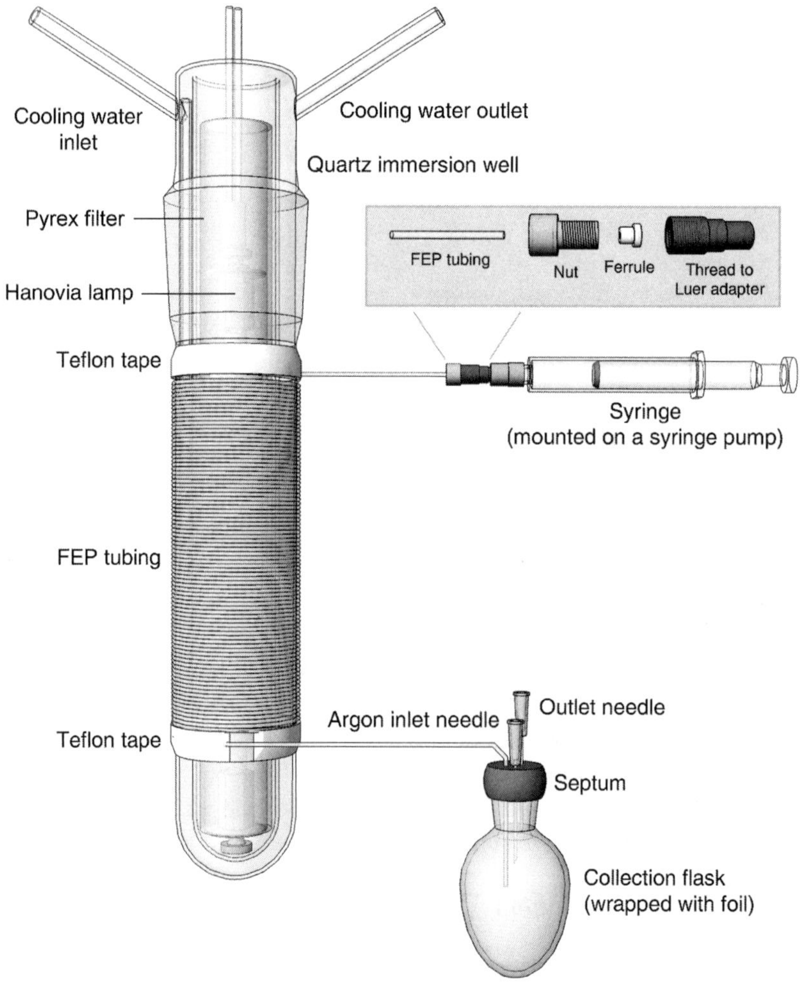

Figure 2. Continuous Flow Reactor for Photochemical Benzannulation

through a rubber septum to a 100-mL, round-bottomed flask equipped with an argon inlet needle and a needle vent. The receiving flask is wrapped in aluminum foil (Figure 3). The immersion well is connected to recirculating tap water via PVC tubing. A Pyrex filter and a Hanovia 450 W medium-pressure mercury lamp are placed inside the immersion well (Note 14).

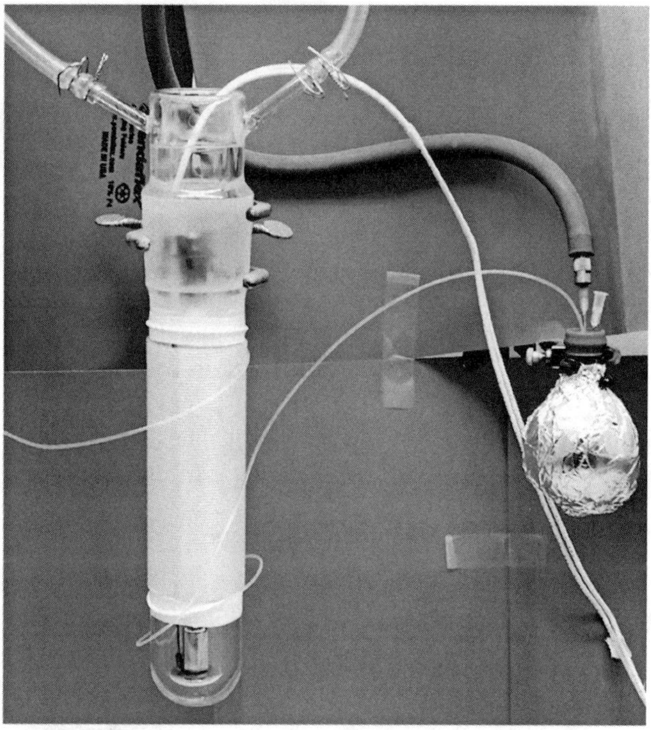

Figure 3. Photograph of Continuous Flow Reactor

A 100-mL pear flask equipped with a rubber septum and argon inlet needle is charged with 2-diazo-1-(2,4-dimethylphenyl)ethan-1-one (**1**) (1.56 g, 8.90 mmol, 1.2 equiv), diethyl benzyl(oct-1-yn-1-yl)phosphoramidate[2] (2.63 g, 7.48 mmol, 1.0 equiv), and CH_2Cl_2 (30 mL) (Notes 15 and 16). The yellow solution is degassed via three freeze-pump-thaw (–196 °C, 0.05 mmHg) cycles. A 20-mL plastic syringe is charged with 5 mL of CH_2Cl_2, the syringe is connected to the Luer adapter, the lamp is turned on (Note 17), and the 5 mL of CH_2Cl_2 is pumped through the system at a rate of 0.58 mL/min (Note 18). Approximately one-half of the reaction mixture is transferred to a 20-mL plastic syringe and pumped through the system at a rate of 0.58 mL/min (calculated residence time = 10.4 min). The syringe is charged with the other half of the reaction solution which is then pumped through the system. The total time required for the injection of the 30-mL reaction solution is 50 min. The 100-mL pear flask is then rinsed with two 0.6-mL portions and one 9 mL portion of CH_2Cl_2 with each portion pumped through the tubing at a rate of 0.58 mL/min. The lamp is turned off, the aluminum foil on the 100-mL collection flask is removed, and the orange solution is concentrated by rotary evaporation (20 °C, 20 mmHg) to afford 3.97 g of dark orange oil (Note 19). This material is dissolved in 50 mL of toluene (Note 20) and the flask is equipped with a 20 x 10 mm, Teflon-coated, oval magnetic stir bar and an 11-cm Liebig condenser fitted with an argon inlet adapter. The orange solution is heated at reflux in an oil bath for 5 h and then concentrated to afford 3.70 g of orange solid (Note 21). This material is dissolved in 15 mL of CH_2Cl_2 and concentrated onto 10 g of silica gel. The resulting free-flowing powder is deposited onto a column (80 mm diameter) of 300 g of silica gel prepared as a slurry in 1:2 ethyl acetate-hexanes. Elution is carried out with a gradient of 1:2 to 4:1 ethyl acetate-hexanes, with a ratio of 1:2 EtOAc-hexanes used for fractions 1-81, 1:1 for fractions 82-121, 2:1 for fractions 122-151, and 4:1 for the remaining fractions. Fractions (35 mL) are collected in 20 x 150 mm test tubes. The desired product is obtained in fractions 70-164. These fractions are combined and the solvent is removed by rotary evaporation (20 °C, 20 mmHg). Further concentration at room temperature, 0.1 mmHg for 10 h provides 3.15 g (88%) of **2** as a pale yellow solid (Notes 22 and 23) (Figure 4).

Figure 4. Product of Step B

Notes

1. Glassware was flame-dried under vacuum (0.1 mmHg), back-filled with argon while hot, and then maintained under the inert atmosphere during the course of the reaction.
2. Solid LiHMDS (98%) was purchased from Aldrich-Fine Chemicals and used as received.
3. Tetrahydrofuran (low water HPLC grade) was purchased from J. T. Baker and purified by pressure filtration through activated alumina immediately prior to use.
4. The submitters prepared a solution of LiHMDS from hexamethyldisilazane (98+%, purchased from Alfa Aesar and distilled at 124-125 °C from CaH_2 prior to use) and butyllithium (purchased from Alfa Aesar and titrated according to the method of Watson and Eastham[3]). The reaction flask was charged via syringe with hexamethyldisilazane (2.81 mL, 2.17 g, 13.5 mmol, 1.1 equiv) and 35 mL of THF. The flask was cooled in an ice/water bath and a solution of *n*-butyllithium (5.20 mL, 2.59 M in hexane, 13.5 mmol, 1.1 equiv) was added dropwise via syringe over 10 min. The reaction mixture was stirred for an additional 10 min and then cooled to -78 °C.
5. 2′,4′-Dimethylacetophenone (97%) was purchased from Alfa Aesar and was used as received.
6. 2,2,2-Trifluoroethyl trifluoroacetate (98%) was purchased from Alfa Aesar and distilled (bp 53-55 °C) from CaH_2 prior to use.
7. Acetonitrile was purchased from Aldrich Chemical Company and distilled (bp 80-81 °C) from CaH_2 prior to use. The submitters employed a 250-mL, two-necked flask for this stage of the reaction.
8. Triethylamine was purchased from J. T. Baker and distilled (bp 87-88.5 °C) from CaH_2 prior to use.
9. 4-Acetamidobenzenesulfonyl azide was purchased from Alfa Aesar and used without purification.
10. The submitters employed silica gel (40-63 μm) that was purchased from Sorbent Technologies and was used as received. The checkers used high purity grade silica gel with a pore size of 60 Å and a 230-400 mesh particle size purchased from Fluka.
11. 2-Diazo-1-(2,4-dimethylphenyl)ethan-1-one (**1**) has the following physical and spectroscopic properties: R_f = 0.20 (1:7 ethyl acetate-

hexanes; Agela silica gel plate); mp 40–41 °C; ^{1}H NMR ($CDCl_3$, 400 MHz) δ: 2.36 (s, 3H), 2.52 (s, 3H), 5.66 (s, 1H), 7.08 – 7.03 (m, 2H), 7.31 (d, J = 7.8 Hz, 1H); ^{13}C NMR ($CDCl_3$, 101 MHz) δ: 20.2, 21.2, 55.9, 126.3, 127.4, 132.3, 134.5, 136.9, 141.2, 189.6; HRMS (EI) calculated for $C_{10}H_{10}N_2O$ $[M]^+$ 174.0788, found 174.0788; IR (film): 3108, 2924, 2095, 1613, 1448, 1343, 1220, 1150, 1008, 853, 824, 753 cm^{-1}; Anal. calcd for $C_{10}H_{10}N_2O$: C, 68.95; H, 5.79; N, 16.08. Found: C, 68.85; H, 5.70; N, 15.92.

12. A second run of Step A provided 1.83 g (86%) of the product.
13. FEP tubing (Catalog No. 1520xL), nut (P-235), ferrule (P-200Nx), and thread to Luer adapter (P-678) were obtained from IDEX Health and Science. The tubing was dried in an oven at 160 °C overnight, allowed to cool in a dessicator, and flushed with argon prior to the assembly of the photochemical flow reactor. It is essential to thoroughly dry the FEP tubing since any residual water will intercept the ketene intermediates resulting in a lower yield of the benzannulation product.
14. The Hanovia 450 W medium-pressure mercury lamp was obtained from Ace Glass Inc. (Catalog No. 7825-34). A double-walled quartz immersion well (Catalog No. 7854-25) was employed with the dimensions o.d. of outer joint = 53 mm, length above joint = 76 mm, distance from bottom of well to joint = 220 mm. The Pyrex filter sleeve was obtained from Ace Glass Inc. (Catalog No. 7835-44).
15. Dichloromethane (HPLC grade) was purchased from J. T. Baker and purified by pressure filtration through activated alumina immediately prior to use.
16. Water-free starting materials are necessary for the reaction; thus the checkers azeotropically dried the diazo ketone with benzene (evaporated 3 x 50 mL from benzene at 20 °C/20 mmHg). The submitters did not report this precaution as being necessary, although they have on occasion observed related diazo ketones to pick up small amounts of water during purification by column chromatography on silica gel.
17. UV light can be damaging to eyesight. The photochemical reactor should be surrounded by an appropriate shield to protect researchers from exposure to UV radiation. The Submitters carried out the reaction in a fume hood with windows covered in aluminum foil secured by tape. UVEX safety glasses were worn for eye protection as an additional precaution when the mercury lamp was in operation.

18. The Submitters employed a Harvard Apparatus PHD 2000 Programmable syringe pump for the reaction. It is advisable to use a plastic syringe since glass syringes sometimes may leak. If a glass syringe is used, then a matching set of syringe barrel and plunger should be chosen and the syringe pump should be kept level to minimize leaking.
19. Alternatively, the next stage of the reaction can be carried out the following day. In this case, the solution from the photolysis should be degassed with a stream of argon for 5 min and then stored under argon at 5 °C.
20. Toluene (Baker analyzed, ACS reagent) was purchased from J. T. Baker and purified by pressure filtration through activated alumina immediately prior to use.
21. TLC analysis was performed with silica gel plates (2 cm x 5 cm, glass backed, purchased from Agela Technologies) with 1:2 ethyl acetate-hexanes as eluent and visualization with Seebach's stain (2.5 g of phosphomolybdic acid, 1 g of $Ce(SO_4)_2$, 6 mL of conc H_2SO_4, and 94 mL of H_2O). Diazo ketone **1**: R_f = 0.58, diethyl benzyl(oct-1-yn-1-yl)phosphoramidate: R_f = 0.40, vinylcyclobutenone intermediate: R_f = 0.49, naphthol benzannulation product **2**: R_f = 0.26.
22. Diethyl benzyl(3-hexyl-4-hydroxy-6,8-dimethylnaphthalen-2-yl)phosphoramidate (**2**) has the following physical and spectroscopic properties: mp 121–123 °C; ^{1}H NMR ($CDCl_3$, 400 MHz) δ: 0.90 (t, *J* = 6.8 Hz, 3H), 1.25–1.39 (m, 13H), 1.69 (bs, 1H), 2.33 (s, 3H), 2.40 (s, 3H), 2.66– 2.81 (m, 2H), 4.05–4.26 (m, 5H), 4.77–4.82 (m, 1H), 6.29 (s, 1H), 6.98 (d, *J* = 8.3 Hz, 2H), 7.13–7.21 (m, 5H), 7.66 (s, 1H); ^{13}C NMR ($CDCl_3$, 101 MHz) δ: 14.3, 16.4 (d, *J* = 7.2 Hz), 19.3, 21.9, 22.9, 26.2, 29.2, 30.4, 31.9, 55.8 (*J* = 5.7 Hz), 63.0 (d, *J* = 13.9 Hz), 118.4, 118.6 (d, *J* = 2.9 Hz), 123.0 (d, *J* = 3.8 Hz), 124.2, 127.6, 128.0, 128.5, 129.6 (d, *J* = 1.6 Hz), 130.0, 133.6, 134.2, 137.0 (d, *J* = 3.3 Hz), 137.8 (d, *J* = 2.4 Hz), 150.4; HRMS (ESI) calculated for $C_{29}H_{40}NO_4P$ $[M+H]^+$ 498.2768, found 498.2765; IR (film): 3271, 2954, 2930, 2870, 2850, 1605, 1571, 1498, 1456, 1413, 1394, 1367, 1334, 1248, 1182, 1154, 1126, 1050, 1023, 962, 915, 891, 864, 817, 807, 738, 719, 702, 599 cm^{-1}; Anal. calcd for $C_{29}H_{40}NO_4P$: C, 70.00; H, 8.10; N, 2.81. Found: C, 69.75; H, 8.12; N, 2.75.
23. A second run of Step B provided 3.16 g (88%) of the product.

Working with Hazardous Chemicals

The procedures in *Organic Syntheses* are intended for use only by persons with proper training in experimental organic chemistry. All hazardous materials should be handled using the standard procedures for work with chemicals described in references such as "Prudent Practices in the Laboratory" (The National Academies Press, Washington, D.C., 2011; the full text can be accessed free of charge at http://www.nap.edu/catalog.php?record_id=12654). All chemical waste should be disposed of in accordance with local regulations. For general guidelines for the management of chemical waste, see Chapter 8 of Prudent Practices.

In some articles in *Organic Syntheses*, chemical-specific hazards are highlighted in red "Caution Notes" within a procedure. It is important to recognize that the absence of a caution note does not imply that no significant hazards are associated with the chemicals involved in that procedure. Prior to performing a reaction, a thorough risk assessment should be carried out that includes a review of the potential hazards associated with each chemical and experimental operation on the scale that is planned for the procedure. Guidelines for carrying out a risk assessment and for analyzing the hazards associated with chemicals can be found in Chapter 4 of Prudent Practices.

The procedures described in *Organic Syntheses* are provided as published and are conducted at one's own risk. *Organic Syntheses, Inc.*, its Editors, and its Board of Directors do not warrant or guarantee the safety of individuals using these procedures and hereby disclaim any liability for any injuries or damages claimed to have resulted from or related in any way to the procedures herein.

Discussion

Benzannulation is an important ring-forming process in which a benzenoid ring system is assembled from two or more precursors in a single step with all or most substituents in place.[4] For many highly substituted aromatic and heteroaromatic target molecules, benzannulation provides the most efficient and attractive synthetic route.

Previously we have developed a benzannulation strategy based on the reaction of activated alkynes with vinylketenes[5] and this chemistry has been applied by our laboratory and others for the synthesis of a number of biologically interesting natural products.[6] A recent variant of this method employs ynamide derivatives as the alkyne reaction partner,[7] enabling the regioselective synthesis of highly substituted anilines which are versatile precursors to multiply substituted indoles and other benzofused nitrogen heterocyclic systems.[8]

The key transformation reported in the present article is an example of the "second-generation" version of the benzannulation developed in our laboratory. In the original version of the benzannulation, vinylketenes are generated via the 4-electron electrocyclic ring opening of cyclobutenones. For the synthesis of polycyclic aromatic and heteroaromatic compounds, however, a superior approach employs α-diazo ketones as the vinylketene precursors. Upon irradiation or thermolysis, α-diazo ketones undergo Wolff rearrangement[9] to generate the key vinylketene intermediates, thereby triggering a cascade involving a series of pericyclic transformations. As outlined in Scheme 1, the highly reactive vinylketene first undergoes rapid [2 + 2] cycloaddition with the ynamide benzannulation partner to yield a vinylcyclobutenone intermediate. Reversible 4-electron electrocyclic cleavage initiated by heat or light then furnishes a dienylketene intermediate, and 6-π electrocyclic ring closure and tautomerization finally generates the highly substituted benzenoid aromatic product.

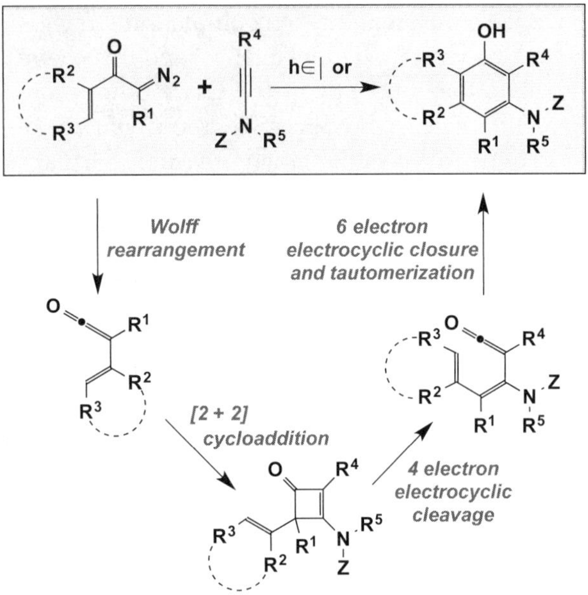

Scheme 1. Mechanism of Benzannulations Based on Diazo Ketones with Ynamides as Reaction Partners

An attractive feature of this "second-generation" version of the benzannulation is that a variety of alkenyl, aryl, and hetaryl α-diazo ketones are readily available, providing access to a wide range of polycyclic aromatic and heteroaromatic compounds. Representative examples of the "second generation" benzannulation with ynamides as reaction partners are presented in Table 1.[10] The benzannulations shown here are best conducted as a two-stage process. As in the case of a number of other "second generation" benzannulations we have studied, a mixture of the desired aromatic product and the intermediate vinylcyclobutenone is isolated after the photochemical phase of the benzannulation. In such cases, heating the crude reaction mixture for several hours serves to complete the conversion of the vinylcyclobutenone to phenol.

Table 1. Selected Examples of Photochemical Benzannulations

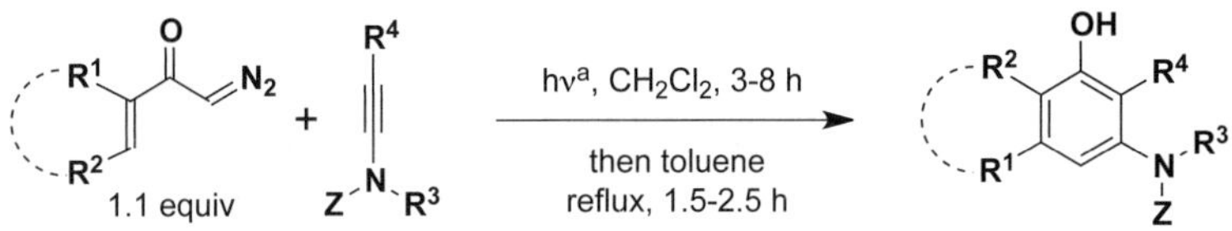

Entry	Diazo ketone	Ynamide	Product	Yield (%)[b]
1	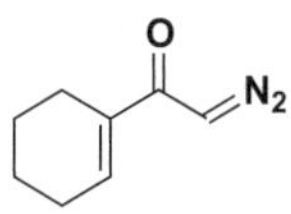	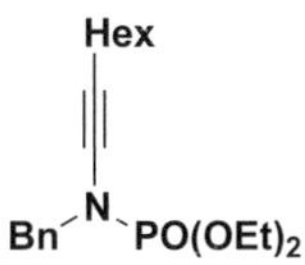	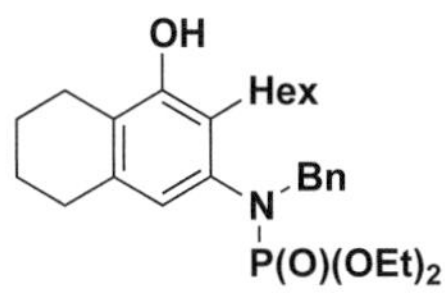	66
2	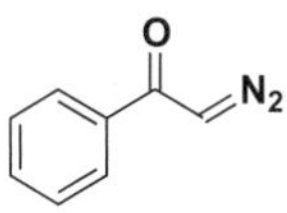	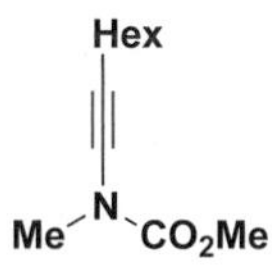	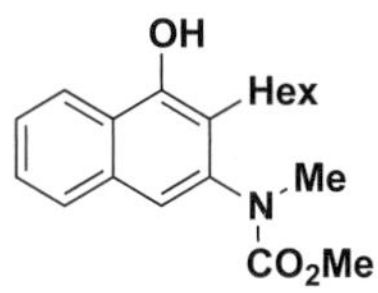	64-76
3	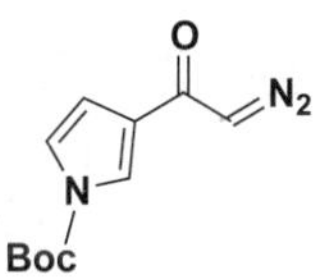	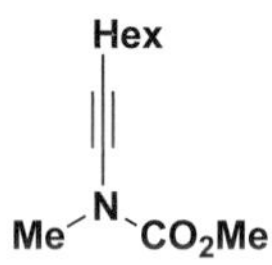	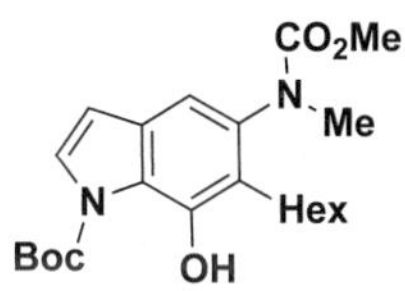	40
4	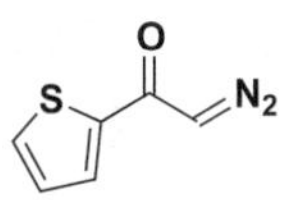	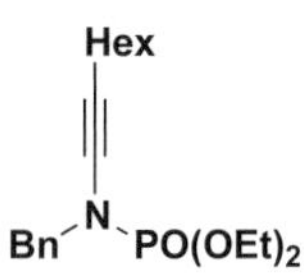	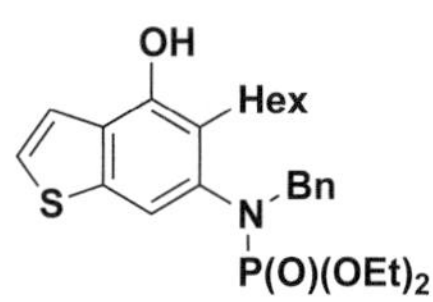	78

[a] Batch processes with irradiation using a Hanovia 450 W medium-pressure mercury lamp. [b] Isolated yield of products purified by column chromatography.

The benzannulation described in the present article features the use of an *N*-phosphoryl ynamide as the alkyne reaction partner. We have reported that *N*-phosphoryl ynamides exhibit significantly greater ketenophilicity than *N*-sulfonyl and *N*-carbonyl ynamides, with [2 + 2] cycloadditions taking place nearly an order of magnitude faster with *N*-phosphoryl ynamides as compared to the corresponding carbamates.[10] *N*-Phosphoryl ynamides are conveniently prepared via the copper-catalyzed alkynylation method reported by Hsung and coworkers,[11] and we have reported a detailed procedure for the synthesis of the ynamide employed in the present article.[2]

α-Diazo ketones function as the vinylketene precursor in the "second generation" version of the benzannulation. Many α-diazo ketones are conveniently prepared by the reaction of diazomethane (or TMS-diazomethane) with carboxylic acid derivatives such as acyl chlorides and certain mixed anhydrides. Alternatively, diazo ketones can also be generated from the corresponding ketones by reaction with sulfonyl azides ("diazo transfer").[12] For the benzannulation described in this article, diazo ketone **1** was synthesized via the detrifluoroacetylative diazo transfer developed in our laboratory.[13] The use of 4-acetamidobenzenesulfonyl azide[14] as the diazo transfer agent in this reaction enables the separation of most of sulfonamide byproduct by filtration and simplifies the purification of the diazo ketone via column chromatography.

Recently we reported that the "second generation" photo-Wolff-based benzannulation can be successfully accomplished under conditions of continuous flow.[10] Continuous flow photochemical reactors provide significant advantages over traditional "batch" reactors.[15] Photochemical reactions carried out in batch are relatively inefficient because light does not penetrate into the bulk of the solution. In contrast, flow reactions can be carried out in narrow gauge tubing with a relatively high surface-to-volume ratio allowing for efficient and uniform irradiation of the reaction mixture. A second important advantage is that photochemical flow reactions often proceed in higher yield. This is due to the fact that side reactions involving the photochemical decomposition of the desired products is minimized with the continuous removal of products from further irradiation in the reactor. Finally, reactions conducted in continuous flow reactors are independent of scale and thus can be scaled up without the need for re-optimization of reaction parameters. In our previous study, we found that the benzannulations shown in Table 1 could each be carried out under conditions of continuous flow with yields equal or superior to those obtained in batch reactions.[10]

The flow photochemical reactor used in this article is based on the design described by Booker-Milburn and coworkers.[16,17] A very attractive feature of this design is its simplicity and the fact that all of the necessary equipment is generally already available in laboratories set up for performing conventional photochemical reactions. For the flow reaction described in the present article, transparent fluorinated ethylene-propylene (FEP) tubing (0.76 mm i.d., 1.59 mm o.d.) is tightly wound around a quartz immersion well cooled with circulating tap water as shown in Figure 1 above. A 450-W medium pressure mercury lamp is placed inside the immersion well with a Pyrex filter sleeve to suppress polymer buildup inside the tubing. A syringe pump is used to drive the reaction mixture (ca. 0.25 M in ynamide and diazo ketone) through the tubing. We have found that this simple apparatus allows preparative photochemical benzannulations to be achieved in excellent yield and with results sometimes superior to that observed in batch reactions.

A key reaction parameter in optimizing photochemical flow reactions is residence time in the irradiation zone of the reactor. We employed a reactor with a shorter length of tubing (internal volume ca. 1.2 mL) for optimization of the residence time for this continuous flow benzannulation. Experiments were conducted at one-tenth of the projected full scale using a small (20%) excess of the diazo ketone. Aliquots of the reaction mixture were analyzed by ^{1}H NMR against 1,3,5-trimethoxybenzene as internal standard to determine the amount of unreacted diazo ketone after irradiation for different periods of time. After heating for 5 h to complete the benzannulation, the product of each run was purified by column chromatography and analyzed by ^{1}H NMR vs. 1,3,5-trimethoxybenze to determine the yield of naphthol produced in the reaction.

Table 2 summarizes the results of our optimization experiments. As expected, faster flow rates (shorter residence times) result in the recovery of larger amounts of unreacted diazo ketone. For our optimal procedure, we chose to employ 1.2 equiv of diazo ketone which allows for efficient reaction even with residence times on the order of 10 min. This residence time corresponds to an excellent throughput of 10.5 mmol/h for the "preparative scale" photochemical flow reactor employed in the procedure described in this article. It should be noted, however, that if desired, the benzannulation can be carried out in comparable overall yield using only 1.05 equiv of diazo ketone simply by increasing the residence time as indicated in Table 2.

Table 2. Optimization of Residence Time for Flow Benzannulation

flow rate (mL/min)	residence time (min)[a]	yield (%)	unreacted diazo ketone (%)
0.057	21	85[b]	6.3
0.086	14	84[b]	7.4
0.115	10	83	18
0.144	8.3	81	17
0.188	6.4	72	21

[a] Residence time = volume of irradiated tubing (mL) / flow rate (mL/min). [b] Average of two runs.

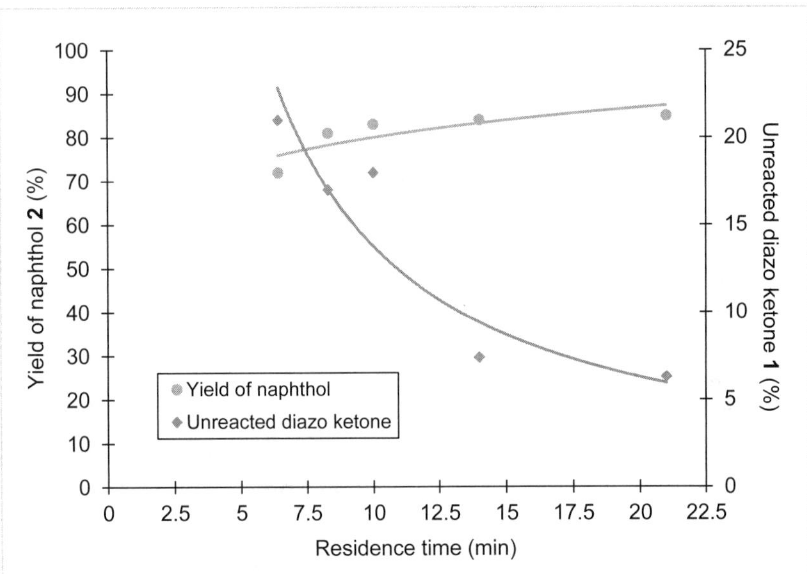

In the case of the benzannulation described in this article, a residence time of 10.4 min was employed with 50 min then required to pump the reaction solution containing 7.3 mmol of ynamide through the photochemical reactor. For comparison, irradiation for 8 h was necessary to achieve complete photo-Wolff rearrangement when the same benzannulation was carried out in batch mode, though the yield in this case (81-90% on 0.3 mmol scale) was comparable to that obtained under condtions of continuous flow. We have previously reported that it is also possible to effect the entire benzannulation cascade by heating, i.e. without the need for irradiation.[10] In this case, however, reaction of the ynamide with 1.2 equiv of diazo ketone **1** in toluene at reflux required 7.5 for complete reaction and produced the desired naphthol **2** in only moderate (55%) yield.

In summary, the "second-generation" benzannulation based on vinylketenes can be conveniently performed under conditions of continuous flow using readily available photochemical equipment and a syringe pump. The high throughput of this reactor allows the benzannulation to be carried out on a multigram scale with reaction times on the order of several hours.

References

1. Department of Chemistry, Massachusetts Institute of Technology, Cambridge, MA 02139. Email: danheisr@mit.edu. We thank the National Science Foundation (CHE-1111567) for generous financial support.
2. Read, J. M.; Wang, Y-P.; Danheiser, R. L. *Org. Synth.* **2015**, *92*, 156–170.
3. Watson, S. C.; Eastham, J. F. *J. Organomet. Chem.* **1967**, *9*, 165–168.
4. For reviews, see: (a) Kotha, S.; Misra, S.; Halder, S. *Tetrahedron* **2008**, *64*, 10775–10790. (b) Saito, S.; Yamamoto, Y. *Chem. Rev.* **2000**, *100*, 2901–2916.
5. Original reports: (a) Danheiser, R. L.; Gee, S. K. *J. Org. Chem.* **1984**, *49*, 1672–1674. (b) Danheiser, R. L.; Brisbois, R. G.; Kowalczyk, J. J.; Miller, R. F. *J. Am. Chem. Soc.* **1990**, *112*, 3093–3100.
6. For selected examples, see (a) Kowalski, C. J.; Lal, G. S. *J. Am. Chem. Soc.* **1988**, *110*, 3693–3695. (b) Smith, A. B., III; Adams, C. M.; Kozmin, S. A.; Paone, D. V. *J. Am. Chem. Soc.* **2001**, *123*, 5925–5937. (c) Dudley, G. B.; Takaki, K. S.; Cha, D. D.; Danheiser, R. L. *Org. Lett.* **2000**, *2*, 3407–3410 and references cited therein.
7. Mak, X. Y.; Crombie, A. L.; Danheiser, R. L. *J. Org. Chem.* **2011**, *76*, 1852–1873.
8. (a) Lam, T. Y.; Wang, Y.-P.; Danheiser, R. L. *J. Org. Chem.* **2013**, *78*, 9396–9414. (b) Willumstad, T. P.; Bordreau, P. D.; Danheiser, R. L. *J. Org. Chem.* **2015**, *80*, 11794–11805.

9. For reviews of the Wolff rearrangement, see: (a) Kirmse, W. *Eur. J. Org. Chem.* **2002**, 2193–2256. (b) Meier, H.; Zeller, K.-P. *Angew. Chem., Int. Ed.* **1975**, *14*, 32–43.
10. Willumstad, T. P.; Haze, O.; Mak, X. Y.; Lam, T. Y.; Wang, Y.-P.; Danheiser, R. L. *J. Org. Chem.* **2013**, *78*, 11450–11469.
11. DeKorver, K. A.; Walton, M. C.; North, T. D.; Hsung, R. P. *Org. Lett.* **2011**, *13*, 4862–4865.
12. For reviews on the synthesis of diazo ketones, see: (a) Maas, G. *Angew. Chem., Int. Ed.* **2009**, *48*, 8186–8195. (b) Zhang, Y.; Wang, J. *Chem. Commun.* **2009**, 5350–5361. (c) Zhang, Y.; Wang, J. *Tetrahedron* **2008**, *64*, 6577–6605. (d) Doyle, M. P.; McKervey, M. A.; Ye, T. *Modern Catalytic Methods for Organic Synthesis of Diazo Compounds: from Cyclopropanes to Ylides*; Wiley & Sons: New York, 1998; pp 1–60. (e) Regitz, M.; Maas, G. *Diazo Compounds: Properties and Synthesis*; Academic Press: Orlando, Fl, 1986.
13. (a) Danheiser, R. L.; Miller, R. F.; Brisbois, R. G.; Park, S. Z. *J. Org. Chem.* **1990**, *55*, 1959–1964. (b) Danheiser, R. L.; Miller, R. F.; Brisbois, R. G. *Org. Synth.* **1996**, *73*, 134–140.
14. (a) Baum, J. S.; Shook, D. A.; Davies, H. M. L.; Smith, H. D. *Synth. Commun.* **1987**, *17*, 1709–1716. (b) Davies, H. M. L.; Cantrell, W. R. Jr.; Romines, K. R.; Baum, J. S. *Org. Synth.* **1992**, *70*, 93. This sulfonyl azide is reported to exhibit no impact sensitivity and is a safer reagent for diazo transfer reactions as compared to mesyl azide and tosyl azide.
15. For reviews of continuous flow photochemical reactions, see: (a) Gilmore, K.; Seeberger, P. H. *Chem. Rec.* **2014**, *14*, 410–418. (b) Schuster, E. M.; Wipf, P. *Isr. J. Chem.* **2014**, *54*, 361–370. (c) Knowles, J. P.; Elliott, L. D.; Booker-Milburn, K. I. *Beilstein J. Org. Chem.* **2012**, *8*, 2025–2052. (d) Oelgemöller, M.; Shvydkiv, O. *Molecules* **2011**, *16*, 7522–7550.
16. Hook, B. D. A.; Dohle, W.; Hirst, P. R.; Pickworth, M.; Berry, M. B.; Booker-Milburn K. I. *J. Org. Chem.* **2005**, *70*, 7558–7564.
17. For an early example of a similar photochemical flow reactor, see Birr, C.; Lochinger, W.; Stahnke, G.; Lang, P. *Liebigs Ann. Chem.* **1972**, *763*, 162–172.

Appendix
Chemical Abstracts Nomenclature (Registry Number)

4-Acetamidobenzenesulfonyl azide; (2158-14-7)
2,2,2-Trifluoroethyl trifluoroacetate; (407-38-5)
LiHMDS: Lithium bis(trimethylsilyl)amide; (4039-32-1)
2′,4′-Dimethylacetophenone; (89-74-7)
Diethyl benzyl(oct-1-yn-1-yl)phosphoramidate: Phosphoramidic acid, *N*-1-octyn-1-yl-*N*-(phenylmethyl)-, diethyl ester; (1332480-36-0)
2-Diazo-1-(2,4-dimethylphenyl)ethan-1-one; (23722-51-2)

John M. Read was born in Laredo, Texas, in 1994. He is currently an undergraduate at the Massachusetts Institute of Technology and expects to receive his B.S. degree in chemistry in 2016. John joined the laboratory of Professor Rick Danheiser in 2013, and his research has focused on the synthesis of highly substituted polycyclic compounds and requisite precursors

Yu-Pu Wang was born in Taipei, Taiwan. He received a B.S. degree in Chemistry in 2009 from Rice University working in the laboratory of Professor James M. Tour. He received his Ph.D. degree at the Massachusetts Institute of Technology in 2015 working in the research group of Professor Rick L. Danheiser on the development of new methods for the synthesis of highly substituted indoles and their application to the synthesis of natural products and polycyclic systems with interesting electronic properties.

Rick L. Danheiser received his undergraduate education at Columbia where he carried out research in the laboratory of Professor Gilbert Stork. He received his Ph.D. at Harvard in 1978 working under the direction of E. J. Corey on the total synthesis of gibberellic acid. Dr. Danheiser is the A. C. Cope Professor of Chemistry at MIT where his research focuses on the design and invention of new annulation and cycloaddition reactions, and their application in the total synthesis of biologically active compounds.

Christian Ebner was born in southern Germany. During his undergraduate studies at ETH Zurich, Switzerland, he moved to the Scripps Research Institute in La Jolla, USA to conduct his master thesis in the laboratory of Professor K. C. Nicolaou involving the development of a novel macroheterocyclization. Since June 2012, he has been a doctoral student in the group of Professor Erick M. Carreira, working on the total synthesis of terpenoid natural products.

Organic Syntheses

Trifluoromethylation of Aryl Iodides Catalyzed by the Copper(I)-Phen Complex

Naoto Shimizu, Hideaki Kondo, Masahiro Oishi, Kenichi Fujikawa, Kazuki Komoda, and Hideki Amii[1]*

Division of Molecular Science, Faculty of Science and Technology, Gunma University, 1-5-1 Tenjin-cho, Kiryu, Gunma, 376-8501, Japan

Checked by Mina Nakhla and John L. Wood

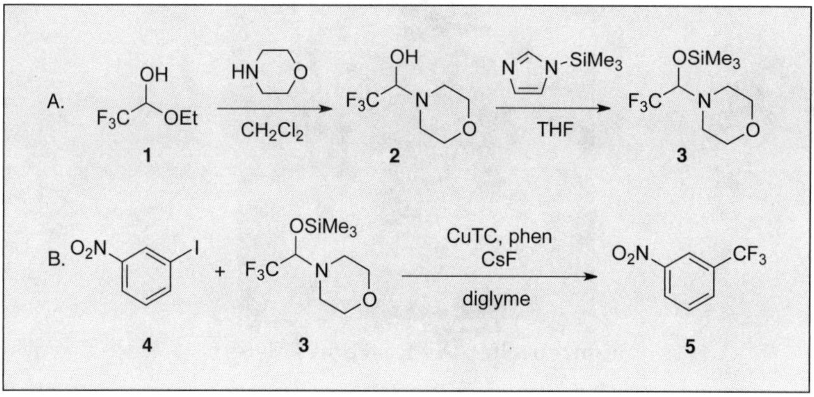

Procedure

A. 4-[2,2,2-Trifluoro-1-(trimethylsilyloxy)ethyl]morpholine (**3**). A 1-L, three-necked, round-bottomed flask is equipped with a 4-cm Teflon-coated magnetic stir bar and the center neck is fitted with a glass gas inlet adapter which is connected to a nitrogen/vacuum double manifold. The other two necks are capped with rubber septa. The flask is charged with molecular sieves 4 Å (40 g) (Note 1). The system is flame-dried under vacuum and backfilled with nitrogen (Note 2). After flame drying the right neck of the flask is equipped with a temperature probe. To the flask are added anhydrous dichloromethane (300 mL, via cannula) (Note 3), morpholine (19.8 g, 19.9 mL, 227 mmol, via syringe) (Note 4), and fluoral ethyl hemiacetal (32.7 g, 26.8 mL, 227 mmol, via syringe) (Note 5). After being

Published on the Web 6/7/2016

stirred at room temperature for 5 min, the resulting mixture is heated in a 30 °C oil bath for 48 h (Notes 6 and 7) (Figure 1). The flask containing the

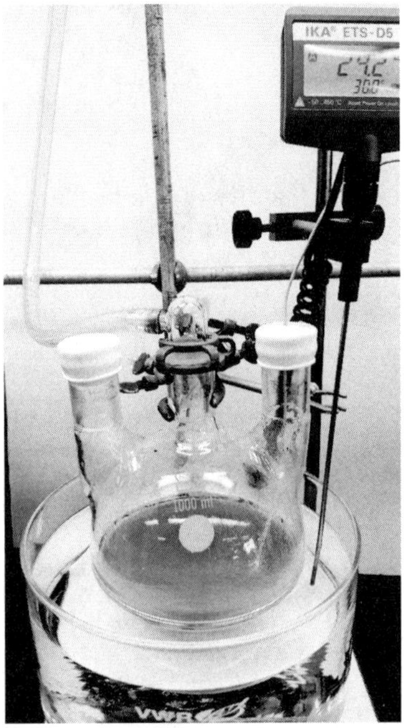

Figure 1. Step A - Reaction 1 Set-up

mixture is then removed from the oil bath and allowed to cool to room temperature. The mixture is then vacuum filtered through a 350 mL fine-fritted glass funnel, which is packed with a 20g pad of Celite (Note 8). The Celite® pad and the reaction flask are washed with CH_2Cl_2 (3 x 100 mL) and the filtrate is transferred to 1 L recovery flask rinsing with dichloromethane (2 x 50 mL). The solvent is removed by rotary evaporation to afford 40.6 g of trifluoroacetaldehyde hemiaminal **2** as a pale yellow oil, which is used in the next step (silylation) without purification (Note 9).

For the silylation step, a 1-L, three-necked, round-bottomed flask is equipped with a 4-cm Teflon-coated magnetic stir bar and the center neck is fitted with a glass gas inlet adapter which is connected to a nitrogen/vacuum double manifold. The other two necks are capped with rubber septa. The flask is flame-dried under vacuum and backfilled with nitrogen. After flame drying the right neck of the flask is equipped with a temperature probe. A solution of trifluoroacetaldehyde hemiaminal **2**

(40.6 g, 219 mmol) in anhydrous tetrahydrofuran (75 mL) (Notes 10 and 11) is prepared and added to the dried flask via cannula. Additional THF (75 mL) is to rinse residual hemiaminal **2** into the three-necked flask. The resultant solution is cooled in an ice bath for 30 min (internal temperature : 2.7 °C).

Figure 2. Step A - Reaction 2 Set-up

To the THF solution of hemiaminal **2** is added *N*-(trimethylsilyl)imidazole (30.8 g, 220 mmol) (Note 12) slowly at a rate of about 2 mL per minute (Note 13) (Figure 2). After the addition is complete the resulting mixture is allowed to warm to room temperature. After being stirred at room temperature for 6 h (Note 14), a saturated aqueous solution of $NaHCO_3$ (150 mL) is added to the reaction and the resultant mixture. transferred to a 1L separatory funnel. Separation of the phases is followed by extraction of the aqueous phase with hexanes (3 x 200 mL) (Note 15). The combined hexanes extracts are washed successively with brine (2 x

100 mL), and then dried over anhydrous sodium sulfate (130 g) (Note 16). The sodium sulfate was removed by vacuum filtration using a glass fritted filter. After rinsing the filter cake with hexane (2 x 50 mL) the filtrate is poured into a 1 L recovery flask rinsing with hexanes (1 x 25 mL). The solvent is then removed under reduced pressure by rotary evaporation to yield 4-[2,2,2-trifluoro-1-(trimethylsilyloxy)ethyl]morpholine (**3**, 46.6–48.7 g, 181–189 mmol, 80–83% over two steps) (Notes 17 and 18) as a pale colorless liquid that is used for step B without further purification (Note 19).

B. 3-Nitro-1-(trifluoromethyl)benzene (**5**). A 1-L, three-necked round-bottomed flask is equipped with a 4-cm Teflon-coated magnetic stir bar and the center neck is fitted with a glass gas inlet adapter which is connected to a nitrogen/vacuum double manifold. The other two necks are capped with rubber septa. The flask is charged with CsF (29.3 g, 193 mmol) (Note 20), flame-dried under vacuum and then backfilled with nitrogen. After flame drying is complete, the right neck of the flask is equipped with a thermometer. To the flask are added copper(I) thiophene-2-carboxylate (CuTC, 1.84 g, 9.65 mmol) (Note 21), 1,10-phenanthroline (phen, 1.74 g, 9.65 mmol) (Note 22), *m*-iodonitrobenzene (24.0 g, 96.5 mmol) (Note 23) and anhydrous diglyme (180 mL) (Note 24). Next, 4-[2,2,2-trifluoro-1-(trimethylsilyloxy)ethyl] morpholine (**3**, 49.6 g, 193 mmol) in 10 mL anhydrous diglyme is transferred by cannula to the reaction flask rinsing with an additional 10 mL of anhydrous diglyme (Note 25). The resulting mixture is heated in a 80 °C oil bath for 24 h (Notes 26 and 27) (Figure 3). After cooling to room temperature the reaction is quenched by the addition of deionized water (250 mL) and the resultant mixture poured into a 2L separatory funnel and extracted with diethyl ether (3 x 150 mL). The combined extracts are washed successively with water (5 x 300 mL) and brine (1 x 300 mL) (Note 28), and then dried over anhydrous sodium sulfate (90 g). Removal of the sodium sulfate by filtration through a fritted filter with rinsing (Et_2O, 3 x 50 mL) was followed by solvent removal on a rotary evaporator (Note 29). The resultant residue is purified by column chromatography on silica gel (Note 30). The fractions containing the product are combined and concentrated under reduced pressure to provide pure 3-nitro-1-(trifluoromethyl)benzene (**5**) as a yellow liquid (14.3 g, 78 %) (Notes 31 and 32).

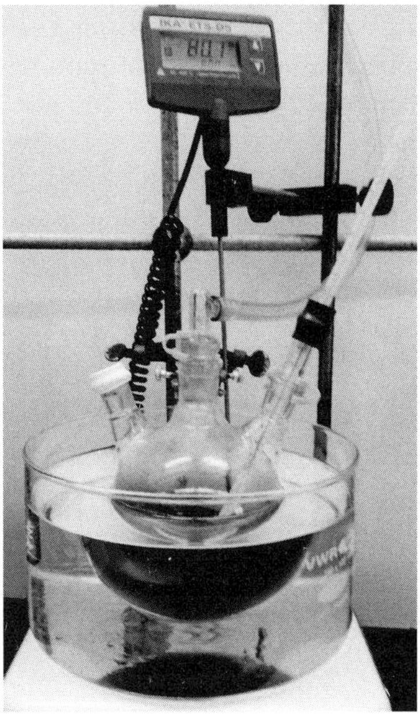

Figure 3. Step B Reaction Set-up

Notes

1. Submitters purchased 4 Å molecular sieves from Wako Pure Chemical Industries. The checkers purchased molecular sieves 4 Å (8–12 mesh) from Acros Organics. The sieves are activated by initially heating in a microwave oven (1100 W general electric model :JES1142 WD 04) for one min and then flame drying under vacuum. After cooling and flushing with nitrogen, this procedure is repeated twice before the sieves are added to the reaction flask.
2. Including the evacuation prior to flame drying, the reaction flasks used in each step of this procedure are evacuated and backfilled with dry nitrogen three times for every step of this procedure.
3. The submitters purchased dehydrated dichloromethane (>98.0%) from Kishida Chemical Co.,Ltd. The checkers purchased non-stabilized DCM from J.T. Baker and passed through two packed columns of neutral alumina in a solvent purification system manufactured by S.G. Waters U.S.A., LLC.

4. The submitters purchased morpholine from Kanto Chemical Co., Inc. and used it as supplied. The checkers purchased morpholine (≥ 99%) from Sigma-Aldrich and also used it as obtained.
5. The submitters purchased trifluoroacetaldehyde ethyl hemiacetal (contains ca. 10% ethanol) from Tokyo Chemical Industry Co., Ltd. and used it as received. The checkers purchased trifluoroacetaldehyde ethyl hemiacetal (>80% , contains ca. 10% ethanol) from TCI America and also used it as received.
6. The reaction mixture turns light brown due to the molecular sieves being crushed over the course of the reaction.
7. Although the submitters and checkers allowed the reaction to proceed for 48 h, the checkers monitored the reaction by NMR and found it to be complete after 2.25 h and undergo no further change during the 48h run.
8. The submitters purchased Celite® 545 from Kishida Chemical Company, Ltd. The checkers purchased Celite® S from Sigma Aldrich. The 350 mL fine-fritted glass funnel had a diameter of 9 cm and a height of 7.5 cm when measured from the top of the frit.
9. The solvent was removed under reduced pressure (160 mmHg – 60 mmHg) on a rotary evaporator (90 rpm, 32–35 °C)
10. The submitters purchased dehydrated tetrahydrofuran from Kishida Chemical Co., Ltd. The checkers purchased non-stabilized THF from Fisher Scientific and passed the solvent through two packed columns of neutral alumina in a solvent purification system manufactured by S.G. Waters U.S.A., LLC.
11. Crude hemiaminal **2** is obtained as a highly viscous oil.
12. The submitters purchased trimethylsilylimidazole (>98.0%) from Tokyo Chemical Industry Co., Ltd. and used it as received. The checkers purchased trimethylsilylimidazole (>98.0%) from Sigma-Aldrich and used it as received.
13. After addition of the trimethylsilylimidazole was complete the internal temperature rose to 5 °C.
14. The measuring of the 6 h reaction time began when the internal thermometer reached room temperature (which for the checkers was 24 °C).
15. The first 200 mL of hexanes is used to wash the reaction flask and then transferred to the separatory for use in the extraction.
16. After extraction the separatory funnel is washed with 50 mL hexanes.

17. Solvent is removed under reduced pressure (140 torr – 70 torr) by rotary evaporation (90 rpm, 28 °C – 39 °C)
18. A reaction run on half scale yielded 23.8 g (92.5 mmol, 80%) of **3**.
19. Although **3** can be carried on crude, for characterization purposes the checkers found that it can be purified via silica gel chromatography eluting with 10% ethyl acetate in hexanes: ^{1}H NMR (400 MHz, $CDCl_3$) δ: 0.18 (s, 9H), 2.67–2.82 (m, 4H), 3.69 (t, J = 4.8 Hz, 4H), 4.38 (q, J = 5.7 Hz, 1H); ^{19}F NMR (376 MHz, $CDCl_3$, C_6F_6) δ: 79.57 (3F, s). ^{13}C NMR (75.6 MHz, $CDCl_3$) δ: 0.04, 48.07, 67.2, 85.62 (q, $^{2}J_{CF}$ = 32.5 Hz), 123.47 (q, $^{1}J_{CF}$ = 286.7 Hz); IR (neat): 2962.3, 2857, 1455.6, 1354.7, 1334.2, 1298.6, 1278.2, 1114.5, 1069.4, 1016.7, 873.1, 839.1 cm^{-1}; +ESI-HRMS *m/z* : calcd for (M+H$^+$) $C_9H_{19}NO_2F_3SiH^+$ = 258.11371, found $C_9H_{19}NO_2F_3SiH^+$ = 258.11328; Anal. calcd. for $C_9H_{18}F_3NO_2Si$: C, 42.01; H, 7.05; N, 5.44; F, 22.15 found: C, 42.08; H, 7.05; N, 5.46; F, 21.98.
20. The submitters purchased cesium fluoride (>97.0%) from Wako Pure Chemical Industries, Ltd. The checkers purchased cesium fluoride (≥ 98%) from Sigma-Aldrich.
21. The submitters purchased copper(I) thiophene-2-carboxylate (>96.0%) from Wako Pure Chemical Industries, Ltd. and used it without further purification. The checkers purchased copper(I) thiophene-2-carboxylate (98.0%) from Oakwood Chemical and used it without further purification.
22. The submitters and checkers purchased anhydrous 1,10-phenanthroline (>99.0%) from Sigma-Aldrich and used it as received.
23. The submitters purchased *m*-iodonitrobenzene from Wako Pure Chemical Industries, Ltd. and used it without further purification. The checkers purchased *m*-iodonitrobenzene from Sigma-Aldrich and used it without further purification.
24. The submitters and checkers purchased anhydrous diethylene glycol dimethyl ether (99.5%) from Sigma-Aldrich and used it as received.
25. The addition of the 180 mL of diglyme was quickly and immediately followed by the rapid addition by cannula of compound **3**. The addition of **3**, including the washing, is complete within 2 min and the flask is then immediately placed in a preheated oil bath. The checkers noted that the oil bath must be preheated to 80 °C, since slow heating of the mixture results in incomplete consumption of starting material.
26. The color of the solution turns dark brown.
27. Monitoring of the reaction by NMR showed that the starting material is consumed and that the reaction is complete within 2.5 h.

28. Each of the 150 mL portions of diethyl ether is used to rinse the reaction flask prior to use in the extraction. The checkers noted a slight exotherm during the water washings and thus to avoid pressure build-up, the two phases were typically swirled together vigorously until the exotherm subsided. At this point the separatory funnel was capped and the extraction continued.
29. The solvent was removed under reduced pressure (210 mmHg – 80 mmhg) using a rotary evaporator. (90 rpm, 27–28 °C).
30. The checkers purchased silica gel SilicaFlash® F60 (40-63 μm / 230 – 500 mesh) from Silicycle. Column chromatography was performed using a glass column 8.5-cm in diameter packed with 1.1 cm sand and 5.7 cm SiO_2 (150 g). The silica was dry packed and wetted with 600 mL hexanes. The compound was loaded neat onto the column and the flask which contained it was washed three times with 5 mL hexanes and these washings were also loaded on to the column. The column was then topped with 1.2 cm sand. The compound was eluted using a gradient solvent system which began with pure hexanes (300 mL), progressed to 0.75% ethyl acetate in hexanes (400 mL) and then progressed to and was held at 1.15% ethyl acetate in hexanes (2600 mL). The checkers found that failing to use a gradient elution and employing 2% ethyl acetate in hexanes results in the co-elution of a minor impurity. The checkers also found that residual diglyme in the crude reaction mixture can cause the compound to elute quickly but is not detrimental to obtaining pure product under the gradient elution conditions. TLC (UV visualization) is used to follow the chromatography. Glass-backed extra hard layer TLC plates, 60 Å (250 μm thickness) containing F-254 indicator were purchased from Silicycle. The R_f value of the title compound is 0.2 (2% EtOAc/hexanes). Fractions 17-74 (30 mL each) are concentrated by rotary evaporation (160 mmHg–125 mmHg, 35–40 °C bath, 90 rpm). To fully remove residual solvent the checkers continued the rotary evaporation beyond the point of apparent solvent removal (about 40 min, 50 mmHg, 90 rpm , 35–42 °C) followed by holding under reduced pressure (30 mmHg) for 1 h.
31. 3-Nitro-1-(trifluoromethyl)benzene (**5**) has the following physical and spectroscopic data: Rf = 0.66 (hexane); ^{1}H NMR (400 MHz, $CDCl_3$) □: 7.74 (t, J = 8.0 Hz, 1H, 7.98 (d, J = 7.8 Hz, 1H), 8.45 (d, J = 8.4 Hz, 1H), 8.52 (s, 1H); ^{19}F NMR (376 MHz, $CDCl_3$, C_6F_6) δ: 66.05 (3 F, s); ^{13}C NMR

(101 MHz, $CDCl_3$) δ: 121.02 (q, J = 4.0 Hz), 122.98 (q, J = 272.9 Hz), 126.82, 130.48, 131.28 (q, J = 3.5 Hz), 132.48 (q, J = 34.3 Hz); IR (Neat): 3098.9, 1625.3, 1537.1, 1487.3, 1445.7, 1358.1, 1319.3, 1281.1, 1173.5, 1128.5, 1108.5, 1065.7, 909.8, 882.9, 812.7, 784.5, 739.6, 697.6 cm^{-1}; MS (EI) *m/z* (%): 191 (M^+, 28), 172 (12), 145 (100), 125 (22), 95 (34), 75 (33), 69 (10); Anal. calcd. for $C_7H_4F_3NO_2$: C, 43.99; H, 2.11; N, 7.33; F, 29.82; found: C, 43.94; H, 2.11; N, 7.27;F, 29.62.

32. A second reaction on the same scale provided 14.5 g (79%) of the product.

Working with Hazardous Compounds

The procedures in *Organic Syntheses* are intended for use only by persons with proper training in experimental organic chemistry. All hazardous materials should be handled using the standard procedures for work with chemicals described in references such as "Prudent Practices in the Laboratory" (The National Academies Press, Washington, D.C., 2011; the full text can be accessed free of charge at http://www.nap.edu/catalog.php?record_id=12654). All chemical waste should be disposed of in accordance with local regulations. For general guidelines for the management of chemical waste, see Chapter 8 of Prudent Practices.

In some articles in *Organic Syntheses*, chemical-specific hazards are highlighted in red "Caution Notes" within a procedure. It is important to recognize that the absence of a caution note does not imply that no significant hazards are associated with the chemicals involved in that procedure. Prior to performing a reaction, a thorough risk assessment should be carried out that includes a review of the potential hazards associated with each chemical and experimental operation on the scale that is planned for the procedure. Guidelines for carrying out a risk assessment and for analyzing the hazards associated with chemicals can be found in Chapter 4 of Prudent Practices.

The procedures described in *Organic Syntheses* are provided as published and are conducted at one's own risk. *Organic Syntheses, Inc.*, its Editors, and its Board of Directors do not warrant or guarantee the safety of individuals using these procedures and hereby disclaim any liability for any injuries or damages claimed to have resulted from or related in any way to the procedures herein.

Discussion

The structural motifs of trifluoromethylated aromatics (Ar-CF_3) have been widely employed in various industrial fields.[2,3] The high lipophilicity, strong electron-withdrawing ability, and characteristic size of the trifluoromethyl group is a key influence in the design of pharmaceuticals, agrochemicals, dyes, liquid crystals, and polymers. Therefore, a great deal of attention has been paid to convenient protocols of introducing trifluoromethyl groups into aromatic rings.[4] With respect to the high regiochemical fidelity in aromatic substitution, the cross-coupling of aryl halides with CF_3Cu is one of the most powerful and versatile methods to construct trifluoromethylated aromatics.[5] Fluoral (trifluoroacetaldehyde) and its derivatives are readily available organofluorine compounds. Hemiaminals of fluoral are known to be convenient sources of trifluoromethyl anion.[6] Previously, we reported the catalytic procedure for aromatic trifluoromethylation by the use of trifluoroacetaldehyde hemiaminal derivative **2** as a cross-coupling partner in the presence of 10 mol% of CuI and 1,10-phenanthroline (phen).[7,8] After screening of Cu catalysts, we found that for the cross-coupling reactions of aryl iodides with 2, the use of copper thiophene-2-carboxylate (CuTC) [9] as a source of copper(I) affords the best results. Representative examples of the selective formation of trifluoromethyl arenes are summarized in Table 1 .

Table 1. Cu-Catalyzed Cross-Coupling for Aromatic Trifluoromethylation

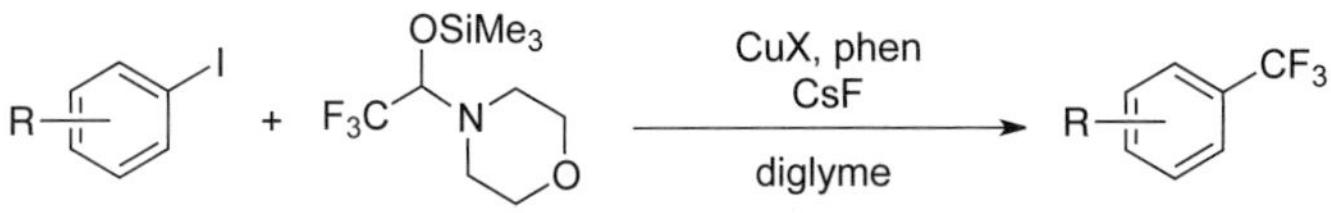

entry	iodoarene	product	yield/% X = I	yield/% X = TC
1	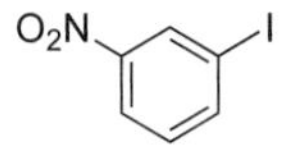	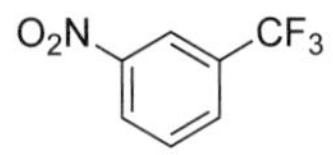	90	>95
2	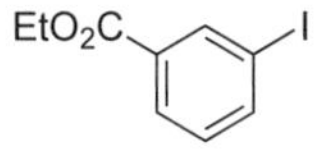	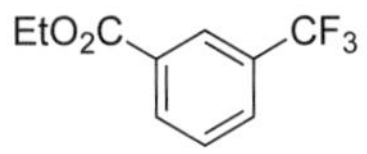	60	75
3	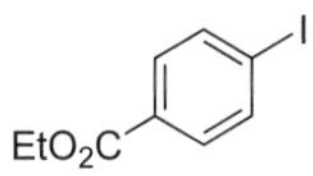	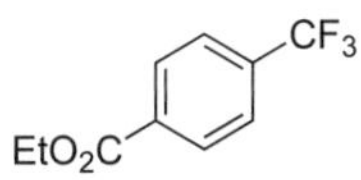	30	91
4	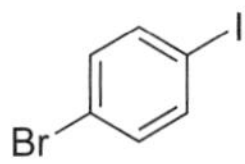	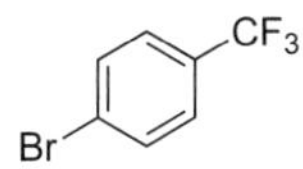	53	83
5	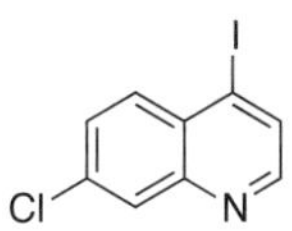	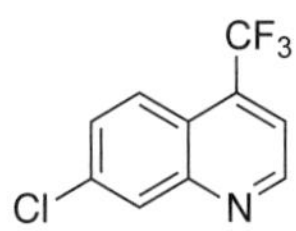	42	91

[a] Determined by ^{19}F NMR analysis (CF_3CH_2OH was used as internal standard).

References

1. Division of Molecular Science, Faculty of Science and Technology, Gunma University, 1-5-1 Tenjin-cho, Kiryu, Gunma, 376-8501, Japan. E-mail: amii@gunma-u.ac.jp. This work was financially supported by the Ministry of Education, Culture, Sports, Science and Technology of Japan and Japan Science and Technology Agency (JST) (ACT-C: Creation of Advanced Catalytic Transformation for the Sustainable Manufacturing at Low Energy, Low Environmental Load.
2. (a) Hiyama, T.; Kanie, K.; Kusumoto, T.; Morizawa, Y.; Shimizu, M. *Organofluorine Compounds: Chemistry and Application*; Springer-Verlag: Berlin, 2000. (b) Kirsch, P. *Modern Fluoroorganic Chemistry: Synthesis,*

Reactivity, Applications; Wiley-VCH: Weinheim, Germany, 2004. (c) Chambers, R. D. *Fluorine in Organic Chemistry*; Blackwell: Oxford, 2004. (d) Uneyama, K. *Organofluorine Chemistry*; Blackwell: Oxford, 2006. (e) Bégué, J.-P.: Bonnet-Delpon D. *Bioorganic and Medicinal Chemistry of Fluorine*; John Wiley & Sons, Inc.: Hoboken, NJ, 2008. (f) Ojima. I. *Fluorine in Medicinal Chemistry and Chemical Biology*; Wiley-Blackwell: Chichester, West Sussex, 2009. (g) Gouverneur, V.; Muller, K. *Fluorine in Pharmaceutical and Medicinal Chemistry: From Biophysical Aspects to Clinical Applications*; World Scientific Publishing Company: London, 2012. (h) Reddy, V. P. *Organofluorine Compounds in Biology and Medicine*; Elsevier: Amsterdam, 2015.

3. Wang, J.; Sánchez-Roselló, M.; Aceña, J. L.; del Pozo, C.; Sorochinsky, A. E.; Fustero, S.; Soloshonok, V. A.; Liu, H. *Chem. Rev.* **2014**, *114*, 2432–2506.
4. (a) Grushin, V. V. *Acc. Chem. Res.* **2010**, *43*, 160–171. (b) Lundgren, R. J.; Stradiotto, M. *Angew. Chem. Int. Ed.* **2010**, *49*, 9322–9324. (c) Furuya, T.; Kamlet, A. S.; Ritter, T. *Nature* **2011**, *473*, 470–477. (d) Roy, S.; Gregg, B. T.; Gribble, G. W.; Le, V.-D.; Roy, S. *Tetrahedron* **2011**, *67*, 2161–2195. (e) Tomashenko, O. A.; Grushin, V. V. *Chem. Rev.* **2011**, *111*, 4475–4521. (f) Amii, H. *J. Synth. Org. Chem. Jpn.* **2011**, *69*, 752–762. (g) Besset, T.; Schneider, C.; Cahard, D. *Angew. Chem. Int. Ed.* **2012**, *51*, 5048–5050. (h) Ye, Y.; Sanford, M. *Synlett* **2012**, *23*, 2005–2013. (i) Wu, X, F.; Neumann, H.; Beller, M. *Chem. Asian J.* **2012**, *7*,1744–1745. (j) Jin, Z.; Hammond, G. B.; Xu, B. *Aldrichimica Acta* **2012**, *45*, 67–83. (k) Studer, A. *Angew. Chem. Int. Ed.* **2012**, *51*, 8950–8958. (l) Qing, F.-L. *Chin. J. Org. Chem.* **2012**, *32*, 815–824. (m) Macé, Y.; Magnier, E. *Eur. J. Org. Chem.* **2012**, 2479–2494. (n) García-Monforte, M. A.; Martínez-Salvador, S.; Menjón B. *Eur. J. Inorg. Chem.* **2012**, 4945–4966. (o) Liu, T.; Shen, Q. *Eur. J. Org. Chem.* **2012**, 6679–6687. (p) Liu, H.; Gu, Z.; Jiang, X. *Adv. Synth. Catal.* **2013**, **355**, 617–626. (q) Liang, T.; Neumann, C. N.; Ritter, T. *Angew. Chem. Int. Ed.* **2013**, *52*, 8214–8264. (r) Wang, H.; Vicic, D.A. *Synlett* **2013**, *24*, 1887–1898. (s) Browne, D. L. *Angew. Chem. Int. Ed.* **2014**, *53*, 1482–1484. (t) Zhang, C. *Org. Biomol. Chem.* **2014**, *12*, 6580–6589. (u) Sodeoka, M.; Egami, H. *Pure Appl. Chem.* **2014**, *86*, 1247–1256. (v) Xu, J.; Liu, X.; Fu, Y. *Tetrahedron Lett.* **2014**, *55*, 585–594. (w) Charpentier, J.; Fruh, N.; Togni, A. *Chem. Rev.* **2015**, *115*, 650–682. (x) Liu, X.; Xu, C.; Wang, M.; Liu, Q. *Chem. Rev.* **2015**, *115*, 683–730. (y) Ni, C.; Hu, M.; Hu, J. *Chem. Rev.* **2015**, *115*, 765–825. (z) Alonso, C.; Martinez de Marigorta, E.; Rubiales, G.; Palacios, F. *Chem. Rev.* **2015**, *115*, 1847–1935.

5. Jouvin, K.; Guissart, C.; Theunissen, C.; Evano, G. Emerging Areas in Copper-Mediated Trifluoromethylations of Aryl Derivatives: Catalytic and Oxidative Cross-Coupling Process. In *Copper-Mediated Cross-Coupling Reactions*; Evano, G.; Blanchard, N., Eds.; John Wiley & Sons, Inc.: Hoboken, NJ, 2013.; pp 515-530.
6. (a) Billard, T.; Langlois, B. R.; Blond, G. *Tetrahedron Lett.* **2000**, *41*, 8777–8780. (b) Billard, T.; Bruns, S.; Langlois, B. R. *Org. Lett.* **2000**, *2*, 2101–2103.
7. Kondo, H.; Oishi, M.; Fujikawa, K.; Amii, H. *Adv. Synth. Cat.* **2011**, *353*, 1247–1252.
8. (a) Oishi, M.; Kondo, H.; Amii, H. *Chem. Commun.* **2009**, 1909–1911. (b) Chu, L.; Qing, F.-L. *Org. Lett.* **2010**, *12*, 5060–5063. (c) Knauber, T.; Arikan, F.; Röschenthaler., G.-V.; Goossen, L. J. *Chem. Eur. J.* **2011**, *17*, 2689–2697. (d) Popov, I.; Lindeman, S.; Daugulis, O. *J. Am. Chem. Soc.* **2011**, *133*, 9286–9289. (e) Liu, T.; Shen, Q. *Org. Lett.* **2011**, *13*, 2342–2345. (f) Morimoto, H.; Tsubogo, T.; Litvinas, N. D.; Hartwig, J. F. *Angew. Chem. Int. Ed.* **2011**, *50*, 3793–3798. (g) Senecal, T. D.; Parsons, A. T.; Buchwald, S. L. *J. Org. Chem.* **2011**, *76*, 1174–1176. (h) Chu, L.; Qing, F.-L. *J. Am. Chem. Soc.* **2012**, *134*, 1298–1304. (i) Nakamura, Y.; Fujiu, M.; Murase, T.; Itoh, Y.; Serizawa, H.; Aikawa, K.; Mikami, K. *Beilstein J. Org. Chem.* **2013**, *9*, 2404–2409. (j) Ogawa, T.; Kumagai, N.; Shibasaki, M. *Angew. Chem. Int. Ed.* **2013**, *52*, 6196–6201. k) Kato, H.; Hirano, K.; Kurauchi, D.; Toriumi, N.; Uchiyama, M. *Chem. Eur. J.* **2015**, *21*, 3895–3900.
9. Miyake, Y.; Ota, S.; Nishibayashi, Y. *Chem. Eur. J.* **2012**, *18*, 13255–13258.

Appendix
Chemical Abstracts Nomenclature (Registry Number)

Morpholine: 1-Oxa-4-azacyclohexane; (110-91-8)
Fluoral ethyl hemiacetal: 1-Ethoxy-2,2,2-trifluoroethanol; (433-27-2)
Trifluoroacetaldehyde hemiaminal: 4-Morpholinemethanol, α-(trifluoromethyl)-; (321573-94-8)
N-(Trimethylsilyl)imidazole: Imidazole, 1-(trimethylsilyl)-; (18156-74-6)
[2,2,2-Trifluoro-1-(trimethylsilyloxy)ethyl]morpholine: 4-[2,2,2-Trifluoro-1-(Trimethylsilyloxy)ethyl]morpholine; (289706-46-3)
Copper(I) thiophene-2-carboxylate: 2-Thiophenecarboxylic acid, copper(1+) salt (1:1); (1292766-17-6)

1,10-Phenanthroline: 4,5-Diazaphenanthrene; (66-71-7)
m-Iodonitrobenzene: 3-Iodo-1-nitrobenzene; (645-00-1)
3-Nitro-1-(trifluoromethyl)benzene: Toluene, α, α, α-trifluoro-*m*-nitro-; (98-46-4)

Naoto Shimizu was born in Gunma, Japan in 1990. He received his B. Eng. in 2012 from Gunma University, where he carried out graduate research in the laboratories of Prof. Hideki Amii. His research focused on the catalytic aromatic trifluoromethylation via β-carbon elimination. After receipt of his M. Eng. degree (2014), he joined Nihon Nohyaku Co.,Ltd.

Hideaki Kondo was born in 1985 in Gifu, Japan. He received his B. Sci. (2008) and M. Sci. (2010) from Kobe University under the direction of Dr. Hideki Amii, working on the development of copper-catalyzed aromatic trifluoromethylation via group transfer from fluoral derivatives. Currently he is working for Kao Corporation.

Masahiro Oishi was born in Shizuoka, Japan in 1983. He received his B. Eng. from Shizuoka University under the direction of Professor Kunihiko Takabe in 2006. Then he stayed at Kobe to perform his studies under the supervision of Dr. Hideki Amii, working on the development of catalytic aromatic trifluoromethylation using organosilicon compounds. He currently worksd at ADEKA Corporation.

Kenichi Fujikawa was born in Nara, Japan in 1987. He received his B. Sci. (2009) and M. Sci. (2011) from Kobe University under the direction of Prof. Hideki Amii. He developed a convenient route for aromatic difluoromethylation. Currently he is working for Nitto Denko Corporation.

Kazuki Komoda was born in Gunma, Japan in 1992. He received his B. Eng. in organic chemistry from Gunma University. He is currently a master course student in the group of Prof. Hideki Amii at Gunma University. His research interests include the cross-coupling reactions to introduce fluorine-containing functional groups into organic compounds.

Hideki Amii was born in Hyogo in 1968. He graduated from Kyoto University, where he received his Doctorate degree in 1996 under the direction of Professor Yoshihiko Ito. During 1996–2003, he worked as Research Associate of the Department of Applied Chemistry, Faculty of Engineering, Okayama University (Prof. Kenji Uneyama's group). He carried out postdoctoral work in France with Dr. Guy Bertrand at Université Paul Sabatier during 2000–2001. In 2003, he was appointed to Associate Professor of Kobe University. In 2010, he moved to Gunma University, where he is currently Professor of Chemistry. His research interest focuses in the synthesis of organofluorine compounds by the use of metal reagents.

Mina Nakhla was born in Adelaide, Australia in 1990. He received his B. S. in biochemistry from St. Edward's University. He is currently a doctoral student under the direction of Prof. John L. Wood at Baylor University. His research interests include natural product total synthesis and organometallic chemistry.

Organic Syntheses

Copper-catalyzed Cyanation of Alkenyl Iodides

Antoine Nitelet, Sara Zahim, Cédric Theunissen, Alexandre Pradal, and Gwilherm Evano*[1]

Laboratoire de Chimie Organique, Service de Chimie et PhysicoChimie Organiques, Université Libre de Bruxelles, Avenue F. D. Roosevelt 50, CP160/06, 1050 Brussels, Belgium

Checked by Sidney M. Wilkerson-Hill and Huw M. L. Davies

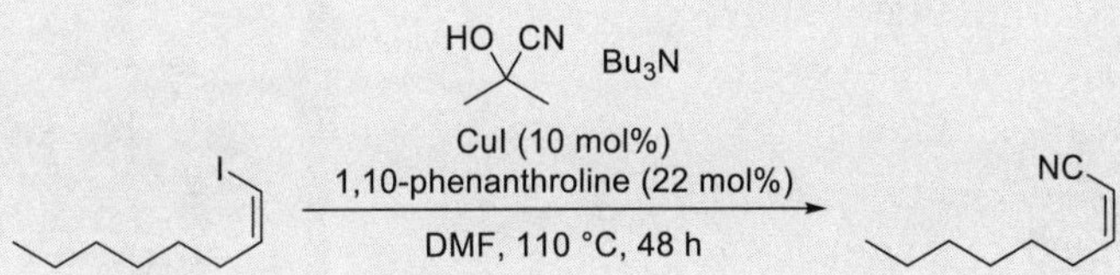

Procedure

Acetone cyanohydrin is highly toxic and must be handled in a well-ventilated fume hood. The aqueous waste solutions should not be in contact with acids and, where permitted, must be treated with excess commercial Clorox® bleach. All glassware and apparatus that comes into contact with it must be washed with commercial Clorox® bleach. All spills around balances or work areas should be immediately cleaned up and the area washed down with bleach.

A. *(Z)-Non-2-enenitrile (**2a**).* A 50-mL, double-necked, oven dried, round-bottomed flask equipped with an octagonal Teflon®-coated magnetic stir bar (0.8 mm x 3.8 cm) is successively charged with (Z)-1-iodooct-1-ene **1a** (13.7 g, 57.5 mmol) (Note 1), copper(I) iodide (1.1 g, 5.8 mmol, 0.1 equiv) (Note 2), and 1,10-phenanthroline (2.3 g, 12.8 mmol, 0.2 equiv) (Note 3). The lateral neck of the flask is equipped with a rubber septum and the top neck of the flask with a water-cooled condenser fitted with a gas inlet (Figure 1).

The flask is then evacuated under vacuum (2 mmHg) and backfilled with argon three times. *N,N*-Dimethylformamide (70 mL) (Note 4) is added via a syringe through the septum and the mixture is stirred for 5 min at room temperature before successively adding tri-*n*-butylamine (17.8 mL, 74.7 mmol, 1.3 equiv) (Note 5) and acetone cyanohydrin (6.3 mL, 69 mmol, 1.2 equiv) (Note 6) via a syringe through the septum. The septum is replaced by a glass stopper and the resulting slurry (Note 7) is heated in a pre-heated oil bath at 110 °C (bath temperature) and kept under vigorous stirring for 48 h (Note 8) (Figure 2). The reaction mixture is then allowed to cool to room temperature, diluted with diethyl ether (150 mL) (Note 9), vacuum filtered through a one-half inch pad of Celite® on a 150-mL medium porosity sintered glass filtration funnel and rinsed with diethyl ether (150 mL) before being diluted with water (300 mL) and transferred to a 1-L separatory funnel. The organic layer is separated and the aqueous layer is extracted with diethyl ether (3 x 100 mL). The combined organic layers are successively washed with water (200 mL) and brine (200 mL), dried over 30 g of anhydrous magnesium sulfate, vacuum filtered through a 150-mL medium porosity sintered glass filtration funnel, and concentrated by rotary evaporation (25 °C, 350 mmHg then 40 mmHg when most of the solvent has been evaporated) to

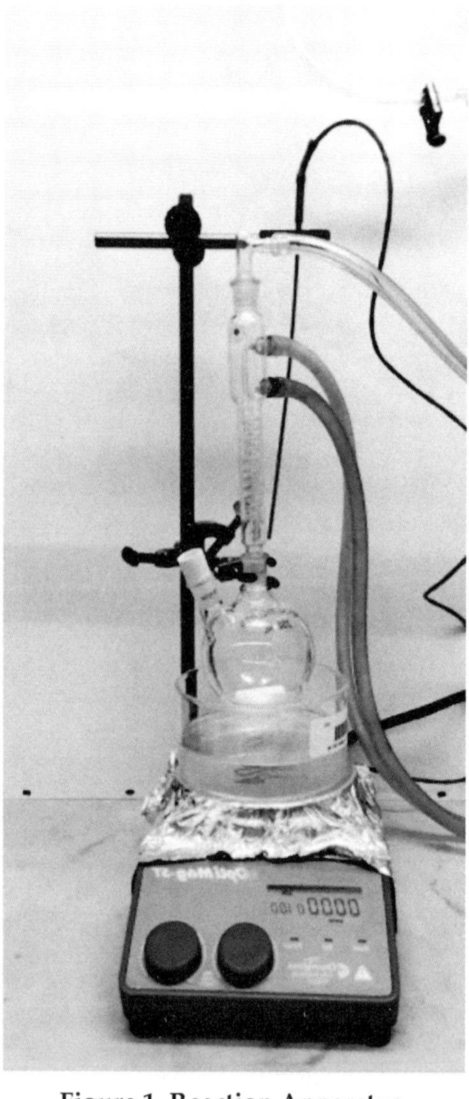

Figure 1. Reaction Apparatus

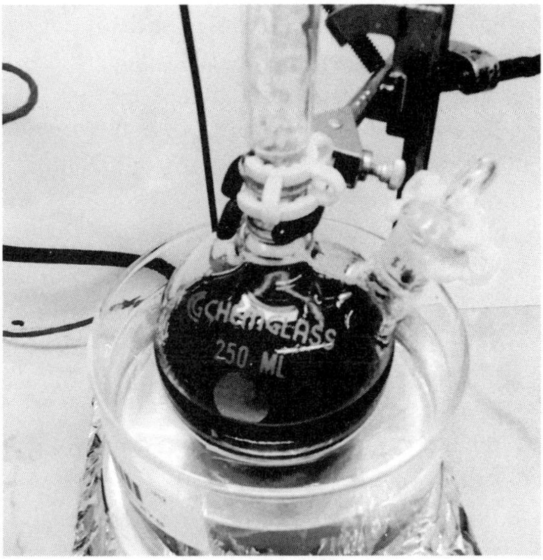

Figure 2. Reaction mixture after heating to 110 °C turns deep red/orange

give 11.2-12.1 g of a brown oil. The product is dissolved in cyclohexane (10 mL) (Note 10) and charged on a column (5 x 40 cm), which is packed with a slurry of 200 g of silica gel (Note 11) in 500 mL of cyclohexane. The column is first eluted with 500 mL of cyclohexane. At that point, fraction collection (~35-mL fractions) is begun, and elution is continued with 500 mL of 2% EtOAc-cyclohexane (98:2 cyclohexane:EtOAc) (Note 12) and then 1250 mL of 3% EtOAc-cyclohexane (97:3 cyclohexane:EtOAc). The desired

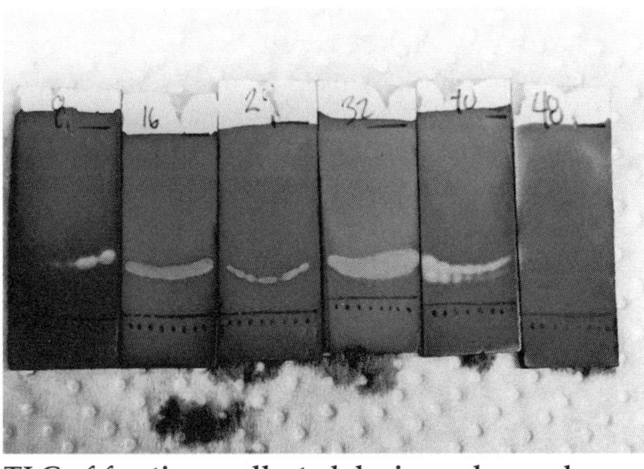

Figure 3. TLC of fractions collected during column chromatography

product is obtained in fractions 3–40 (Notes 13 and 14) (Figure 3), which are concentrated by rotary evaporation (30 °C, 100 mmHg then 15 mmHg when most of the solvent has been evaporated). The resulting oil is dried under vacuum (3 mmHg, 2 h) (Note 15) to afford (Z)-non-2-enenitrile (**2a**) (6.3 g, 80%) as a pale yellow oil (Notes 16, 17, 18, 19, 20, and 21).

Notes

1. Synthesized according to Yang, D.; Cwynar, V. A.; Hart, D. J.; Madanmohan, J.; Lee, J.; Lyons, J.; Caffrey, M. *Org. Synth.* **2012**, *89*, 183–201.
2. Copper iodide (99.999%) was purchased from Sigma-Aldrich and used as received. The yield was not significantly changed when switching to 98% copper iodide (purchased from Sigma-Aldrich and used without purification).
3. Anhydrous 1,10-phenanthroline (99%) was purchased from Alfa-Aesar and used as received.
4. *N,N*-Dimethylformamide (reagent grade, >99%) was purchased from Fisher and purified by distillation over calcium hydride under argon at 0.06 mmHg. The yield was not significantly changed when switching to 99.8% extra dry *N,N*-dimethylformamide stored over molecular sieves (purchased from Acros Organics and used without purification). The Checkers used the latter source of *N,N*-dimethylformamide.
5. Tri-*n*-butylamine (99%) was purchased from Sigma-Aldrich. It was purified by distillation over calcium hydride under argon at 0.06 mmHg and stored under argon in a flask covered with aluminum foil to protect it from the light. The Checkers used 20 mL of the amine over 2.8 g of CaH for the distillation.
6. Acetone cyanohydrin (99%) was purchased from Sigma-Aldrich and used as received.
7. The reaction mixture turns to dark orange at this stage and then back to brown upon heating.
8. TLC analysis was performed on Merck aluminum silica gel plates 60 F_{254}. Reaction conversion was ascertained using the following procedure. The reaction mixture was spotted directly on the TLC plate, which was eluted with 4% EtOAc-cyclohexane (96:4 cyclohexane:EtOAc), and visualized with $KMnO_4$ stain after heating:

starting material **1a** (R_f = 0.7, a $KMnO_4$ active spot); product **2a** (R_f = 0.22, a $KMnO_4$ active spot). The $KMnO_4$ stain was prepared using 1.5 g of $KMnO_4$ and 10 g of K_2CO_3 dissolved in 150 mL of deionized water and 5.0 mL of a 5% weight NaOH solution. The Checkers prepared their $KMnO_4$ stain according to the procedure of Pirrung, M. C. "The Synthetic Organic Chemist's Companion." **2007**, Wiley, p. 171.

9. Diethyl ether (reagent grade, >99%) was purchased from VWR and used as received.
10. Cyclohexane (analytical reagent grade, >99.9%) was purchased from Fisher and distilled before use to remove non-volatile residues.
11. Silica gel Geduran® Si60 particle size 40-63 μm was purchased from Merck.
12. Ethyl acetate (analytical reagent grade, >99.9%) was purchased from Fisher and distilled before use to remove non-volatile residues.
13. The presence of the product **2a** is visualized by TLC analysis on Merck aluminum silica gel plates 60 F_{254} (R_f = 0.22, 96:4 cyclohexane:EtOAc, a $KMnO_4$ active spot).
14. Minor amounts of the unconjugated isomer, (*E*)-non-3-enenitrile, can be detected by TLC analysis on Merck aluminum silica gel plates 60 F_{254} (R_f = 0.20, 96:4 cyclohexane:EtOAc, a $KMnO_4$ active spot). All fractions for which TLC analyses indicate that the intensity of the spot corresponding to this byproduct was less important than the intensity of the spot corresponding to **2a** were collected. The Checkers found that collecting *only* the spots containing product **2a** gives 68% isolated yield of product without the undesired unconjugated product.
15. The product being volatile, drying it under higher vacuum or for longer times results in significant losses of product. During one run, the Checkers rinsed the test tubes containing product with ethyl acetate (5 mL each) to ensure complete product collection. Upon concentration, the Checkers found that this added ethyl acetate could not be removed completely without significant loss of the product on high vacuum. Rinsing the test tubes is not recommended.
16. (Z)-Non-2-enenitrile **2a** has the following physical properties: IR (ATR) 2957, 2930, 2859, 2220, 1621, 1465, 739 cm^{-1}; ^1H NMR (400 MHz, $CDCl_3$) δ: 0.88 (t, *J* = 6.9 Hz, 3H), 1.23–1.39 (m, 6H), 1.40–1.52 (m, 2H), 2.42 (dq, *J* = 7.6, 1.3 Hz, 2H), 5.30 (dt, *J* = 10.9, 1.3 Hz, 1H), 6.48 (dt, *J* = 10.9, 7.7 Hz, 1H); ^{13}C NMR (100 MHz, $CDCl_3$) δ: 14.0, 22.5, 28.2, 28.7, 31.5, 31.9, 99.4, 116.1, 155.3; ESIHRMS *m*/*z* calcd for $C_9H_{16}N^+$ $[M+H]^+$ 138.1277, found 138.1278.

17. Purity 97.1% *wt.* by quantitative ^{1}H NMR assay using ethylene carbonate as an internal standard. The Checkers found that the product could be obtained without contamination from the unconjugated alkene byproduct in chromatographic fractions 3 – 32 accounting for 68% yield. The purity refers to the total combined yield (79-80%) of product **2a** plus its unconjugated isomer.
18. The product contains 2-3% *wt.* of the unconjugated isomer, (*E*)-non-3-enenitrile, as determined by ^{1}H NMR analysis.
19. The product has a rather unpleasant smell and should therefore be handled in a well-ventilated fume hood.
20. The product was stored under argon at –20 °C and found to be stable for more than a year at this temperature.
21. A second run of the reaction provided 6.2 g (79%) of the product in 97.7% purity as assessed by Q NMR with ethylene carbonate as the internal standard.

Working with Hazardous Chemicals

The procedures in *Organic Syntheses* are intended for use only by persons with proper training in experimental organic chemistry. All hazardous materials should be handled using the standard procedures for work with chemicals described in references such as "Prudent Practices in the Laboratory" (The National Academies Press, Washington, D.C., 2011; the full text can be accessed free of charge at http://www.nap.edu/catalog.php?record_id=12654). All chemical waste should be disposed of in accordance with local regulations. For general guidelines for the management of chemical waste, see Chapter 8 of Prudent Practices.

In some articles in *Organic Syntheses*, chemical-specific hazards are highlighted in red "Caution Notes" within a procedure. It is important to recognize that the absence of a caution note does not imply that no significant hazards are associated with the chemicals involved in that procedure. Prior to performing a reaction, a thorough risk assessment should be carried out that includes a review of the potential hazards associated with each chemical and experimental operation on the scale that is planned for the procedure. Guidelines for carrying out a risk assessment

and for analyzing the hazards associated with chemicals can be found in Chapter 4 of Prudent Practices.

The procedures described in *Organic Syntheses* are provided as published and are conducted at one's own risk. *Organic Syntheses, Inc.*, its Editors, and its Board of Directors do not warrant or guarantee the safety of individuals using these procedures and hereby disclaim any liability for any injuries or damages claimed to have resulted from or related in any way to the procedures herein.

Discussion

Alkenyl nitriles, commonly called acrylonitriles, are important building blocks and intermediates in organic synthesis where they have been used in an impressive number of transformations involving the activated alkene, the nitrile or both. They are in addition found in various natural products, the most famous examples probably being hemi-phorboxazole A[2] or calyculin A,[3] and compounds from the pharmaceutical industry such as trilostane (Modrenal®),[4] a steroidal alkenyl nitrile used for the treatment of Cushing's disease and breast cancer, or rilpivirine (Edurant®), a non-nucleoside reverse transcriptase inhibitor used for the treatment of HIV infections.[5]

The classical synthesis of acrylonitriles typically rely on Wittig,[6] Horner-Wadsworth-Emmons[7] or Peterson[8] olefination reactions. While these reactions are in general efficient and high yielding, they are often restricted to the preparation of one of the *E* or *Z* isomer and/or yield mixtures of these isomers whose separation is in general troublesome. Other standard techniques available to access these building blocks include the dehydration of conjugated amides or oximes,[9] Heck-type reactions[10] or cross-metathesis,[11] just to mention the methods that are the most commonly used for the synthesis of conjugated nitriles.[12] While extensively used, these routes however still have major limitations in terms of scope, yields and *E*/*Z* selectivity, the last point being in general the main issue. An interesting solution to this problem lies in the metal-catalyzed cyanation of alkenyl halides. Indeed, their *E* and *Z* isomers can be easily prepared by a range of robust and reliable methods such as the Takai[13] or Stork-Zhao[14] olefinations of aldehydes, Brown's hydroboration of iodoalkynes followed by protonolysis[15] or the homologation/elimination sequence from benzyl halides and diiodomethane.[16] Provided that the stereochemistry of the double bond can be preserved during the cross-coupling, the cyanation of alkenyl halides could provide an interesting, general and stereocontroled route to acrylonitriles.

While this approach had already been investigated before we started our studies, most catalytic systems based on nickel,[17] cobalt[18] or palladium[19] are often limited to the preparation of aryl-substituted alkenyl nitriles, lack generality and/or require expensive complexes and ligands. A general, stereospecific and robust cyanation of alkenyl halides was therefore still an unmet challenge, which prompted us to develop the copper-catalyzed cyanation we reported in 2014 and that is discussed below.[20]

The cyanation of alkenyl iodides was found to be best performed using a catalytic system similar to the one developed by Taillefer for the cyanation of aryl halides[21,22] based on the combination of copper(I) iodide (10 mol%) and 1,10-phenanthroline (20 mol%) in DMF at 110°C for 16-48 hours. The best source of cyanide was found to be acetone cyanohydrin,[23] which decomposes quantitatively to acetone and cyanide in the presence of a base, tributylamine in our case. A slight excess (1.2 and 1.3 equivalents of the former and the latter, respectively) was shown to be sufficient for full conversion and, while a slow addition of the base is usually required in metal-catalyzed cross-coupling reactions involving acetone cyanohydrin in order to minimize the amount of cyanide present in solution and therefore avoid catalyst poisoning, all the base could be added at the beginning of the reaction in our case without significant catalyst deactivation.

Representative examples of this copper-catalyzed cyanation of alkenyl iodides are shown in Table 1.[20] As evidenced by these results, the procedure was found to be suitable for the preparation of a wide range of alkenyl nitriles, substituted not only with aryl groups but also with alkyl chains. The presence of substituents close to the reacting center was not detrimental since 1-substituted and 2,2-disubstituted 1-alkenyl-iodides were conveniently transformed to the corresponding conjugated nitriles under the reaction conditions. Importantly, the configuration of the double bond was fully retained, which therefore enabled the cyanation of Z-iodoalkenes without noticeable isomerization of the double bond. Benzyl and silyl ethers, acetals and acetates were shown to be stable under the reaction conditions, and an aromatic bromide was also found to be compatible. While most alkenyl iodides were found to be reactive under our standard reaction conditions, the corresponding bromide were, however, found to be totally inert and recovered at the end of the reaction.

Table 1. Copper-catalyzed cyanation of alkenyl iodides

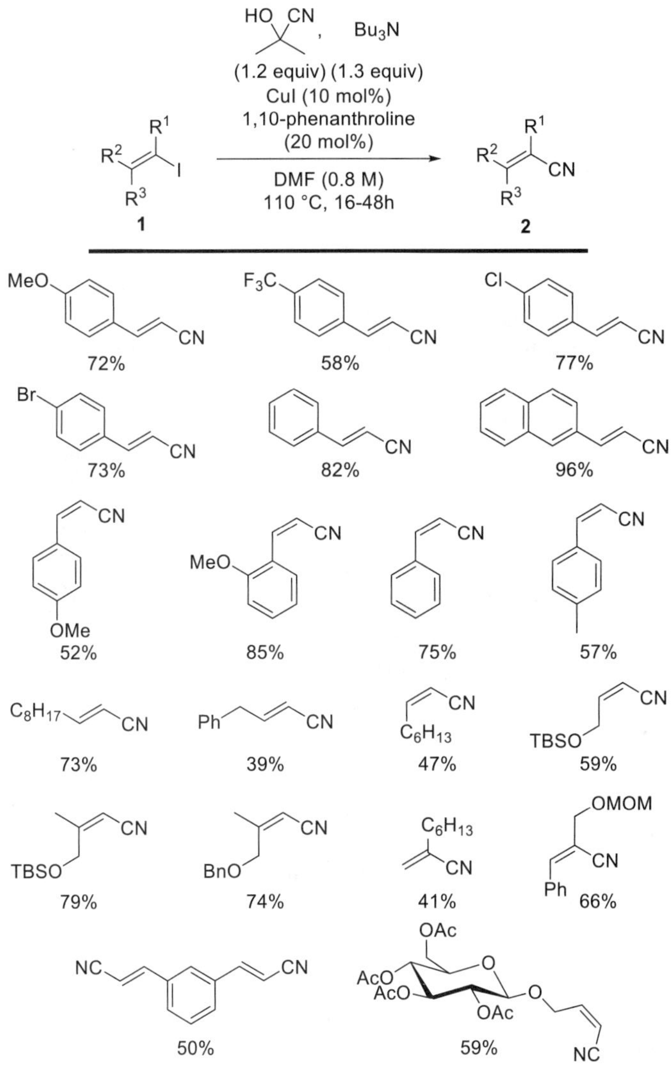

In summary, a general and practical copper-catalyzed vinylic version of the Rosenmund-von Braun cyanation reaction was developed and provides an efficient entry to both *E*- and *Z*- polysubstituted alkenyl nitriles. The cyanation of (Z)-1-iodooct-1-ene described in this procedure is illustrative of this reaction which can be conveniently performed on a multi-gram scale.

References

1. Laboratoire de Chimie Organique, Service de Chimie et PhysicoChimie Organiques, Université Libre de Bruxelles, Avenue F. D. Roosevelt 50, CP160/06, 1050 Brussels, Belgium. E-mail: gevano@ulb.ac.be. This research was supported by the FNRS (Incentive Grant for Scientific Research n° F.4530.13) and the Fonds pour la formation à la Recherche dans l'Industrie et dans l'Agriculture.
2. Dalisay, D. S.; Molinski, T. F. *Org. Lett.* **2009**, *11*, 1967–1970.
3. Kato, Y.; Fusetani, N.; Matsunaga, S.; Hashimoto, K.; Fujita S; Furuya, T. *J. Am. Chem. Soc.* **1986**, *108*, 2780–2781.
4. (a) Komanicky, P.; Spark, R. F.; Melby, J. C. *J. Clin. Endocrinol. Metab.* **1978**, *47*, 1042–1051. (b) Puddefoot, J. R.; Barker, S.; Glover, H. R.; Malouitre, S. D. M.; Vinson, G. P. *Int. J. Cancer* **2002**, *101*, 17–22.
5. Goebel, F.; Yakovlev, A.; Pozniak, A. L.; Vinogradova, E.; Boogaerts, G.; Hoetelmans, Ri.; de Béthune, M.-P. P.; Peeters, M.; Woodfall, B. *AIDS* **2006**, *20*, 1721–1726.
6. (a) Schiemenz, G. P.; Engelhard, H. *Chem. Ber.* **1961**, *94*, 578–585. (b) Bestmann, H. J.; Pfohl, S. *Justus Liebigs Ann. Chem.* **1974**, 1688–1693. (c) Kojima, S.; Kawaguchi, K.; Matsukawa, S.; Uchida, K.; Akiba, K.-y. *Chem. Lett.* **2002**, 170–171. For a recent, highly *Z*-selective variant involving *N*-tosyl-imines, see: (d) Fang, F.; Li, Y.; Tian, S.-K. *Eur. J. Org. Chem.* **2011**, 1084–1091.
7. (a) Zhang, T. Y.; O'Toole, J. C.; Dunigan, J. M. *Tetrahedron Lett.* **1998**, *39*, 1461-1464. (b) Ando, K.; Okumura, M.; Nagaya, S. *Tetrahedron Lett.* **2013**, *54*, 2026-2028.
8. (a) Yamakado, Y.; Ishiguro, M.; Ikeda, N.; Yamamoto, H. *J. Am. Chem. Soc.* **1981**, *103*, 5568–5570. (b) Haruta, R.; Ishiguro, M.; Furuta, K.; Mori, A.; Ikeda, N.; Yamamoto, H. *Chem. Lett.* **1982**, 1093–1096. (c) Furuta, K.; Ishiguro, M.; Haruta, R.; Ikeda, N.; Yamamoto, H. *Bull. Chem. Soc. Jpn.* **1984**, *57*, 2768–2776. (d) Palomo, C.; Aizpurua, J. M.; Aurrekoetxea, N. *Tetrahedron Lett.* **1990**, *31*, 2209–2210. (e) Sato, Y.; Niinomi, Y. *J. Chem. Soc., Chem. Commun.* **1982**, 56–57. (f) Matsuda, I.; Okada, H.; Izumi, Y.

Bull. Chem. Soc. Jpn. **1983**, *56*, 528–532. (g) Kojima, S.; Fukuzaki, T.; Yamakawa, A.; Murai, Y. *Org. Lett.* **2004**, *6*, 3917–3920.

9. For examples, see: (a) Ishihara, K.; Furuya, Y.; Yamamoto, H. *Angew. Chem., Int. Ed.* **2002**, *41*, 2983–2986. (b) Yamaguchi, K.; Fujiwara, H.; Ogasawara, Y.; Kotani, M.; Mizuno, N. *Angew. Chem., Int. Ed.* **2007**, *46*, 3922–3925. (c) Zhou, S.; Addis, D.; Das, S.; Junge, K. Beller, M. *Chem. Commun.* **2009**, 4883–4885.
10. For examples, see: (a) Bumagin, N. A.; More, P. G.; Beletskaya, I. P. *J. Organomet. Chem.* **1989**, *371*, 397–401. (b) Ruan, J.; Li, X.; Saidi, O.; Xiao, J. *J. Am. Chem. Soc.* **2008**, *130*, 2424–2425.
11. (a) Crowe, W. E.; Goldberg, D. R. *J. Am. Chem. Soc.* **1995**, *117*, 5162–5163. (b) Randl, S.; Gessler, S.; Wakamatsu, H.; Blechert, S. *Synlett* **2001**, 430–432. (c) Love, J. A.; Morgan, J. P.; Trnka, T. M.; Grubbs, R. H. *Angew. Chem., Int. Ed.* **2002**, *41*, 4035–4037. (d) Michrowska, A.; Bujok, R.; Harutyunyan, S.; Sashuk, V.; Dolgonos, G.; Grela, K. *J. Am. Chem. Soc.* **2004**, *126*, 9318–9325.
12. For other representative routes to alkenyl nitriles, see: (a) Nakao, Y.; Yada, A.; Ebata, S.; Hiyama, T. *J. Am. Chem. Soc.* **2007**, *129*, 2428–2429. (b) Cheng, Y.-n.; Duan, Z.; Yu, L.; Li, Z.; Zhu, Y.; Wu, Y. *Org. Lett.* **2008**, *10*, 901–904. (c) Oishi, T.; Yamaguchi, K.; Mizuno, N. *Angew. Chem., Int. Ed.* **2009**, *48*, 6286–6288. (d) Hirata, Y.; Yukawa, T.; Kashihara, N.; Nakao, Y.; Hiyama, T. *J. Am. Chem. Soc.* **2009**, *131*, 10964–10973. (e) Nakao, Y.; Yada, A.; Hiyama, T. *J. Am. Chem. Soc.* **2010**, *132*, 10024–10026. (f) Qin, C.; Jiao, N. *J. Am. Chem. Soc.* **2010**, *132*, 15893–15895. (g) Zhou, W.; Xu, J.; Zhang, L.; Jiao, N. *Org. Lett.* **2010**, *12*, 2888–2891. (h) Rokade, B. V.; Prabhu, K. R. *J. Org. Chem.* **2012**, *77*, 5364–5370. (i) Rokade, B. V.; Malekar, S. K.; Prabhu, K. R. *Chem. Commun.* **2012**, *48*, 5506–5508. (j) Minami, Y.; Yoshiyasu, H.; Nakao, Y.; Hiyama, T. *Angew. Chem., Int. Ed.*, **2013**, *52*, 883–887. (k) Yang, X.; Arai, S.; Nishida, A. *Adv. Synth. Catal.* **2013**, *355*, 2974–2985. (l) Wang, Z.; Chang, S. *Org. Lett.* **2013**, *15*, 1990–1993.
13. (a) Takai, K.; Nitta, K.; Utimoto, K., *J. Am. Chem. Soc.* **1986**, *108*, 7408–7410. (b) Okazoe, T.; Takai, K.; Utimoto, K., *J. Am. Chem. Soc.* **1987**, *109*, 951–953.
14. Stork, G.; Zhao, K. *Tetrahedron Lett.* **1989**, *30*, 2173–2174.
15. (a) Brown, H. C.; Blue, C. D.; Nelson, D. J.; Bhat, N. G. *J. Org. Chem.* **1989**, *54*, 6064–6067. Also see the following procedures from *Organic Syntheses*: (b) Denmark, S. E.; Wang, Z. *Org. Synth.* **2005**, *81*, 42–53. (c)

Yang, D.; Cwynar, V. A.; Hart, D. J.; Madanmohan, J.; Lee, J.; Lyons, J.; Caffrey, M. *Org. Synth.* **2012**, *89*, 183–201.

16. (a) Bull, J. A.; Mousseau, J. J.; Charette, A. B. *Org. Lett.* **2008**, *10*, 5485–5488. (b) Bull, J. A.; Mousseau, J. J.; Charette, A. B. *Org. Synth.* **2010**, *87*, 170–177.
17. (a) Sakakibara, Y.; Yadani, N.; Ibuki, I.; Sakai, M.; Uchino, N. *Chem. Lett.* **1982**, 1565–1566. (b) Sakakibara, Y.; Enami, H.; Ogawa, H.; Fujimoto, S.; Kato, H.; Kunitake, K.; Sasaki, K.; Sakai, M. *Bull. Chem. Soc. Jpn.* **1995**, *68*, 3137–3143.
18. (a) Funakibi, T.; Hosomi, H.; Yoshida, S.; Tarama, K. *J. Am. Chem. Soc.* **1982**, *104*, 1560–1568. (b) Funakibi, T.; Kishi, H.; Sato, Y.; Yoshida, S. *Bull. Chem. Soc. Jpn.* **1983**, *56*, 649–650.
19. (a) Yamamura, K.; Murahashi, S.-I. *Tetrahedron Lett.* **1977**, *18*, 4429–4430. (b) Alterman, M.; Hallberg, A. *J. Org. Chem.* **2000**, *65*, 7984–7989. (c) Li, L.-H.; Pan, Z.-L.; Duan, X.-H.; Liang, Y.-M. *Synlett* **2006**, 2094–2098. (d) Powell, K. J.; Han, L.-C.; Sharma, P.; Moses, J. E. *Org. Lett.* **2014**, *16*, 2158–2161.
20. Pradal, A.; Evano, G. *Chem. Commun.* **2014**, *50*, 11907–11910.
21. Cristau, H.-J.; Ouali, A.; Spindler, J.-F. ; Taillefer, M. *Chem. Eur. J.* **2005**, *11*, 2483–2492.
22. For other examples of the copper-catalyzed cyanation of aryl halides, see: (a) Zanon, J.; Klapars, A.; Buchwald, S. L. *J. Am. Chem. Soc.* **2003**, *125*, 2890–2891. (b) Schareina, T.; Zapf, A.; Beller, M. *Tetrahedron Lett.* **2005**, *46*, 2585–2588. (c) Schareina, T.; Zapf, A.; Mägerlein, W.; Müller, N.; Beller, M. *Chem. Eur. J.* **2007**, *13*, 6249–6254. (d) Schareina, T.; Zapf, A.; Mägerlein, W.; Müller, N.; Beller, M. *Synlett* **2007**, 555–558. (e) Schareina, T.; Zapf, A.; Cotté, A.; Müller, N.; Beller, M. *Synthesis* **2008**, 3351–3355. For general references on the copper-mediated cyanation of aryl halides or related reagents, see: (f) Schareina, T.; Beller, M. In *Copper-Mediated Cross-Coupling Reactions*, Evano, G.; Blanchard, N. Eds.; Wiley: Hoboken, 2013, pp. 313–334. (g) Wen, Q.; Jin, J.; Zhang, L.; Luo, Y.; Lu, P.; Wang, Y. *Tetrahedron Lett.* **2014**, *55*, 1271–1280.
23. For the use of acetone cyanohydrin in selected metal-catalyzed cyanations, see: (a) Sundermeier, M.; Zapf, A.; Beller, M. *Angew. Chem., Int. Ed.* **2003**, *42*, 1661–1664. (b) Schareina, T.; Zapf, A.; Cotte, A.; Gotta, M.; Beller, M. *Adv. Synth. Catal.* **2011**, *353*, 777–780. Also see refs 19d, 21 and 22f.

Appendix
Chemical Abstracts Nomenclature (Registry Number)

(Z)-Non-2-enenitrile: 2-Nonenenitrile, (2Z)-; (40856-15-3)
(Z)-1-Iodooct-1-ene: 1-Octene, 1-iodo-, (1Z)-; (52356-93-1)
Copper(I) iodide: Copper iodide; (7681-65-4)
1,10-Phenanthroline; (66-71-7)
Tri-*n*-butylamine: 1-Butanamine, *N,N*-dibutyl-; (102-82-9)
Acetone cyanohydrin: Propanenitrile, 2-hydroxy-2-methyl-; (75-86-5)

Antoine Nitelet was born in Charleroi in 1990 and studied chemistry at the Université Libre de Bruxelles. In 2012, he joined the group of Prof. Gwilherm Evano as a master student and obtained a F.R.I.A. Ph.D. fellowship in 2013. His work focuses on the development of new copper-catalyzed cross-coupling reactions involving alkenyl halides and their application to the synthesis of natural products.

Sara Zahim was born in Tanger in 1986 and studied chemistry at the Université Libre de Bruxelles. In 2011, she joined the groups of Profs. Gwilherm Evano and Ivan Jabin as a master student and obtained a F.R.I.A. Ph.D. fellowship in 2012. Her work focuses on the development of new methods for the selective functionalization of calix[6]arenes and their application to the design of new molecular receptors and new ligands for metal-catalyzed reactions.

Cédric Theunissen was born in Brussels in 1989 and studied chemistry at the Université Libre de Bruxelles. In 2012, he obtained his master thesis, under the supervision of Prof. Cécile Moucheron, which focused on the synthesis of new ruthenium complexes designed to interact with DNA in an anti-cancer approach. He then joined the group of Prof. Gwilherm Evano and obtained a F.R.I.A. Ph.D. fellowship. His work focuses on the development of new copper-mediated transformations and on the study of the reactivity of ynamides.

Alexandre Pradal was born in 1987 in Maisons-Laffitte (France) and received his Ph.D. in 2012 under the supervision of Drs. Véronique Michelet and Patrick Toullec with whom he worked on electrophilic carbocyclization and asymmetric cycloisomerization reactions of 1,n-enynes in the presence of platinum and gold catalysts. He then joined the group of Prof. Gwilherm Evano at the Université Libre de Bruxelles as a postoctoral researcher where he worked on the copper-catalyzed cyanation of alkenyl iodides. He next moved to the University of Nottingham as a postdoctoral researcher in the group of Prof. Christopher Moody, working on the total synthesis of diazonamide A and is currently working with Prof. Vincent Dalla at the Université du Havre.

Gwilherm Evano studied chemistry at the Ecole Normale Supérieure in Paris and received his Ph.D. from the Université Pierre et Marie Curie in 2002 under the supervision of Profs. François Couty and Claude Agami. After postdoctoral studies with Prof. James S. Panek at Boston University, he joined the CNRS at the University of Versailles in 2004. He then moved to the Université Libre de Bruxelles as associate professor in 2012. He has coauthored 80 publications and 10 book chapters with research interests in developing copper-mediated transformations and the chemistry of heteroatom-substituted alkynes as well as the total synthesis of natural and/or biologically relevant products.

Sidney Wilkerson-Hill was born in Kinston, North Carolina in 1988 and studied chemistry at North Carolina State University. In 2015, Sidney received his Ph.D. under the supervision of Prof. Richmond Sarpong at the University of California, Berkeley where his research focused on using cycloisomerization reactions to access natural product scaffolds. He is currently a UNCF-Merck postdoctoral fellow in the group of Prof. Huw Davies at Emory University in Atlanta, GA where his research focuses on developing novel reactions using *N*-sulfonyltriazoles and rhodium tetracarboxylate catalysts.

Organic Syntheses

Visible Light Photocatalysis of Radical Cation Diels–Alder Cycloadditions: Preparation of Tris(2,2'-bipyrazyl) Ruthenium(II) Bis(tetrakis(3,5-bis(trifluoromethyl)phenyl)borate)

Shane D. Lies, Shishi Lin, and Tehshik P. Yoon[1]*

Department of Chemistry, University of Wisconsin-Madison, 1101 University Avenue, Madison, Wisconsin 53706, United States

Checked by Stanley Chang and Richmond Sarpong

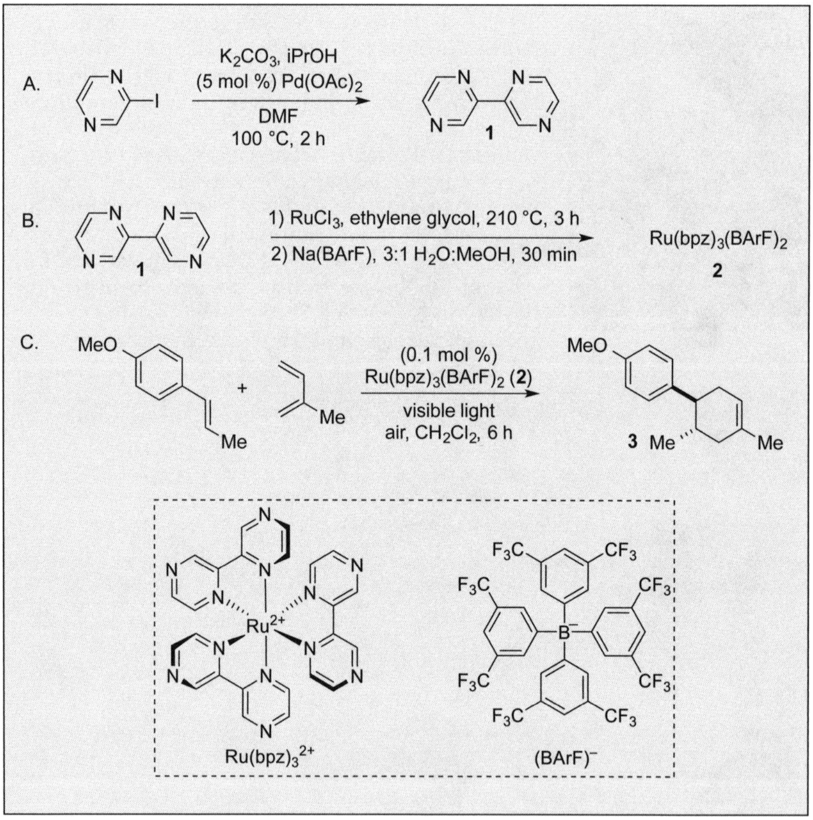

DOI: 10.15227/orgsyn.093.0178
Published on the Web 6/28/2016

Procedure

A. *2,2′-Bipyrazine* (**1**). A 250-mL, three-necked, round-bottomed flask open to air is equipped with a magnetic stir bar (15 x 31 mm, egg-shaped, Teflon-coated), glass stopper, and thermometer. The reaction vessel is charged with 2-iodopyrazine (4.01 g, 19.5 mmol, 1.0 equiv) (Note 1) and *N,N*-dimethylformamide (55.0 mL) (Note 2). To this solution are added potassium carbonate (4.03 g, 29.1 mmol, 1.5 equiv) (Note 3), palladium (II) acetate (218 mg, 0.973 mmol, 0.05 equiv) (Note 4), and 2-propanol (3.0 mL, 39.2 mmol, 2.0 equiv) (Note 5), resulting in a heterogeneous light-orange mixture. A reflux condenser is attached (Figure 1), and the mixture is heated to 100 °C with a silicone oil bath under nitrogen atmosphere and stirred (500 rpm) for 2 h (internal temperature = 101–103 °C) (Note 6). The oil bath is removed, and the reaction is allowed to cool to 30 °C (internal

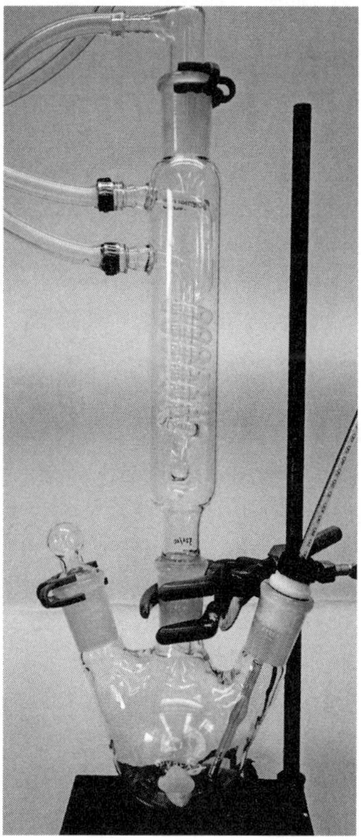

Figure 1. Reaction Apparatus used in Step A (provided by Checker)

temperature) over 60 min. A saturated aqueous solution of ammonium chloride (40 mL) is added drop-wise using a glass pipette to the reaction mixture over 5 min (Note 7). The reaction mixture is again cooled to 30 °C (internal temperature) over 15 min, then the mixture is transferred to a 500-mL separatory funnel and diluted with ethyl acetate (75 mL) and water (10 mL). After extraction and separation, the aqueous phase is extracted with ethyl acetate (3 x 75 mL). The combined organic fractions are washed with saturated aqueous solution of NH_4Cl (3 x 75 mL), and dried over Na_2SO_4 (15 g). The brown solution is filtered through a 150-mL coarse-porosity sintered glass funnel into a 500-mL recovery flask, and the solvent removed by rotary evaporation (35 °C, *ca.* 200 mmHg) in two portions to furnish 1.26 g of light brown solid. The residue is recrystallized from EtOAc (30 mL) (Note 8) to afford 1.11–1.17 g (mp 186–188 °C, 7.00–7.39 mmol, 72–76%) of 2,2′-bipyrazine as light brown flakes (Note 9).

B. *Ru(bpz)$_3$(BArF)$_2$ (**2**)*. A 100 mL, three-necked, round-bottomed flask is equipped with magnetic stir bar (17 x 7 mm, egg-shaped, Teflon-coated), glass stopper, thermometer, and charged with ruthenium(III) chloride (294 mg, 1.42 mmol, 1.0 equiv) (Note 10). 2,2′-Bipyrazine (**2**) (1.00 g, 6.33 mmol, 4.5 equiv) and ethylene glycol (25.0 mL) (Note 11) are added to the flask, resulting in a turbid dark-brown mixture. A reflux condenser is attached (Figure 2), and the mixture is heated under nitrogen atmosphere to 214 °C with a silicone oil bath (Note 12). The stirred solution (500 rpm) is refluxed for 3 h (internal temperature = 195–197 °C), after which time the oil bath is removed and the reaction is allowed to cool to 60 °C (internal temperature) (Note 13). Upon further cooling over 30 min, the reaction mixture is added drop-wise using glass pipette over 15 min to a 500-mL Erlenmeyer flask containing acetone (200 mL) (Note 14), which results in the precipitation of an orange solid. The round-bottomed flask is rinsed with acetone (2 x 25 mL) and sonicated (15 s) each time, and the combined mixtures are stored at 3 °C for 14 h. The resulting precipitates are collected by vacuum filtration (Note 15) to afford tris(2,2′-bipyrazyl)ruthenium(II) chloride as a dark-red powder (Note 16).

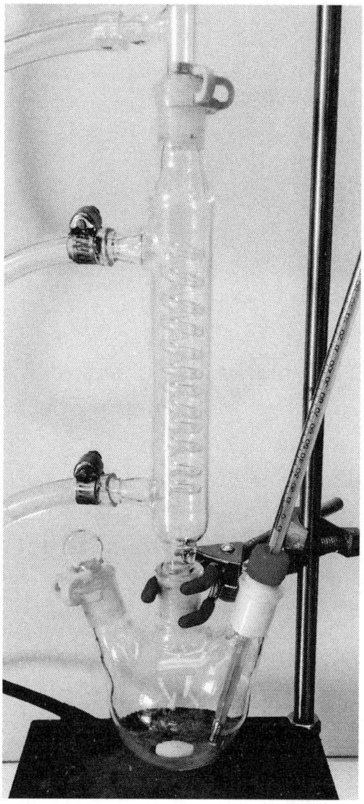

Figure 2. Reaction Apparatus used in Step B (provided by Checker)

The precipitate is transferred to a 250-mL, single-necked, round-bottomed flask equipped with a magnetic stir bar (37 x 17 mm, egg-shaped, Teflon-coated) and water (52.5 mL) (Note 17) is added to afford an opaque dark-red solution. A 24-mL vial is charged with NaBArF (2.48 g, 2.80 mmol, 2.0 equiv) (Note 18) and methanol (17.5 mL) (Note 19), capped (Teflon lining), and inverted until fully mixed. The resulting cloudy white mixture is added to the round-bottomed flask using a glass pipette over 1.5 min, causing a light-brown semisolid precipitate to form. The mixture is stirred for 30 min, and the solid is collected by vacuum filtration (Note 20) to afford a brown powder. The powder is dissolved in acetone (10 mL) and loaded onto a column (4.5 x 18.5 cm) of alumina (250 g) (Note 21) that has been conditioned with dichloromethane. The top of the alumina column is protected with a thin layer of sand (1.5 cm), and the loaded material is eluted by gravity elution with CH_2Cl_2 (450 mL) (Note 22). The eluent is collected in 17.5 mL fractions, and the dark colored product is obtained in fractions 12–18. Concentration of these fractions by rotary evaporation (35 °C, *ca.* 200 mmHg) furnishes a dark-brown slurry. The slurry is dissolved in acetone (8 mL) and the chromatography procedure is repeated. The eluent is collected in 17.5 mL fractions, and the dark-brown product is obtained in fractions 12–18. Concentration of these fractions by rotary evaporation (35 °C, *ca.* 200 mmHg) furnishes a dark-brown slurry (Note 23).

The slurry is dissolved in dichloromethane (75 mL) (Note 24), transferred to a 250-mL Erlenmeyer flask, and layered with benzene (60 mL) (Note 25) by slow pouring from a 50-mL graduated cylinder (2 x 30 mL portions) over 5 min. The solution is cooled to 3 °C for 16 h, and the resulting crystals are collected by vacuum filtration (air aspirator, *ca.* 300 mmHg) on a 60-mL medium-porosity sintered glass funnel. The crystals are washed with ice-cold benzene (2 x 30 mL) and air-dried for 1 h. Residual dichloromethane (*ca.* 7-8%) is observed from recrystallization (Note 26). The recrystallized product is dried under vacuum (0.3 mmHg) for 4 h to furnish 1.47 g (0.64 mmol, 45%) of $Ru(bpz)_3(BArF)_2$ as orange needles (Notes 27, 28, and 29).

C. *1-(4,6-Dimethyl-3-cyclohexen-1-yl)-4-methoxybenzene* (**3**). A 1 L, three-necked, round-bottomed flask open to air, equipped with a magnetic stir bar (20 x 40 mm, egg-shaped, Teflon-coated) and a thermometer, is charged with *trans*-anethole (5.06 mL, 5.01 g, 33.8 mmol, 1.0 equiv) (Note 30). Dichloromethane (420 mL) (Note 22), $Ru(bpz)_3(BArF)_2$ (**2**) (78.9 mg, 0.034 mmol, 0.001 equiv), and isoprene (9.9 mL, 6.73 g, 98.8 mmol, 2.9 equiv) (Note 31) are transferred into the flask under air, resulting in a clear yellow solution. The mixture is placed 5 cm in front of a 26 W compact fluorescent light bulb (Note 32), stirred (500 rpm) and irradiated open to air (Figure 3) (Note 33). The progress of the reaction is monitored by gas chromatography (Note 34). Upon completion (6 h), the stir-bar and thermometer are removed and rinsed with dichloromethane (5 mL), and the clear dark-orange mixture is transferred to a 500-mL recovery flask, and the three-necked flask is rinsed with dichloromethane (15 mL). The clear, dark-orange mixture is concentrated by rotary evaporation (35 °C, *ca.* 200 mmHg) to afford a brown oil containing dark brown precipitates. The residue is eluted through a column (4.5 x 4.0 cm) of silica (25 g) (Note 35) using a 50:1 mixture of pentanes:ethyl acetate (350 mL) (Note 36) into a 500-mL recovery flask. The clear solution is concentrated by rotary evaporation (35 °C, *ca.* 200 mmHg) to provide a colorless oil. The crude product is transferred to a 25-mL, round-bottomed flask equipped with a magnetic stir bar (12.5 x 9 mm,

Teflon-coated). The desired product is obtained by distillation through a short-path distillation apparatus at 108–110 °C (0.5 mmHg) to afford 6.67 g (31 mmol, 91%, run 1) and 6.78 g (31 mmol, 93%, run 2) of the title compound as a clear, colorless oil (Note 37).

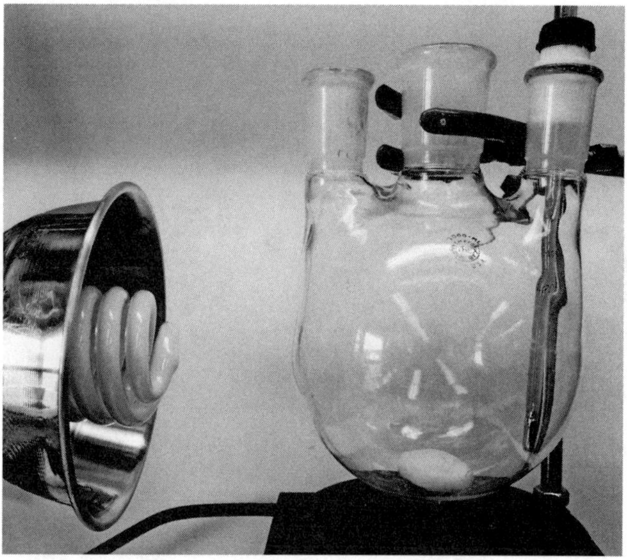

Figure 3. Reaction Apparatus used in Step C (provided by Checker)

Notes

1. 2-Iodopyrazine (97%) was purchased from Sigma-Aldrich and distilled immediately prior to use. The Checker also performed one run using 2-iodopyrazine (97%+) that was purchased from Ark Pharm and distilled immediately prior to use. The starting material is distilled under vacuum through a short-path distillation apparatus at 41–42 °C (1.2 mmHg) and provided a light yellow oil.
2. *N,N*-Dimethylformamide (≥ 99.8%) was purchased from Sigma-Aldrich and used as received.
3. Potassium carbonate (≥ 99.0%) was purchased from Sigma-Aldrich and used as received.
4. Palladium(II) acetate (99+% [99.95+% Pd]) was purchased from Strem Chemicals Inc. and used as received.

5. 2-Propanol (≥ 99.8%) was purchased from Sigma-Aldrich and used as received.
6. The Submitter and Checker observed that during heating, the heterogeneous mixture turns orange, then dark reddish-purple, and the solid changes from white to brown.

Figure 4. Color changes observed in Step A (provided by Submitter)

The Checker follows progress of the reaction by TLC analysis on silica gel (first placed under 0.3 mmHg vacuum for 5 min) with 50% EtOAc-hexane as eluent. Visualization is performed with UV irradiation, where the 2-iodopyrazine starting material has R_f = 0.52 and the 2,2'-bipyrazine product has $R_f = 0.27$.

7. An exotherm of *ca.* 12 °C was observed upon the addition of saturated aqueous solution of NH_4Cl.
8. The Checker performed recrystallization as follows. The residue is suspended in EtOAc (25 mL) and transferred open to air into a 100-mL round-bottomed flask. Following removal of EtOAc *in vacuo* (35 °C, *ca.* 200 mmHg), a magnetic stir bar (14 x 27.5 mm, egg-shaped, Teflon-coated) and EtOAc (30 mL) are added to the flask. A reflux condenser is attached and the solution is heated to reflux in a silicone oil bath (95 °C). After dissolution of the solid and 15 min at reflux, the flask was removed from the oil bath and the stir bar was removed using a magnetic rod. The flask is allowed to cool under ambient conditions over 1 h, the condenser replaced with a glass stopper, and then placed in a refrigerator at 7 °C for 13 h. The resulting crystals are collected by vacuum filtration (air aspirator, *ca.* 300 mmHg) through a Büchner

funnel (68 mm diameter; fitted with filter paper) with a 250-mL Erlenmeyer flask with side-arm, washed with hexanes (4 x 5 mL), and air-dried by vacuum (air aspirator, *ca.* 300 mmHg) for 30 min. The mother liquor is concentrated by rotary evaporation (35 °C, *ca.* 200 mmHg) to furnish a light-brown solid. The residue is transferred to a 100-mL round-bottomed flask with magnetic stir bar (14 x 27.5 mm, egg-shaped, Teflon-coated). Crystallization was carried out as described for the first crop, using 5 mL EtOAc. The two crops are combined and used in the next step. The Submitter performed recrystallization using a 50-mL Erlenmeyer flask, heated to 75 °C on a hot plate with occasional swirling, and using the same volume of EtOAc for each recrystallization.

9. Characterization data for **1** obtained by Checker: ^{1}H NMR (400 MHz, $CDCl_3$) δ: 8.66 (s, 4H), 9.60 (s, 2H); ^{13}C NMR (101 MHz, $CDCl_3$) δ: 143.6, 143.9, 145.3, 149.4; FT-IR (ATR) 3077, 3051, 3015, 1465, 1383, 1092, 1028, 1019, 846 cm^{-1}; Exact mass EI-MS: calculated *m/z* for $[C_8H_6N_4]^+$: 158.0592, found: 158.0591; Anal. calcd for $C_8H_6N_4$: C, 60.75; H, 3.82; N, 35.42. Found: C, 60.60; H, 3.94; N, 35.15. Picture provided by Checker.

Figure 5. Product of Step A (provided by Checker)

10. Ruthenium(III) chloride hydrate (40-43% Ru [99.9%-Ru]) was purchased from Strem Chemicals and used as received.
11. Ethylene glycol (≥ 99%) was purchased from Sigma-Aldrich and used as received.
12. The oil bath and reaction vessel was wrapped with aluminum foil during the course of the reaction to prevent radiative heat loss.

13. The Submitter reported isolation of 681 mg of crude $Ru(bpz)_3Cl_2$ (~2.5% by mass of residual ethylene glycol) following the submitted procedure for precipitation. The Checker observed that precipitation following Submitter's procedure consistently led to low isolated mass (251 mg and 248 mg for 2 separate runs).
   The Checker observed that consistent yield of crude $Ru(bpz)_3Cl_2$ may be obtained when excess ethylene glycol is removed at this stage by distillation using the procedure as follows. The condenser of the three-necked round-bottomed flask is replaced with a short-path distillation apparatus, and the thermometer is replaced with a glass stopper. The distillation apparatus is fitted with a 100-mL round-bottomed flask (with marking at 20 mL volume). Ethylene glycol is distilled under vacuum by heating in a silicone oil bath to 125 °C. Following removal of 20 mL of ethylene glycol at 70–78 °C (2.9 mmHg), the reaction mixture was cooled again prior to precipitation. Note: the distilled ethylene glycol also contains white precipitates that are identified as 2,2′-bipyrazine by analysis of the corresponding ^{1}H NMR spectrum.

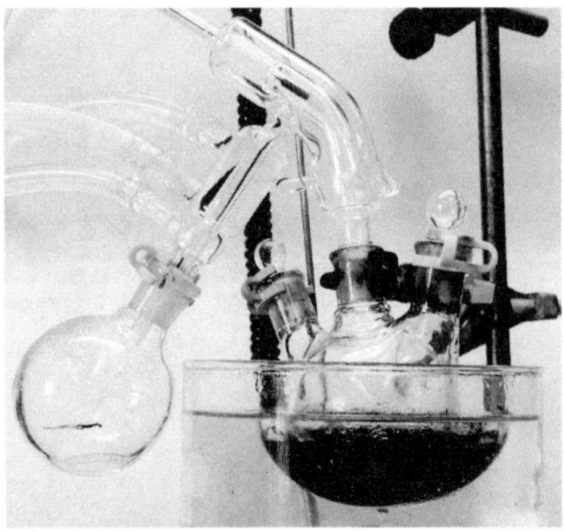

Figure 6. Distillation Apparatus for Removal of Ethylene Glycol (provided by Checker)

14. Acetone (≥ 99.9%) was purchased from Sigma-Aldrich and used as received.
15. The Checker performed isolation of tris(2,2′-bipyrazyl)ruthenium(II) chloride as follows. The 500-mL Erlenmeyer flask is first sonicated for 15 s with swirling. The precipitates are collected by vacuum filtration

(air aspirator, *ca.* 300 mmHg) using a 60-mL medium-porosity sintered glass funnel with a 500-mL Erlenmeyer flask with side-arm. The 500-mL Erlenmeyer flask containing crude material is rinsed with ice-cold acetone (2 x 15 mL). Any solid deposited on the sides of the flask can be collected by scraping during rinsing. Vacuum is turned off and the collected solid is stirred in ice-cold acetone (2 x 15 mL), then vacuum is reapplied, resulting in a dark-red paste. The solid is air dried for 30 min, agitating by scraping (once every 10 min) to furnish red-brown powder (670 and 658 mg for 2 separate runs).

Figure 7. Tris(2,2′-bipyrazyl)ruthenium(II) chloride (provided by Checker)

16. Data for crude tris(2,2′-bipyrazyl)ruthenium(II) chloride: ^{1}H NMR (400 MHz, DMSO-d_6) δ 10.21 (m, 6H), 8.71 (d, J = 3.0 Hz, 6H), 8.03 (d, J = 3.0 Hz, 6H); ^{13}C NMR (101 MHz, DMSO-d_6) δ 150.3, 148.5, 147.0, 145.5; FT-IR (ATR) 2999, 2974, 1585, 1407 (s), 1269, 1157 (s), 828, 660 (s), 468 cm^{-1}; Exact mass ESI-MS: calculated *m/z* for $[C_{24}H_{18}N_{12}ClRu]^+$: 611.0504, found: 611.0500.
17. The Checker used water (HPLC grade, submicron filtered) that was purchased from Fisher Scientific and used as received.
18. Sodium tetrakis[bis(3,5-trifluoromethyl)phenyl]borate (97%) was purchased from Alfa Aesar and used as received.
19. Methanol (absolute) was purchased from Macron Fine Chemicals and used as received.
20. The Checker performed vacuum filtration of the crude solid product as follows. The crude reaction mixture is transferred to a 150-mL medium-porosity sintered glass funnel with a 250-mL Erlenmeyer flask with side-arm. Vacuum (air aspirator, *ca.* 300 mmHg) is applied for 16 h until clumps of catalyst are obtained, and the crude solid is agitated then

rinsed with water (10 mL). After drying for 12 h, the crude solid is again agitated, then rinsed with water (10 mL). Additional drying for 12 h affords 2.49 g of a brown powder (picture provided by Checker).

Figure 8. $Ru(bpz)_3(BArF)_2$ as isolated after filtration (provided by Checker)

The Submitter performed vacuum filtration using procedure as follows. The crude reaction mixture is transferred to a 150-mL medium-porosity sintered glass funnel with a 250-mL round-bottomed receiving flask. Vacuum (water aspirator) is applied for 24–48 h (filtration time varies by aspirator pressure), and the crude solid is rinsed with water (2 x 10 mL) then dried to afford a dark brown powder.

21. Alumina (puriss., ≥ 98% Al_2O_3 basis) was purchased from Sigma-Aldrich and used as received.
The Submitter dissolved the powder in acetone (10 mL) and loaded onto a column (4.5 x 20 cm) of alumina (200 g). The Submitter and Checker performed the column by gravity elution. The Submitter suggests running columns using nitrogen or air positive pressure may reduce the resolution between the product and the brown impurity.
22. The Submitter used dichloromethane (≥ 99.5%) that was purchased from Sigma-Aldrich and used as received. The Checker used dichloromethane (≥ 99.5%) that was purchased from Fisher Scientific and used as received.
23. The Checker observed oiling out of product during recrystallization if only two alumina columns were performed. In these cases, the dark-brown slurry is dissolved in acetone (6 mL) and loaded onto a column (4.5 x 18.5 cm) of alumina (250 g) that had been conditioned with dichloromethane. The top of the alumina column is protected with a thin layer of sand (1.5 cm), and the loaded material is eluted by gravity

filtration with CH_2Cl_2 (450 mL). The eluent is collected in 17.5 mL fractions, and the colored product in fractions 12–18 are collected (2.2-2.3 g) (Note 24). For improved reproducibility, a minimum total of four alumina columns may be necessary.

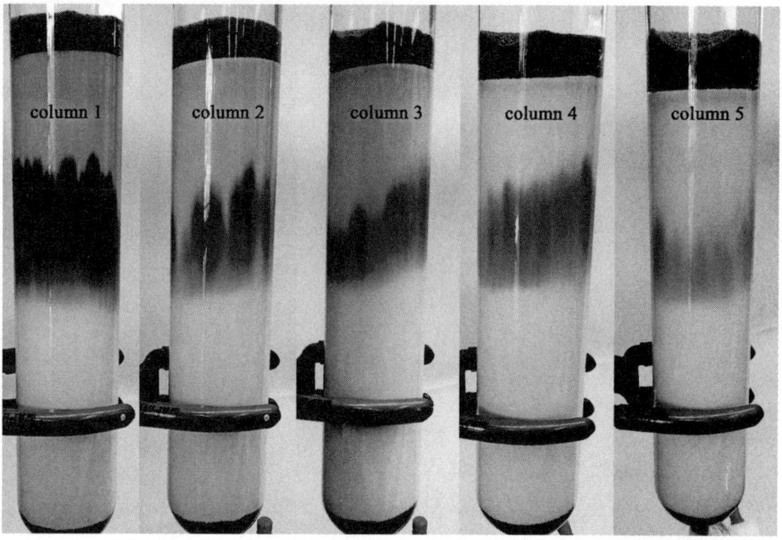

Figure 9. Alumina Columns (provided by Checker)

24. The Checker observed that the crude product may begin to precipitate during dilution with dichloromethane when using a crude product that has been concentrated to complete dryness. This leads to problematic recrystallization, which traps dichloromethane that is difficult to remove at a later stage. To prevent unwanted precipitation, the Checker recommends concentrating to a viscous oil, and immediately adding dichloromethane. In cases where precipitation occurs, the Checker adds acetone (2 mL) (Note 14) to re-dissolve precipitates, then re-attempts concentrating to a viscous oil by rotary evaporation (35 °C, *ca.* 200 mmHg).
25. Benzene (≥ 99.0%) was purchased from Sigma-Aldrich and used as received.
26. The Submitter's ^1H NMR spectrum indicates that the Submitter's sample may contain a non-negligible amount of dichloromethane and a small amount of benzene. The Checker used 10 mg dimethyl fumarate (Sigma-Aldrich, TraceCERT®) as internal standard with 48 mg

recrystallized product **2** to calculate weight percent of desired product by Q NMR.

27. Q NMR results for run 1: 1.58 g isolated mass (92.9% pure by Q NMR; 10.2 mg dimethyl fumarate, 47.8 mg analyte), 1.47 g adjusted mass (45% adjusted yield); run 2: 1.70 g isolated mass (92.3% pure by Q NMR; 10.2 mg dimethyl fumarate, 47.7 mg analyte), 1.57 g adjust mass (48% adjusted yield).
28. Characterization data for **2** (anhydrous catalyst, based on EA result and absence of OH absorption in IR spectrum) obtained by Checker: mp = 273–276 °C; ^{1}H NMR (400 MHz, Acetone-d_6) δ: 7.67 (br s, 8H), 7.78 (m, 16H), 8.47 (dd, J = 3.1, 0.8 Hz, 6H), 8.79 (d, J = 3.2 Hz, 6H), 10.21 (br s, 6H); ^{13}C NMR (101 MHz, Acetone-d_6) δ: 118.5 (app. p, $^{3}J_{C\text{-}F}$ = 3.8 Hz), 125.4 (q, $^{1}J_{C\text{-}F}$ = 272.0 Hz), 130.0 (qq, $^{2}J_{C\text{-}F}$ = 31.4 Hz, $^{3}J_{C\text{-}B}$ = 2.9 Hz), 135.6, 146.6, 148.2, 149.8, 151.9, 162.6 (1:1:1:1 q, $^{1}J_{C\text{-}B}$ = 49.8 Hz); ^{19}F NMR (376 MHz, Acetone-d_6) δ −62.1; UV (CH_2Cl_2); λ_{max}, nm (ε): 294 (5.71 x 10^4); FT-IR (ATR) 1698, 1409, 1354 (s), 1275 (s), 1117 (br, s), 932, 887, 744, 712, 682 cm^{-1}; Exact mass ESI-MS: calculated *m/z* for $[C_{24}H_{18}N_{12}Ru]^{2+}$: 288.0405, found: 288.0409; Exact mass ESI-MS: calculated *m/z* for $[C_{32}H_{12}BF_{24}]^{-}$: 863.0654, found: 863.0627; Anal. calcd for $C_{88}H_{42}B_2F_{48}N_{12}Ru$: C, 45.92; H, 1.84; N, 7.30. Found: C, 45.71; H, 1.53; N, 7.26. Note: elemental analysis obtained using solvent-free sample of **2**.

Figure 10. Color of Product Formed in Step B (provided by Checker)

The Submitter reported that the product is isolated as the trihydrate, but can be dried *in vacuo* over P_2O_5 for 48 h. For the subsequent procedure, drying the catalyst is not necessary. μ

29. A second run by the Checkers provided 1.57 g (0.68 mmol) of Ru(bpz)$_3$(BArF)$_2$ as orange needles with mp = 274–277 °C.

30. *trans*-Anethole (≥ 99%, FCC, FG) was purchased from Sigma-Aldrich and purified by silica gel flash chromatography and vacuum distillation. The Submitter observed generation of an oxidized impurity (*p*-anisaldehyde) upon ambient storage conditions, and performed flash chromatography due to similar boiling points of *trans*-anethole and *p*-anisaldehyde. Distillation further removes residual solvent following flash chromatography. Purified *trans*-anethole can be stored for several days when stored in a refrigerator. Purification is performed as follows: 10 g of *trans*-anethole is charged onto a column (4.5 x 12.5 cm) of silica gel (75 g) that had been conditioned with a 50:1 mixture of hexanes:ethyl acetate. The top of the silica column is protected with a thin layer of sand (1.5 cm) and the loaded material is eluted with a 50:1 mixture of hexanes:ethyl acetate (510 mL). The eluent is collected in 17.5 mL fractions and the UV-active fractions (fractions 9–19) are combined into a 500-mL recovery flask and concentrated by rotary evaporation (35 °C, *ca.* 200 mmHg). The residue is transferred to a 50 mL round-bottomed flask equipped with a magnetic stir bar (12.5 x 9 mm, Teflon-coated) and distilled immediately prior to use. The starting material is distilled at 59–61 °C under vacuum (0.7 mmHg) through a short-path distillation apparatus to provide a colorless oil.
31. The Checker used isoprene (≥ 99%) that was purchased from Sigma-Aldrich and distilled under a prior to use. Isoprene (30 g) is distilled through a short-path distillation apparatus at 34–35 °C (760 mmHg) to provide a colorless liquid. The Submitter distilled isoprene under nitrogen prior to use.
32. Both the Submitter and Checker used 26 W compact fluorescent light bulb with the following specifications: GE Reveal® (product code 67454, model number FLE26HT3) 1570-lumen T3 spiral CFL bulb.
33. The Checker observed that during the first 10 min of reaction, the mixture turns from yellow to dark orange.

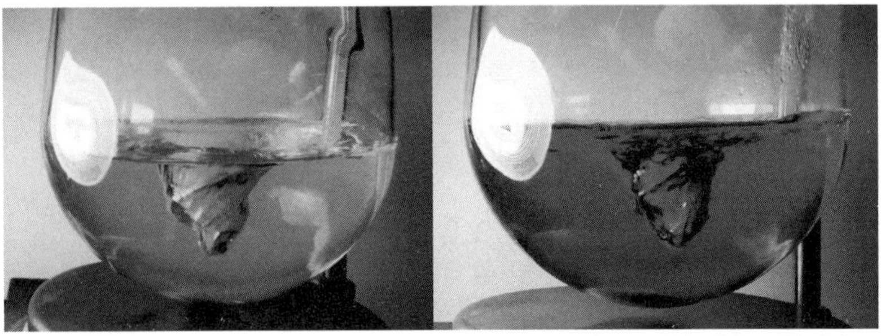

Figure 11. Color Change Observed in Step C (provided by Checker)

The Checker also observed an increase in internal temperature to 27.5–28.5 °C during the course of the reaction.

34. The Checkers performed gas chromatography using a Shimadzu GCMS-QP2010 SE Gas Chromatograph. Column: Shimadzu SHRXI-5MS (cat no. 220-94764-02), −60 to 330/350 °C; 30 m length x 0.25 mm ID x 0.25 μm thickness; 5% diphenyl 95% dimethyl polysiloxane. Injector: 200 °C; detector: 200 °C. Oven program: 80 °C for 1 min, then 20 °C/min to 250 °C for 8.5 min, then 250 °C for 2.5 min. Retention times: *trans*-anethole (5.8 min), 1-(4,6-Dimethyl-3-cyclohexen-1-yl)-4-methoxybenzene (8.5 min). The Submitters performed gas chromatography using a Shimadzu GC-2010 Gas Chromatograph. Column: Restek RTX-5: 350 °C; 15 m x 250 μm x 0.25 μm; 5% diphenyl 95% dimethyl polysiloxane. Injector: 300 °C; detector FID: 300 °C. Oven program: 50 °C for 0 min then 15 °C/min to 130 °C for 0 min then 30 °C/min to 250 °C for 5.0 min. Retention times: anethole (5.0 min), 1-(4,6-Dimethyl-3-cyclohexen-1-yl)-4-methoxybenzene (7.4 min), anethole homodimer (9.5 min). Reaction is complete when anethole is consumed and ratio of product to anethole homodimer is ≥ 100:1.
35. Silica (60 Å, 230-400 mesh, 40-63 μm particle size) was purchased from Sigma-Aldrich and used as received.
36. Pentanes (98%) was purchased from Sigma-Aldrich and used as received. Ethyl acetate (≥ 99.5%) was purchased from Sigma-Aldrich and used as received. The Submitter notes that hexanes can be used as an alternative to pentanes.
37. Characterization data for **3**: ^1H NMR (400 MHz, $CDCl_3$) δ: 0.70 (d, *J* = 6.3 Hz, 3H), 1.69 (br s, 3H), 1.74–1.94 (m, 2H), 2.03–2.25 (m, 3H), 2.30 (ddd, *J* = 10.5, 10.5, 5.2 Hz, 1H), 3.79 (s, 3H), 5.44 (br s, 1H), 6.84 (d, *J* = 8.5 Hz, 2H), 7.08 (d, *J* = 8.5 Hz, 2H); ^{13}C NMR (101 MHz, $CDCl_3$) δ:

20.4, 23.5, 34.1, 35.4, 40.0, 47.1, 55.3, 113.8, 121.0, 128.6, 133.9, 138.3, 157.9; FT-IR (ATR) 2949, 2888, 2833, 1611, 1511 (s), 1440, 1270 (s), 1242, 1175, 1108, 1038, 823, 801 cm^{-1}; Exact mass EI-MS: calculated *m/z* for $[C_{15}H_{20}O]^+$: 216.1514, found: 216.1518; Anal. calcd for $C_{15}H_{20}O$: C, 83.28; H, 9.32. Found: C, 83.17; H, 9.27.

38. A second run by the Checkers provided 6.78 g (31 mmol, 93%) of the product as a clear, colorless oil.

Working with Hazardous Chemicals

The procedures in *Organic Syntheses* are intended for use only by persons with proper training in experimental organic chemistry. All hazardous materials should be handled using the standard procedures for work with chemicals described in references such as "Prudent Practices in the Laboratory" (The National Academies Press, Washington, D.C., 2011; the full text can be accessed free of charge at http://www.nap.edu/catalog.php?record_id=12654). All chemical waste should be disposed of in accordance with local regulations. For general guidelines for the management of chemical waste, see Chapter 8 of Prudent Practices.

In some articles in *Organic Syntheses*, chemical-specific hazards are highlighted in red "Caution Notes" within a procedure. It is important to recognize that the absence of a caution note does not imply that no significant hazards are associated with the chemicals involved in that procedure. Prior to performing a reaction, a thorough risk assessment should be carried out that includes a review of the potential hazards associated with each chemical and experimental operation on the scale that is planned for the procedure. Guidelines for carrying out a risk assessment and for analyzing the hazards associated with chemicals can be found in Chapter 4 of Prudent Practices.

The procedures described in *Organic Syntheses* are provided as published and are conducted at one's own risk. *Organic Syntheses, Inc.*, its Editors, and its Board of Directors do not warrant or guarantee the safety of individuals using these procedures and hereby disclaim any liability for any injuries or damages claimed to have resulted from or related in any way to the procedures herein.

Discussion

The Diels–Alder reaction ranks among the most important reactions for the synthesis of complex organic molecules. One major limitation of classical Diels–Alder methodology, however, arises from electronic considerations; efficient [4+2] cycloadditions generally require the reaction of one electron-rich reaction component (typically the diene) and one electron-deficient component (typically the dienophile). In this regard, radical cation Diels–Alder reactions initiated by single-electron oxidation of an electron-rich dienophile can overcome this fundamental limitation and enable the efficient [4+2] cycloaddition of two electron-rich reaction partners that would otherwise be electronically mismatched.

A variety of one-electron oxidation methods have been used to initiate radical cation Diels–Alder reactions. The most common classical methods use chemical oxidants such as aminium salts,[2] although somewhat high loadings of the oxidant are typically required to achieve complete conversion. The radical cation intermediates can also be generated by photoinitiated electron transfer (PET). A variety of organic photosensitizers have been utilized for this purpose, although the poor chemical stability and relatively short excited state lifetimes again necessitate somewhat high loadings of the photooxidant.

We have recently described a highly efficient and versatile method for radical cation Diels–Alder reactions. The highly oxidizing $Ru(bpz)_3^{2+}$ photocatalyst utilized in this method absorbs strongly in the visible range, enabling the reaction to be conducted with variety of conveniently available visible light sources, including ambient sunlight.[3] This catalyst also exhibits excellent chemical stability and a long-lived redox-active excited state, enabling loadings of the photocatalyst as low as 0.1 mol%. The procedure described here is adapted from our initial communication.

Table 1 demonstrates that the substrate scope for the photocatalytic radical cation Diels–Alder cycloaddition is quite broad. A variety of structurally diverse dienes react efficiently, including sterically encumbered and cyclic dienes as well as dienes bearing aryl and heteroatom substituents. The regioselectivity in all cases was excellent and consistent with an umpolung polarization of the styryl radical cation, in which the oxidized arene moiety can be considered an electron-withdrawing group. Diastereoselectivity, where relevant, was also excellent and highly *endo* selective. The scope of the reaction with respect to the dienophile is quite broad as well, so long as the alkene is electron-rich enough to be oxidized by the excited photocatalyst. Thus, a variety of styrenes bearing electron-donating *para* substituents react smoothly. More importantly, non-styrenic alkenes such as enol ethers and enamines are also readily oxidized and are suitable substrates for the radical cation Diels–Alder cycloaddition.

Table 1. Scope of the Radical Cation Diels–Alder Cycloaddition Reaction[a]

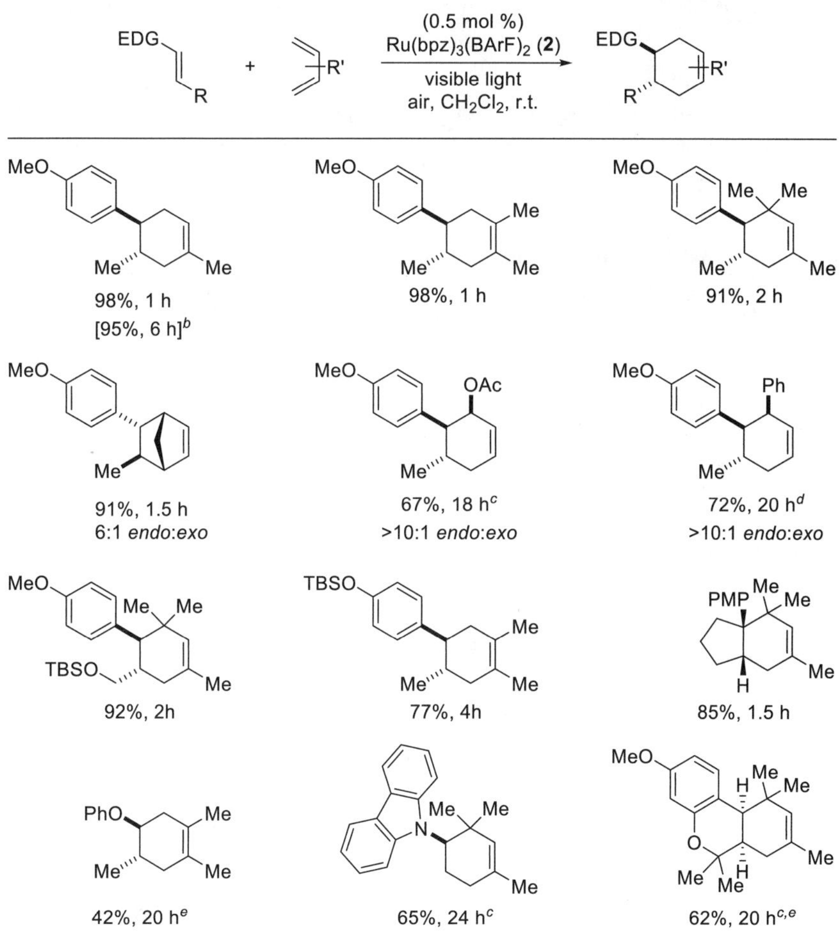

[a]Conditions, unless specified otherwise: 0.5 mol % $Ru(bpz)_3(BArF)_2$, 3 equiv. diene, irradiated with a 23 W compact fluorescent bulb. [b]0.1 mol % $Ru(bpz)_3(BArF)_2$, 25 W CFL, 5 gram scale. [c]3 mol % $Ru(bpz)_3(BArF)_2$ [d]2 equiv. diene. [e]Yield determined by ^1H NMR analysis using CH_2Br_2 as an external standard.

The optimal photocatalyst identified for the radical cation Diels–Alder in our studies was $Ru(bpz)_3(BArF)_2$. The electron-deficient bipyrazyl ligands render this photocatalyst a substantially more powerful photooxidant than the more common parent tris(bipyridyl) complex. Due to the attractive photoredox properties of $Ru(bpz)_3^{2+}$ coupled with its chemical stability and long excited state lifetime, this photocatalyst has proven to be powerfully enabling in a variety of photooxidative transformations (Table 2). These have included oxidative [3+2] cycloadditions of phenols,[4] thiol–ene reactions,[5] and aerobic cyclooxygenation reactions,[6] among others.[7] However, previously reported methods for the synthesis of the bipyrazyl ligand are not easily scalable or provide poor yields. The improved method described in this manuscript is based upon optimized conditions reported recently by our group.[8]

Table 2. Photooxidative transformations using $Ru(bpz)_3^{2+}$

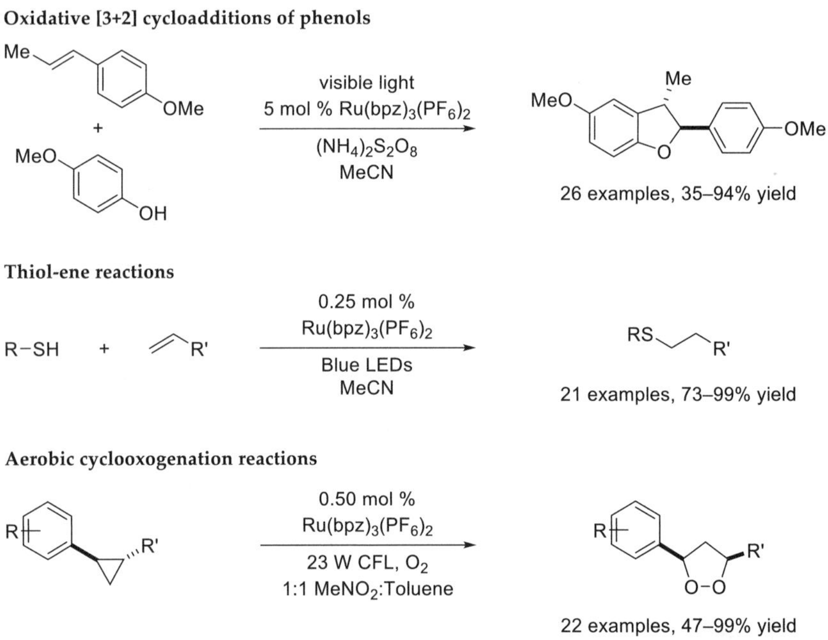

References

1. Department of Chemistry, University of Wisconsin–Madison, 1101 University Avenue, Madison, Wisconsin, 5370, United States. E-mail: tyoon@chem.wisc.edu. This publication is based on work supported by NIH-NIGMS (GM095666), the Sloan Foundation, the Beckman Foundation, and Research Corporation. The authors also thank Sigma–Aldrich for a gift of ruthenium(III) chloride used in these studies.
2. (a) Bellville, D. J.; Wirth, D. W.; Bauld, N. L. *J. Am. Chem. Soc.* **1981**, *103*, 718–720. (b) Bauld, N. L.; Bellville, D. J.; Pabon, R.; Chelsky, R.; Green, G. *J. Am. Chem. Soc.* **1983**, *105*, 2378–2382. (c) Bellville, D. J.; Bauld, N. L.; Pabon, R.; Gardner, S. A. *J. Am. Chem. Soc.* **1983**, *105*, 3584–3588. (d) Pabon, R. A.; Bauld, N. L. *J. Am. Chem. Soc.* **1984**, *106*, 1145–1146. (e) Wiest, O.; Steckhan, E.; Grein, F. *J. Org. Chem.* **1992**, *57*, 4034–4037. (f) Yueh, W.; Bauld, N. L. *J. Phys. Org. Chem.* **1996**, *9*, 529–538. (g) Bauld, N. L.; Yang, J. *Org. Lett.* **1999**, *1*, 773–774. (h) Valley, N. A.; Wiest, O. *J. Org. Chem.* **2007**, *72*, 559–566. (i) Sevov, C. S.; Wiest, O. *J. Org. Chem.* **2008**, *73*, 7909–7915. (j) Lim, H. N.; Parker, K. A. *Org. Lett.* **2013**, *15*, 398–401. (k) Lim, H. N.; Parker, K. A. *J. Org. Chem.* **2014**, *79*, 919–926.
3. Lin, S.; Ischay, M. A.; Fry, C. G.; Yoon, T. P. *J. Am. Chem. Soc.* **2011**, *133*, 19350–19353.
4. Blum, T. R.; Zhu, Y.; Nordeen, S. A.; Yoon, T. P. *Angew. Chem. Int. Ed.* **2014**, *53*, 11056–11059.
5. Tyson, E. L.; Ament, M. S.; Yoon, T. P. *J. Org. Chem.* **2013**, *78*, 2046–2050.
6. Lu, Z.; Parrish, J. D.; Yoon, T. P. *Tetrahedron* **2014**, *70*, 4270–4278.
7. (a) Maity, S.; Zhu, M.; Shinabery, R. S.; Zheng, N. *Angew. Chem. Int. Ed.* **2012**, *51*, 222–226. (b) Maity, S.; Zheng, N. *Angew. Chem. Int. Ed.* **2012**, *51*, 9562–9566.
8. Schultz, D. M.; Sawicki, J. W.; Yoon, T. P. *Beilstein J. Org. Chem.* **2015**, *11* 61–65.

Appendix
Chemical Abstracts Nomenclature (Registry Number)

2-Iodopyrazine: Pyrazine, 2-iodo-; (32111-21-0)
Potassium carbonate: Carbonic acid, potassium salt (1:2); (584-08-7)

Palladium (II) acetate: Acetic acid, palladium(2+) salt (2:1); (3375-31-3)
Ammonium chloride; (12125-02-9)
2,2′-Bipyrazine; (10199-00-5)
Ruthenium (III) chloride: Ruthenium chloride ($RuCl_3$); (10049-08-8)
$Ru(bpz)_3Cl_2$: Ruthenium(2+), tris(2,2'-bipyrazine-□*N*[1],□*N*[1'])-, chloride (1:2), (*OC*-6-11)-; (80925-50-4)
Sodium BArF: Borate(1-), tetrakis[3,5-bis(trifluoromethyl)phenyl]-, sodium (1:1); (79060-88-1)
$Ru(bpz)_3(BArF)_2$: Ruthenium(2+), tris(2,2'-bipyrazine-κ*N*[1],κ*N*[1'])-, (*OC*-6-11)-, tetrakis[3,5-bis(trifluoromethyl)phenyl]borate(1-) (1:2); (1350432-81-3)
trans-Anethole: Benzene, 1-methoxy-4-(1-propen-1-yl)-; (4180-23-8)
Isoprene: 1,3-Butadiene, 2-methyl-; (78-79-5)
1-(4,6-Dimethyl-3-cyclohexen-1-yl)-4-methoxybenzene: Benzene, 1-[(1*R*,6*R*)-4,6-dimethyl-3-cyclohexen-1-yl]-4-methoxy-, *rel*-; (112150-17-1)

Shane Lies was born in Waterloo, Iowa in 1987. He received his B.S. in Biochemistry from the University of Northern Iowa in 2014 where he worked in the laboratory of Prof. Jeffrey Elbert. That year, he began his doctoral studies at the University of Wisconsin–Madison in the laboratories of Prof. Tehshik Yoon. His graduate work has focused on the development of methods for the synthesis of cyclobutane derivatives using photoredox chemistry.

Shishi Lin grew up in sunny Singapore and received her B.Sc. degree in Chemistry from the National University of Singapore in 2009 where she carried out undergraduate research on the enantioselective protonation reactions catalyzed by chiral bicyclic guanidine with Prof. Choon-Hong Tan. She then moved across the globe to the University of Wisconsin-Madison where she completed her Ph.D. degree in Chemistry with Tehshik Yoon. Her graduate research focused on the electronically-mismatched radical cation Diels–Alder cycloaddition reactions by visible light photocatalysis. Currently she is an industrial postdoctoral fellow at Merck Process Chemistry.

Tehshik Yoon received his A.B. in Chemistry from Harvard University in 1996 working with Prof. David Evans. His graduate degrees were received from Caltech, his M.S. under the direction of Prof. Erick Carreira, and his Ph.D. from Prof. David MacMillan. After an NIH postdoctoral fellowship in Prof. Eric Jacobsen's laboratory at Harvard University, Yoon began his independent career at the University of Wisconsin–Madison in 2005. He was promoted to Associate Professor in 2012 and to Professor in 2014.

Stanley Chang received his B.Sc. at the University of British Columbia in 2009 and carried out undergraduate research with Prof. Chris Orvig. Later he studied chemistry at Simon Fraser University, obtaining his Ph.D. in 2015 under the direction of Prof. Robert Britton, investigating the total syntheses of 5,5-spiroacetal natural products. He is currently a Postdoctoral Research Fellow with Prof. Richmond Sarpong at UC Berkeley, exploring the total synthesis of complex terpene natural products.

Organic Syntheses

Preparation of Methyl 1,2,3,4-tetra-*O*-acetyl-β-D-glucopyranuronate

Aisling Ní Cheallaigh,[$] Garrett T. Potter,[$] John M. Gardiner,[$] and Gavin J. Miller[^*1]

[^]Synthesis and Medicinal Chemistry Cluster, School of Physical and Geographical Sciences, Keele University, Keele, Staffordshire, ST5 5BG, U.K.
[$]Manchester Institute of Biotechnology and School of Chemistry, 131 Princess Street, The University of Manchester, Manchester, M1 7DN, U.K.

Checked by Michal Achmatowicz and Margaret Faul

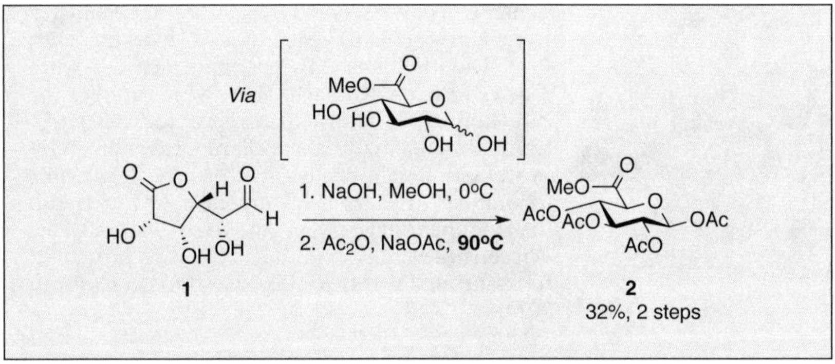

Procedure

A. *Methyl 1,2,3,4-tetra-O-acetyl-β-D-glucopyranuronate* **2.** A three-necked 250-mL round-bottomed flask on a stirrer plate is equipped with a 2-cm Teflon coated magnetic stirrer bar, rubber septum, thermometer (with insert to monitor the internal reaction temperature) and a connection to a nitrogen line (Figure 1).

DOI: 10.15227/orgsyn.093.0200

Figure 1. Reaction set-up

Under inert atmosphere the flask is charged with methanol (MeOH) (50 mL) (Note 1), NaOH (20 mg, 0.5 mmol, 0.02 equiv) (Note 2) and then stirred vigorously for 10 min to complete dissolution of NaOH. The clear solution is cooled to 0 °C with an ice/water bath and to this is added D-glucuronolactone **1** (5.0 g, 28 mmol, 1.0 equiv) portion-wise (Note 3) over 25 min (Note 4). The solution is stirred for an additional 30 min, while maintaining the internal temperature at 0 °C. Upon complete dissolution of **1**, the reaction solution is transferred to a one-necked 250-mL round-bottomed flask and the solvent removed by rotary evaporation (30 °C, 20 mmHg) (Note 5) to yield a pale yellow foam which is dried under vacuum (5 mmHg) for 2 h to remove the majority of residual MeOH (Figure 2).

Figure 2. Pale yellow foam after removal of methanol

The flask (Note 6) containing the foam is charged with a 2-cm Teflon coated magnetic stirrer bar, acetic anhydride (54 g, 50 mL, 530 mmol, 19 equiv) and anhydrous NaOAc (9.5 g, 116 mmol, 4.1 equiv) (Note 7). A reflux condenser is attached, the reaction placed under nitrogen and the mixture heated (Note 8) to 90 °C for 1.5 h (Note 9). The light-brown suspension (Figure 3) is then allowed to cool. The solution is no longer stirred and is left standing overnight.

Figure 3. Light brown suspension

The resulting precipitate is diluted with EtOAc (100 mL) and any solids broken up carefully with a spatula. This enables stirring of the suspension, which is continued vigorously for 5 min. The suspension is then filtered through a 100-mL medium porosity sintered glass funnel into a 0.5 L flask, washing the solids with an additional portion of EtOAc (100 mL). This filtrate is retained. Sodium bicarbonate ($NaHCO_3$) (90 g, 1.1 mol, 38 equiv) is weighed into a 1.0 L beaker on a stirrer plate and distilled water (250 mL) added. A 4-cm Teflon coated stirrer bar is added and stirring commenced to dissolve the majority of the $NaHCO_3$. The reaction filtrate is added slowly to this stirred solution (Note 10) and stirring is continued for >30 min, or until all effervescence had ceased. The solution is then transferred to a 1.0 L separating funnel and the organic/aqueous layers separated. The aqueous phase is extracted once with EtOAc (50 mL) and the organics combined, washed with H_2O (50 mL), dried over $MgSO_4$ (10.0 g) for 10 min, and filtered through a sintered glass funnel into a flask. The solvent is removed on a rotary evaporator (40 °C, 20 mmHg) to reveal a sticky, brown oil mixed with solids. This residue is dissolved in hot (70 °C) EtOH (50 mL) and allowed to cool, which induces crystallisation of **2**. The solids are collected by vacuum filtration (Note 11) and dried under vacuum for 2 h (5 mmHg) furnishing 3.40 g (32%) of **2** as a light tan solid (Figure 4) (Notes 12 and 13).

Figure 4. Product (2) as a light tan solid

Notes

1. Methanol (MeOH, HPLC Grade) was obtained from Sigma-Aldrich and used as received.
2. Sodium hydroxide (NaOH) was obtained from Fisher Scientific in pellet form and was used as received. For this procedure a pellet of NaOH was ground in a pestle with a mortar and the required amount weighed out into a small sample vial for transfer to the stirred MeOH solution.
3. D-Glucuronolactone **1** [D-(+)-glucuronic acid-□-lactone, >99%] was purchased from Sigma Aldrich and used as received.
4. D -Glucuronolactone **1** was added to the solution in 5 x 1.0 g portions in 5 minute intervals. The solution color may turn from clear to very pale yellow during this addition process.
5. Upon complete dissolution of **1** into solution the evaporation of MeOH should be completed with a rotary evaporator bath temperature not exceeding 30 °C as this may lead to decomposition of the intermediate free sugar. As the last traces of MeOH are removed, the residue may begin to foam inside the flask and the intermediate material will appear as a light yellow foam.

6. As necessary and depending on the heating setup in use, the flask can be fitted with a Y-shaped adapter for a flexible thermocouple or a 2-necked flask used.
7. Acetic anhydride (ReagentPlus, ≥99%) and anhydrous NaOAc (ReagentPlus, ≥99%) were both obtained from Sigma Aldrich and used as received.
8. The pale yellow foam is quite sticky and may not immediately dissolve upon addition of the acetic anhydride and NaOAc. Heating from ambient temperature up to 90 °C is required to achieve this. As the solvent warms, the reagents will dissolve and a pale brown suspension results. The mixture was heated using a Drysyn™ heating block and temperature thermocouple.
9. If lower heating temperature of 50 °C is applied for 4 h and then the solution is allowed to cool to room temperature overnight, the reaction delivers an improved yield of **2** (4.8–6.0 g, 45–60%) as white solid, but with a reduced diastereomeric preference for the *β* anomer. Experiments run at 50 °C gave approximately 3/7 (□/*β*) diastereomeric mixtures, as judged by ^{1}H NMR analysis.

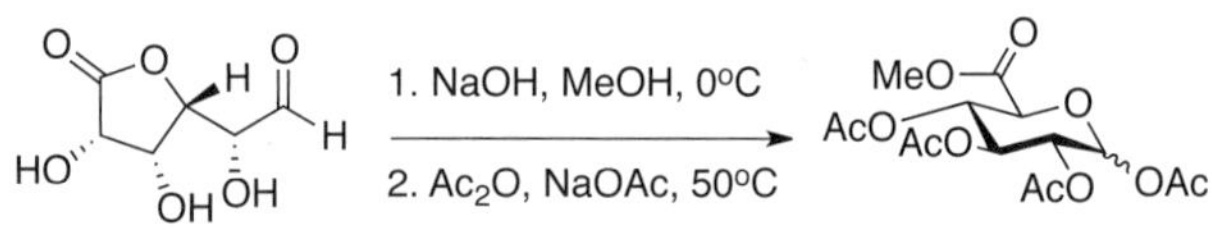

1 **2** α:β (3:7)

Scheme 1. Reaction performed at 50 °C

10. This step requires a careful addition of the filtrate to control the effervescence of CO_2 upon destruction of excess Ac_2O and neutralization of AcOH. Approximately 50 mL of filtrate should be added to the $NaHCO_3$ solution every 5 minutes.
11. Following filtration, the solids should be washed with 10-20 mL of ice-cold diethyl ether.
12. Analytical data for **2**[2]: TLC R_f 0.15 (3/1, hexane:EOAc) visualised with phosphomolybdic acid (PMA) stain. mp 175–177 °C; $[\alpha]_D^{20}$ +0.70° (*c* 0.55, CH_2Cl_2). D-(+)-Glucuronic acid γ-lactone $[\alpha]_D^{20}$ +18.0° (*c* 8.0, H_2O) (lit. 18.8°. IR (film): 2955, 1751, 1371, 1202, 1144, 1115, 1087, 1177, 1037 cm^{-1}; ^{1}H NMR (400 MHz, $CDCl_3$) δ: 2.04 (s, 3H), 2.05 (s, 6H), 2.13 (s, 3H), 3.75 (s, 3H), 4.19 (d, *J* = 9.4 Hz, 1H), 5.15 (t, *J* = 8.3 Hz, 1H), 5.26 (t, *J* = 9.3

Hz, 1H), 5.32 (t, J = 9.1 Hz, 1H), 5.78 (d, J = 7.8 Hz, 1H); ^{13}C NMR (100 MHz, $CDCl_3$) δ: 20.4, 20.5, 20.6, 20.8, 53.0, 68.9, 70.2, 71.8, 73.0, 91.4, 166.8, 168.8, 169.1, 169.4, 169.9. MS *m/z* 399 (100, [$M+Na^+$]); [$M+Na^+$] calcd for $C_{15}H_{20}NaO_{11}$: 399.08978. Found: 399.08980 (0.1 ppm). The β anomer is identified by a large ^{3}J (H_1-H_2) coupling (J = 7.8 Hz). The product can be stored on the shelf for prolonged periods.

13. A second run of the reaction on the same scale provided 3.45 g (32%) of the product.

Working with Hazardous Chemicals

The procedures in *Organic Syntheses* are intended for use only by persons with proper training in experimental organic chemistry. All hazardous materials should be handled using the standard procedures for work with chemicals described in references such as "Prudent Practices in the Laboratory" (The National Academies Press, Washington, D.C., 2011; the full text can be accessed free of charge at http://www.nap.edu/catalog.php?record_id=12654). All chemical waste should be disposed of in accordance with local regulations. For general guidelines for the management of chemical waste, see Chapter 8 of Prudent Practices.

In some articles in *Organic Syntheses*, chemical-specific hazards are highlighted in red "Caution Notes" within a procedure. It is important to recognize that the absence of a caution note does not imply that no significant hazards are associated with the chemicals involved in that procedure. Prior to performing a reaction, a thorough risk assessment should be carried out that includes a review of the potential hazards associated with each chemical and experimental operation on the scale that is planned for the procedure. Guidelines for carrying out a risk assessment and for analyzing the hazards associated with chemicals can be found in Chapter 4 of Prudent Practices.

The procedures described in *Organic Syntheses* are provided as published and are conducted at one's own risk. *Organic Syntheses, Inc.*, its Editors, and its Board of Directors do not warrant or guarantee the safety of individuals using these procedures and hereby disclaim any liability for any injuries or damages claimed to have resulted from or related in any way to the procedures herein.

Discussion

The synthesis of methyl 1,2,3,4-tetra-*O*-acetyl-□-D-glucopyranuronate **2** described builds on an original report by Bollenback *et al.*[2] In their article the authors prepare **2** using acid (periodic acid) or base (pyridine) catalyzed acetylation of the intermediate pyranouronate, obtained from methanolysis of **1**. While this is a useful procedure on large scale (the original scale was >2.3 mol, 400 g of **1**) we did not find it reliably reproducible on small to medium scales (often producing brown tars needing multiple repetitive chromatographic steps to isolate low % yields of **2**), and we also sought to eliminate the use of periodic acid and/or pyridine. The use of NaOAc in acetic anhydride proved a convenient alternative for the acetylation step and the process could be reliably scaled to 10 g of **1**. This method further provides an economic alternative to **2** derived from the commercially available, but more expensive D-glucuronic acid.[3]

D-Glucuronate **2** is a commonly required uronate building block (Scheme 2). For example, it has been used to access non-readily available L-IdoA units for the construction of complex heparin-related disaccharides,[4] *via* an intermediate C5-bromide (Scheme 2, A). It has also seen utility for access to and characterization of the important glucuronide-containing metabolite **5** (Scheme 2, C).[5] Importantly, **2** also serves as a vital precursor for the formation of alternative C-1 anomeric derivatives **4** (complicit to wider glycosylation applications and methodologies) including trichloroacetimidates,[5] hemi-acetals,[6] and glycosyl bromides[7] (Scheme 2, B).

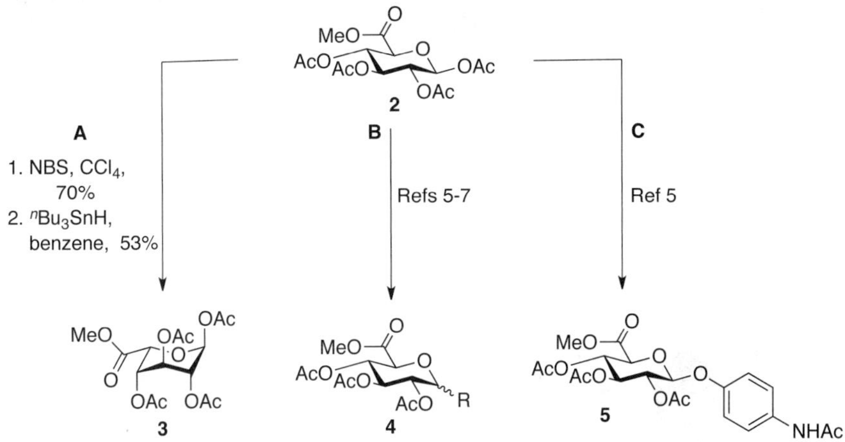

Scheme 2. Synthetic applications of glucuronate building block 2. R = Br (*via* $TiCl_4$, DCM), R = OH (*via* $^{n}Bu_3SnOMe$, benzene), R=$OC(O)CCl_3$ (*via* $ClC(O)CCl_3$, DBU, DCM).

References

1. Email: g.j.miller@keele.ac.uk. This work was supported by Keele University, Keele, Staffordshire, ST5 5BG, U.K.
2. Bollenback, G. N.; Long, J. W.; Benjamin, D. G.; Lindquist, J. A. *J. Am. Chem. Soc.* **1955**, *77*, 3310–3315.
3. Tosin, M.; Murphy, P. V. *Org. Lett.* **2002**, *4*, 3675–3678.
4. Yu, H. N.; Furukawa, J.-I.; Ikeda, T; Wong, C.-H. *Org. Lett.* **2004**, *6*, 723–726.
5. Hayes, J. A.; Eccles, K. S.; Lawrence, S. E.; Moynihan, H. A. *Carbohydr. Res.* **2012**, *349*, 108–112.
6. Nudelman, A.; Herzig, J.; Gottlieb, H. E.; Keinan, E.; Sterling, J. *Carbohydr. Res.* **1987**, *162*, 145–152.
7. Cheng, T.-C.; Roffler, S. R.; Tzou, S.-C.; Chuang, K.-H.; Su, Y.-C.; Chuang, C.-H.; Kao, C.-H.; Chen, C.-S.; Harn, I.-H.; Liu, K.-Y.; Cheng, T.-L.; Leu, Y.-L. *J. Am. Chem. Soc.* **2012**, *134*, 3103–3110.

Appendix
Chemical Abstracts Nomenclature (Registry Number)

D-Glucuronolactone: D-(+)-glucuronic acid-□-lactone; (32449-92-6)
Acetic anhydride: acetic anhydride; (108-24-7)
Sodium acetate: Sodium acetate; (127-09-3)

Gavin Miller was born in Newport, Wales and read a MChem in Chemistry at UMIST. He then undertook a Ph.D. in carbohydrate chemistry at the University of Manchester, followed by PDRA positions at Brown University, USA and at St Andrews, UK. Gavin then moved to industry, firstly at Ferring Pharmaceuticals and secondly at Peakdale Molecular. He returned to the University of Manchester in 2009 as a PDRA in the Manchester Institute of Biotechnology and secondly as a fixed-term lecturer in 2013, within the School of Chemistry. He took up his current Lectureship in Organic Chemistry at Keele University in 2015.

Aisling Ní Cheallaigh undertook a BSc. in Medicinal Chemistry and Chemical Biology at University College Dublin (UCD) and graduated with first class honours. Upon receipt of a prestigious IRCSET scholarship she began her postgraduate studies through which she obtained a Ph.D. in Synthetic Carbohydrate Chemistry. She then held a PDRA position at the University of Manchester carrying out organic synthesis of monoterpenoid derivatives for use in a synthetic biology programme. She currently holds a Senior Scientist position in Chemical Development at AstraZeneca.

Garrett Potter pursued a B.S. in Biochemistry/ Chemistry at The University of California, San Diego (UCSD). He then worked for biopharmaceutical start-up companies Actimis Pharmaceuticals and Axikin Pharmaceuticals prior to pursuing postgraduate studies with a Ph.D. in carbohydrate chemistry at the University of Manchester. He held a PDRA position at the University of Manchester, synthesizing carbohydrate building blocks toward heparan sulfate-derived bioactive scaffolds. Currently, he is a Postdoctoral Research Scholar at Stanford University in the ChEM-H Medicinal Chemistry Knowledge Center.

John Gardiner received his undergraduate education and his Ph.D. (Martin Bryce's lab) working on total synthesis of alkaloids, from Durham University. He joined Mike Jung's group at the University of California, Los Angeles (UCLA) for 2 years working on nucleoside and de novo carbohydrate syntheses. He moved to Manchester (UMIST) then the University of Manchester and Manchester Institute of Biotechnology (MIB) where his group's research has focused on carbohydrate methods, oligosaccharide syntheses, glycosylaminoglycans and mimetics, biocatalytic reductions, synthesis / biotransformations of terpenoids, polyaromatic dendrimers and heterocyclic medicinal chemistry.

Michal Achmatowicz received his undergraduate education at the University of Warsaw, Poland where he completed his undergraduate thesis in Chemistry in 1997. He then joined Professor Janusz Jarczak's research group at the Institute of Organic Chemistry of the Polish Academy of Sciences in Warsaw to pursue his Ph.D. in organic chemistry. From 2001 to 20013 he was a post-doctoral research fellow with Professor Louis S. Hegedus at the Colorado State University. Subsequently he joined the Chemical Process Research and Development at Amgen at Thousand Oaks, California, where he has been developing robust processes toward active pharmaceutical ingredients, coauthoring several publications, and enjoying rock climbing in his spare time.

Organic Syntheses

Preparation of 1,5-Dioxaspiro[5.5]undecan-3-one

Robert A. Craig, II, Russell C. Smith, Beau P. Pritchett, Benzi I. Estipona, and Brian M. Stoltz*[1]

The Warren and Katharine Schlinger Laboratory for Chemistry and Chemical Engineering, Division of Chemistry and Chemical Engineering, California Institute of Technology, 1200 East California Boulevard, MC 101-20, Pasadena, California 91125, United States

Checked by Marcel Weiss and Mohammad Movassaghi

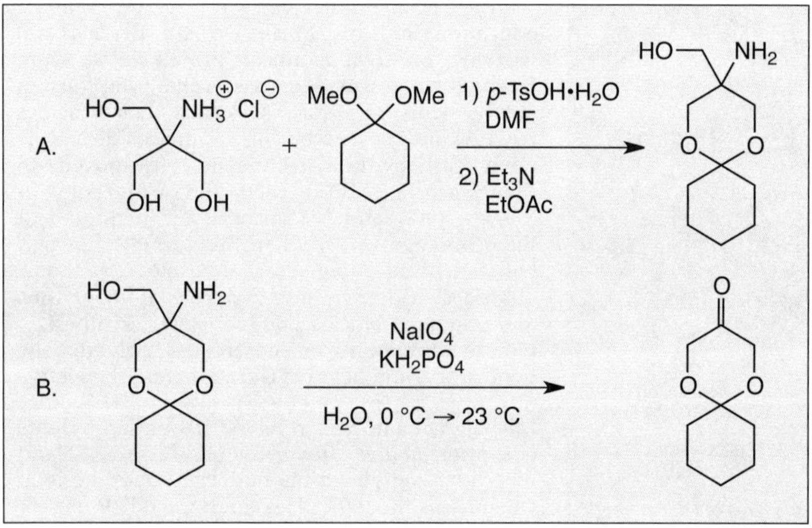

Procedure

A. *3-Amino-3-(hydroxymethyl)-1,5-dioxaspiro[5.5]undecane*. A 1-L single-necked, round-bottomed flask is equipped with an egg-shaped, Teflon®-coated magnetic stirring bar (3.5 cm x 1.5 cm), capped with a rubber septum, flame-dried under vacuum, and cooled under an argon atmosphere (Note 1). After cooling to ambient temperature (21–23 °C), to the flask is added anhydrous *N,N*-dimethylformamide (DMF, 365 mL, 0.78 M) via cannula. Subsequently, tris(hydroxymethyl)aminomethane hydrochloride (45.0 g, 286 mmol, 1.00 equiv) (Note 2 and 3) is added in a single portion as white crystalline solid. The reaction vessel is immediately resealed with a rubber septum under inert atmosphere and stirring is commenced (Figure 1). To this white suspension is added 1,1-dimethoxycyclohexane (50.0 mL, 47.4 g, 329 mmol, 1.15 equiv) via syringe in one portion (Note 4). Lastly, to the off-white slurry is added *para*-toluenesulfonic acid monohydrate (p-TsOH•H_2O, 1.63 g, 8.57 mmol, 0.03 equiv) as a solid in one

Figure 1. Heterogeneous Reaction Mixture at Initiation of Transketalization

portion quickly, immediately replacing the rubber septum to maintain an inert atmosphere. After stirring for 18 h at ambient temperature, the off-white heterogeneous reaction mixture becomes a completely homogenous, pale yellow solution (Figure 2, Note 5). The reaction flask is then placed in an oil bath and the rubber septum is replaced with a short path distillation head with a hollow water-jacketed column 6.0 cm in length, equipped with a thermometer and a 500-mL single-necked, round-bottomed receiving flask being cooled in an ice-water bath (Figure 3, Note 6). The volatiles are distilled off under high vacuum with heating (Note 7) and vigorous stirring (Note 8) to furnish an extremely viscous off-white semi-solid (Figure 4).

Figure 2. Homogeneous Reaction Mixture at Completion of Transketalization

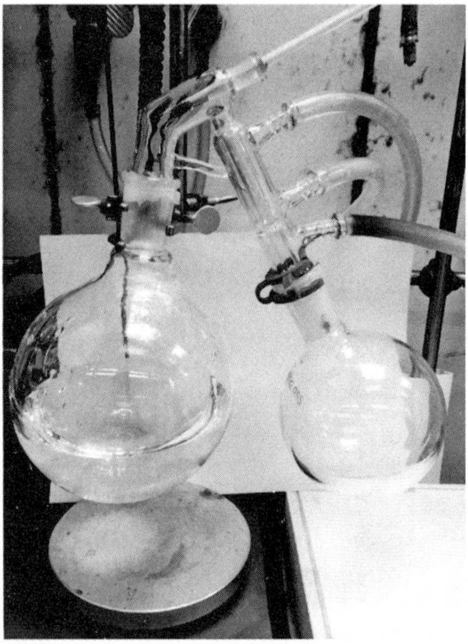

Figure 3. Distillation Apparatus Before Introduction of Heating and Cooling Baths (Baths Omitted for Visual Clarity)

This residue is triturated with diethyl ether (700 mL), resulting in the precipitation of a white solid (Note 9). The heterogeneous suspension is then filtered through a medium porosity sintered glass funnel (Note 10). The solid is then washed with a single 100 mL portion of ethyl acetate. The lumpy white solid is then briefly dried for 2–3 min by pulling air through the sintered glass funnel and then ground into a fine powder using a mortar and pestle (Note 11). Once thoroughly ground, the white solid is triturated with three 100 mL portions of ethyl acetate in a medium porosity sintered glass funnel (Note 10). The resultant white powder is dried under high vacuum (23 °C, 0.40 mmHg) for 12 h to afford an amorphous white solid (63 g).

Figure 4. Semi-Solid Product After Distillation

Following trituration, this white solid (63 g) is split into two equal portions and each portion is added to a separate 1-L single-necked, round-bottomed flask equipped with an egg-shaped, Teflon®-coated magnetic stirring bar (3.5 cm x 1.5 cm) (Note 12). To each of these round-bottomed flasks is added ethyl acetate (EtOAc, 400 mL) (Note 13) and stirring commenced. To each white, heterogeneous suspension is then added triethylamine (Et_3N, 23.7 mL, 179 mmol) (Note 14) dropwise over 6 min. After 12 h, each white, heterogeneous reaction mixture is filtered through a medium porosity sintered glass funnel (Note 10), collecting the filtrate in a

1-L round-bottomed flask. For each portion, the filter cake ("Solid A") is subsequently washed with ethyl acetate (4 x 50 mL). The washes are combined with the filtrate, concentrated by rotary evaporation (23 °C, 13 mmHg), and dried under high vacuum (23 °C, 0.40 mmHg) for 12 h to afford a fluffy white solid (the free-base amino alcohol, "Solid B"). The two filter cakes (Solid A) are then combined in a 1-L single-necked, round-bottomed flask equipped with an egg-shaped, Teflon®-coated magnetic stirring bar (3.5 cm x 1.5 cm) and suspended in ethyl acetate (250 mL) with stirring. Triethylamine (18.0 mL, 129 mmol) (Note 15) is subsequently added dropwise over 3 min. The pure white suspension is stirred vigorously for 24 h and then filtered through a medium porosity sintered glass funnel (Note 10). The filter cake is then washed with ethyl acetate (4 x 75 mL). The filtrate and washes are added to the flask containing Solid B and then concentrated by rotary evaporation (23 °C, 13 mmHg) to furnish 3-amino-3-(hydroxymethyl)-1,5-dioxaspiro[5.5]undecane (42.6 g, 74% combined yield over the two iterations, melting point: 91–92 °C) as an amorphous white solid (Notes 16 and 17).

B. *1,5-Dioxaspiro[5.5]undecan-3-one.* 3-Amino-3-(hydroxymethyl)-1,5-dioxaspiro[5.5]undecane (31.4 g, 156 mmol, 1.00 equiv) (Note 18) is added to a 2-L three-necked, round-bottomed flask equipped with an egg-shaped, Teflon®-coated magnetic stirring bar (3.5 cm x 1.5 cm) and suspended in water (520 mL). The flask is fitted with a pressure-equalizing 250-mL addition funnel capped with a rubber septum in the middle neck, a rubber septum equipped with a thermometer in one of the side necks, and capped with a rubber septum in the third neck (Figure 5). Subsequently, potassium dihydrogen phosphate (KH_2PO_4, 25.5 g, 187 mmol, 1.20 equiv) is added. Stirring is commenced and the heterogeneous white suspension is cooled in an ice-water bath (Note 19). To the resultant stirred heterogeneous reaction mixture is added a solution of sodium (meta)periodate ($NaIO_4$, 33.4 g, 156 mmol, 1.00 equiv) as a pale yellow solution in water (520 mL, 0.30 M) dropwise over 3 h through the addition funnel (Note 20). The reaction is then allowed to stir for 1.0 h before removing the ice-water bath (Note 21) and allowing the homogenous, pale yellow reaction mixture to warm slowly. The starting material is completely consumed 3.0 h after removing the cooling bath as determined by TLC (Note 22, Figure 6). Anhydrous sodium thiosulfate (24.7 g, 156 mmol, 1.00 equiv) (Note 23) is then immediately added in one portion (Note 24). The solution is allowed to stir for 20 min at which time the pale yellow homogenous reaction mixture is transferred to a 2-L separatory funnel. The aqueous reaction mixture is extracted with dichloromethane (12 x 150 mL) (Note 25). The organic layers are combined in a 2-L Erlenmeyer flask equipped with a cylindrical, Teflon®-coated magnetic stirring bar (7.4 cm x 1.2 cm) and dried over 400 g of anhydrous sodium sulfate for 20 min with stirring. The organic solution is subsequently filtered through grade 4 filter paper and concentrated in vacuo (23 °C, 50 mmHg) and under high vacuum (23 °C, 0.4 mmHg, 20 min) to afford 1,5-dioxaspiro[5.5]undecan-3-one (25.5 g, 96% yield, boiling point: 82–88 °C, 0.60 mmHg) as a pale yellow oil (Notes 26, 27, 28, 29, and 30).

Figure 5. Oxidative Cleavage Reaction Set-up Before Cooling

Figure 6. Oxidative Cleavage After Completion Before Quenching

Notes

1. The reaction may alternatively be run under a nitrogen atmosphere with no deleterious effects.
2. Tris(hydroxymethyl)aminomethane hydrochloride (Trizma® hydrochloride, ≥99%, crystalline, Sigma-Aldrich), 1,1-dimethoxycyclohexane (99%, Sigma-Aldrich), *para*-toluenesulfonic acid monohydrate (≥98.5%, ACS reagent grade, Sigma-Aldrich), potassium dihydrogen phosphate (anhydrous, 99.8%, ACS reagent grade, Mallinckrodt), sodium (meta)periodate (≥99%, Sigma-Aldrich), anhydrous sodium thiosulfate (99%, Sigma-Aldrich), and anhydrous sodium sulfate (≥99%, granular (10-60 mesh), Macron; Submitters: ≥99%, powder, Fisher Scientific) are purchased and used as received. *N,N*-Dimethylformamide (DMF, ≥99.9%, HPLC grade, Sigma-Aldrich;

Submitters: ≥99.8%, ACS reagent grade, Mallinckrodt) is degassed with argon and purified by passage through two columns of molecular sieves (9 cm x 60 cm) under argon. Triethylamine (≥99.5%, EMD; Submitters: ≥99%, Sigma-Aldrich) is degassed with argon and purified by passage through two columns of molecular sieves (9 cm x 60 cm) under argon (Submitters: triethylamine distilled from calcium hydride prior to use). Water is distilled and purified with a Barnstead NANOpure Infinity UV/UF system and had a pH of ca. 7 (Submitters: water pH between 5 and 6). Diethyl ether (≥99%, anhydrous, ACS reagent grade, Fisher Scientific; Submitters: ≥99%, ACS reagent grade, EMD), ethyl acetate (≥99.5%, ACS reagent grade, Sigma-Aldrich; Submitters: ≥99.5%, ACS reagent grade, EMD), and dichloromethane (≥99.5%, ACS reagent grade, Sigma-Aldrich; Submitters: ≥99.5%, ACS reagent grade, EMD) are purchased and used as received without needing to remove dissolved gasses or water from the solvent. For the purpose of purity analysis, dimethyl fumarate (TraceCERT® certified reference material, Sigma-Aldrich) is used as an internal standard.

3. Crystalline tris(hydroxymethyl)aminomethane hydrochloride used for this procedure from Sigma-Aldrich arrived with some large chunks of material mixed in with fine white crystals. These large chunks are broken up prior to use, using a mortar and pestle if necessary, to ensure efficient transketalization since the material is not fully soluble in the *N,N*-dimethylformamide reaction solvent.
4. Varying the rate of addition of 1,1-dimethoxycyclohexane to the reaction flask has provided no noticeable effect on the yield of the reaction or the purity of the 3-amino-3-(hydroxymethyl)-1,5-dioxaspiro[5.5]undecane.
5. Extended reaction times up to 30 h have shown no deleterious effects on the overall yield or purity of 3-amino-3-(hydroxymethyl)-1,5-dioxaspiro[5.5]undecane.
6. The internal temperature of the receiving flask is not monitored, however, the temperature of the surrounding ice-water bath is between 3 °C and 5 °C.
7. Reaction flask is initially placed under vacuum CAREFULLY since at ambient temperature (21–23 °C) under reduced pressure (>1 mmHg), the methanol byproduct from the transketalization will boil vigorously. Only after all of the methanol has been distilled off is the reaction flask heated to facilitate the distillation of the *N,N*-dimethylformamide. Typically, the oil bath is heated to between 50 °C to 65 °C depending on

the high vacuum manifold pressure, resulting in the distillation of the *N*,*N*-dimethylformamide at an indicated temperature range from 35 °C to 39 °C in the distillation head.

8. Vigorous stirring is necessary to ensure full removal of the volatiles by distillation. Without vigorous stirring, the pale yellow semi-solid resulting from the distillation process will trap *N*,*N*-dimethylformamide which can lead to the contamination of the aminoalcohol product.
9. This step is *critical* to achieving optimal yield. The trituration process will take some time as the semi-solid pale yellow gel that results from the distillation process hardens when cooled to ambient temperature. The semi-solid should be manually ground and triturated until no pale yellow semi-solid remains either on the walls of the flask or suspended in solution.
10. A medium porosity sintered glass funnel refers to a sintered glass funnel with 10–15 μm sintered glass mesh.
11. In order to ensure complete removal of *N*,*N*-dimethylformamide and *para*-toluenesulfonic acid, the white solid must be ground as finely as possible. If these impurities remain, however, they will not affect the subsequent reactions and can be separated from 1,5-dioxaspiro[5.5]undecan-3-one by filtration through silica gel (see Note 30).
12. No precautions are taken to exclude either water or air from the reaction flask. The flask is neither dried prior to use nor placed under an inert atmosphere. The product has been split into two portions to allow for more vigorous and efficient stirring of the heterogeneous reaction mixture.
13. Assuming quantitative yield after the first step of the crude amorphous white solid intermediate, this volume of ethyl acetate corresponds to a concentration of 0.36 M. As a reminder, the ethyl acetate used as the solvent for this step is purchased and used as received without the need to remove dissolved gasses or water from the solvent.
14. Assuming quantitative yield after the first step of the crude amorphous white solid intermediate, 1.25 equivalents of triethylamine are added.
15. Since the majority of the 3-amino-3-(hydroxymethyl)-1,5-dioxaspiro[5.5]undecane is isolated after the first free base procedure and filtration, a large excess of triethylamine is added here to help drive any remaining hydrochloride salt of 3-amino-3-(hydroxymethyl)-1,5-dioxaspiro[5.5]undecane to the desired product. Smaller amounts of triethylamine may be used here, however, this will result in a slightly

reduced isolated yield of 3-amino-3-(hydroxymethyl)-1,5-dioxaspiro[5.5]undecane.

16. *3-Amino-3-(hydroxymethyl)-1,5-dioxaspiro[5.5]undecane*: R_f = 0.17 (1:4 methanol:dichloromethane eluent); ^{1}H NMR (400 MHz, $CDCl_3$) δ: 1.36–1.45 (*m*, 2H), 1.45–1.60 (*m*, 4H), 1.68–1.81 (*m*, 4H), 2.01 (*bs*, 2H), 2.04 (*bs*, 1H), 3.48 (*s*, 2H), 3.52 (*d*, J = 11.9 Hz, 2H), 3.79 (*d*, J = 11.9 Hz, 2H), ^{13}C NMR (101 MHz, $CDCl_3$) δ: 22.6, 25.7, 31.1, 33.8, 50.5, 64.7, 66.4, 98.6. IR (film): 3350, 3294, 2938, 2860, 1619, 1449, 1331, 1285, 1255, 1160, 1110, 1087, 918 cm^{-1}. HRMS (MM: ESI-APCI) *m/z* calcd for $C_{10}H_{20}NO_3$ $[M+H]^+$: 202.1438, found 202.1431; Elemental anal. calcd for $C_{10}H_{19}NO_3$: C, 59.68; H, 9.52; N, 6.96; O, 23.85, found: C, 59.67; H, 9.62; N, 6.86; O, 23.85. The product is indefinitely stable at ambient temperature stored under an inert atmosphere (argon or nitrogen).
17. A second run on half-scale provided 21.2 g (74%) of the product.
18. Only a portion of the 3-amino-3-(hydroxymethyl)-1,5-dioxaspiro[5.5]undecane from Step A is advanced for ease in using glassware no larger than 2-L in volume.
19. The internal temperature of the reaction mixture is measured between 2 °C and 3 °C after cooling.
20. The internal temperature of the reaction mixture during addition ranges between 1 °C and 3 °C.
21. The internal temperature of the reaction mixture is between 0 °C and 1 °C.
22. Progress of the reaction can be monitored by TLC analysis on silica gel using 20% methanol in dichloromethane eluent with visualization by potassium permanganate. The aminoalcohol starting material has an R_f = 0.17 and the ketone product has an R_f = 0.90. At the time of the reaction quench, 3.0 h after removing the reaction vessel from the cooling bath, the internal temperature is between 16 °C and 17 °C.
23. The internal temperature of the reaction warms to between 18 °C and 19 °C after the addition of anhydrous sodium thiosulfate. Alternatively, sodium thiosulfate pentahydrate may also be used without any deleterious effects on the overall efficacy of the described transformation.
24. This portion of the procedure is mildly *time sensitive*. Stirring overnight should be avoided, as prolonged exposure of 1,5-dioxaspiro[5.5]undecan-3-one to the reaction conditions will result in decomposition by ketal cleavage, as evidenced by the presence of

cyclohexanone, which can be observed in the ^{1}H NMR spectrum of the product.

25. Washes of the aqueous layer with dichloromethane should be followed by TLC analysis on silica gel (20% diethyl ether in hexanes eluent, *p*-anisaldehyde stain, R_f = 0.38) to ensure the product has been completely extracted. The product is persistent in the aqueous layer, and remains visible by TLC until the twelfth extraction.
26. *1,5-Dioxaspiro[5.5]undecan-3-one*: R_f = 0.90 (1:4 methanol: dichloromethane eluent) and 0.38 (1:4 diethyl ether:hexanes eluent); ^{1}H NMR (400 MHz, $CDCl_3$) δ: 1.40–1.49 (*m*, 2H), 1.55–1.64 (*m*, 4H), 1.70–1.77 (*m*, 4H), 4.18 (*s*, 4H), ^{13}C NMR (101 MHz, $CDCl_3$) δ: 22.8, 25.2, 32.5, 66.5, 99.8, 208.0; IR (film): 2938, 2863, 1751, 1448, 1435, 1425, 1369, 1338, 1281, 1264, 1239, 1201, 1162, 1146, 1118, 1079, 1058, 1028, 922, 847, 825 cm^{-1}; HRMS calcd for $C_9H_{15}O_3$ $[M+H]^+$: 171.1016, found 171.1008; Elemental anal. calcd for $C_9H_{14}O_3$: C 63.51, H 8.29, O 28.20, found C 63.53, H 8.49, O 27.98; Quantitative NMR Analysis: title compound (20.4 mg) with dimethylfumarate standard (19.7 mg) in $CDCl_3$, 100.0% pure (see ^{1}H NMR spectrum below). The product may be stored at –20 °C under an inert atmosphere (argon or nitrogen) for >3 months, however, prolonged storage or storage at higher temperature will result in slow decomposition by ketal cleavage.
27. A second run on half-scale provided 12.7 (96%) of the product.
28. After storage overnight in a freezer at –20 °C as a neat oil, 1,5-dioxaspiro[5.5]undecan-3-one solidifies to become a tan solid that does not melt upon warming back to ambient temperature (ca. 23 °C). Melting point of this tan solid: 28–30 °C.
29. Residual dichloromethane can be challenging to remove from the title compound. To aid in the removal of this impurity, the dioxanone product can be diluted with pentane (ca. 1.5 mL pentane/g of dioxanone) and then concentrated in vacuo (23 °C, 13 mmHg) and under high vacuum (23 °C, 0.40 mmHg, 5 min). After repeating this procedure two additional times, the removal of residual dichloromethane will be complete.
30. The procedure, as written, will afford the title compound in analytically pure form. Neither column chromatography nor distillation are necessary. However, if traces of *N,N*-dimethylformamide or *para*-toluenesulfonic acid from Step A remain with 1,5-dioxaspiro[5.5]undecan-3-one after aqueous work-up, the impure product may be filtered through a pad of silica gel (5 cm diameter, 3 cm

depth) in a medium porosity sintered glass funnel (see Note 10) as a solution in dichloromethane (200 mL), rinsing the silica gel with dichloromethane (100 mL) after filtration. This procedure typically provides 1,5-dioxaspiro[5.5]undecan-3-one in a slightly reduced isolated yield from 3-amino-3-(hydroxymethyl)-1,5-dioxaspiro-[5.5]undecane. Attempts to purify 1,5-dioxaspiro[5.5]undecan-3-one by either short-path distillation or bulb-to-bulb distillation (82–88 °C, 0.60–0.70 mmHg) result in partial decomposition of the dioxanone product by ketal cleavage and contamination of the distilled product with cyclohexanone.

Working with Hazardous Chemicals

The procedures in *Organic Syntheses* are intended for use only by persons with proper training in experimental organic chemistry. All hazardous materials should be handled using the standard procedures for work with chemicals described in references such as "Prudent Practices in the Laboratory" (The National Academies Press, Washington, D.C., 2011; the full text can be accessed free of charge at http://www.nap.edu/catalog.php?record_id=12654). All chemical waste should be disposed of in accordance with local regulations. For general guidelines for the management of chemical waste, see Chapter 8 of Prudent Practices.

In some articles in *Organic Syntheses*, chemical-specific hazards are highlighted in red "Caution Notes" within a procedure. It is important to recognize that the absence of a caution note does not imply that no significant hazards are associated with the chemicals involved in that procedure. Prior to performing a reaction, a thorough risk assessment should be carried out that includes a review of the potential hazards associated with each chemical and experimental operation on the scale that is planned for the procedure. Guidelines for carrying out a risk assessment and for analyzing the hazards associated with chemicals can be found in Chapter 4 of Prudent Practices.

The procedures described in *Organic Syntheses* are provided as published and are conducted at one's own risk. *Organic Syntheses, Inc.*, its Editors, and its Board of Directors do not warrant or guarantee the safety of individuals using these procedures and hereby disclaim any liability for any injuries or damages claimed to have resulted from or related in any way to the procedures herein.

Discussion

The procedure for the synthesis of 1,5-dioxaspiro[5.5]undecan-3-one is based on our previous report[2] and was originally adapted from the work of Forbes[3] and inspired by the work of Majewski,[4] Hoppe,[5] and Enders[6] on the synthesis of related ketodioxanone scaffolds. Using the procedure described herein, the synthesis of the analogous diethyl ketal product has been accomplished using 3,3-dimethoxypentane in place of 1,1-dimethoxycyclohexane.[7] This modification provided the corresponding ketodioxanone in a comparable yield to 1,5-dioxaspiro[5.5]undecan-3-one.

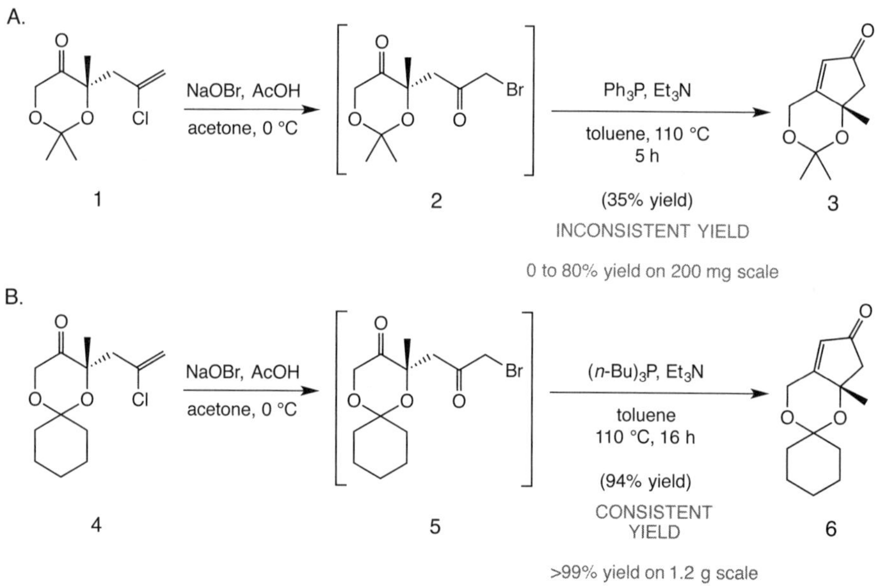

Scheme 1. Effect of Ketal Identity on Oxidative α-Bromination

The title compound has been previously synthesized by another method.[8] It has been employed successfully as a nucleophile in palladium-catalyzed enantioselective allylic alkylation[2] as well as stereoselective aldol condensation.[9] As a synthetic building block, 1,5-dioxaspiro[5.5]undecan-3-one offers the advantage of a ketal that is significantly more stable to acidic reaction conditions in comparison to the acetonide and diethylketal orthologues. For example, exposure of acetonide **1** to acidic oxidative α-bromination conditions furnished α-bromoketone **2** in widely variable yields, most typically between 0% and 20% (Scheme 1.A).[2] Although the conversion of α-bromoketone **2** to cyclopentenone **3** was accomplished, the labile nature of the acetonide contributed to the diminished yield.

Alternatively, employment of cyclohexyl ketal analog **4** enabled consistent quantitative conversion to α-bromoketone **5** on gram scale (Scheme 1.B).[2] Under optimized reaction conditions, α-bromoketone **5** was converted to cyclopentenone **6** in excellent yield. Although the comparatively robust cyclohexyl ketal had enabled the efficient formation of the cyclopentenone scaffold (e.g. **6**), the protecting group was easily removed from benzoyl ester **7** under relatively mild acidic transketalization conditions to provide diol **8** in 97% yield (Scheme 2).

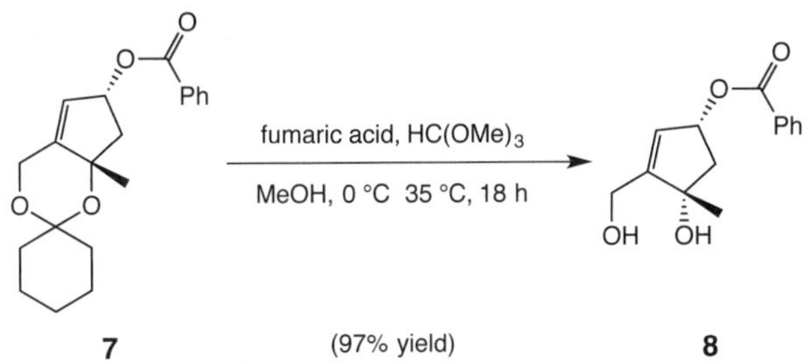

Scheme 2. Removal of Cyclohexyl Ketal Under Mild Acidic Conditions

Ketal orthologues of the title compound have been studied more extensively than the title compound itself, showing the utility of ketodioxanones as nucleophiles in enantioselective palladium-catalyzed allylic alkylations,[2,10] diastereoselective alkylations,[6,11] α,α′-annulations,[12] and organocatalyic stereoselective aldol condensations.[13] The products of these transformations have been successfully applied to the synthesis of a variety of simple and complex carbohydrates[13b,f,g,k] and biologically active natural products.[10a,c,11 a–b,12a,13a,c–d,h,14]

References

1. The Warren and Katharine Schlinger Laboratory for Chemistry and Chemical Engineering, Division of Chemistry and Chemical

Engineering, California Institute of Technology, 1200 East California Boulevard, MC 101-20, Pasadena, California 91125, United States. stoltz@caltech.edu; Ph. (626) 395-6064; Fax (626) 395-8436. The authors wish to thank the NIH NIH-NIGMS (R01M080269—01), Amgen, and Caltech for financial support. R.A.C. gratefully acknowledges the support of this work provided by a fellowship from the National Cancer Institute of the National Institutes of Health under Award Number F31A17435. B.P.P. thanks the NSF for a predoctoral fellowship. B.I.E. thanks the Caltech Summer Undergraduate Research Fellowship (SURF) office for funding.

2. Craig, R. A., II; Roizen, J. L.; Smith, R. C.; Jones, A. C.; Stoltz, B. M. *Org. Lett.* **2012**, *14*, 5716–5719.
3. Forbes, D. C.; Ene, D. G.; Doyle, M. P. *Synthesis* **1998**, 879–882.
4. Majewski, M.; Gleave, D. M.; Nowak, P. *Can. J. Chem.* **1995**, *73*, 1616–1626.
5. Hoppe, D.; Schmincke, H.; Kleemann, H.-W. *Tetrahedron* **1989**, *45*, 687–694.
6. Enders, D.; Bockstiegel, B. *Synthesis* **1989**, 493–496.
7. Craig, R. A., II. Progress Toward the Enantioselective Total Synthesis of Ineleganolide and the Polycyclic Norcembranoid Diterpenes and Construction of the Ineleganoloids. Ph.D. dissertation, California Institute of Technology, Pasadena, CA, 2015.
8. Kim, K. S.; Hong, S. D. *Tetrahedron Lett.* **2000**, *41*, 5909–5913.
9. (a) Suri, J. T.; Mitsumori, S.; Albertshofer, K.; Tanaka, F.; Barbas, C. F., III. *J. Org. Chem.* **2006**, *71*, 3822–3828. (b) Majewski, M.; Niewczas, I.; Palyam, N. *Synlett* **2006**, 2387–2390. (c) Suri, J. T.; Ramachary, D. B.; Barbas, C. F., III. *Org. Lett.* **2005**, *7*, 1383–1385.
10. (a) Liu, Y.; Liniger, M.; McFadden, R. M.; Roizen, J. L.; Malette, J.; Reeves, C. M.; Behenna, D. C.; Seto, M.; Kim, J.; Mohr, J. T.; Virgil, S. C.; Stoltz, B. M. *Beilstein J. Org. Chem.* **2014**, *10*, 2501–2512. (b) Behenna, D. C.; Mohr, J. T.; Sherden, N. H.; Marinescu, S. C.; Harned, A. M.; Tani, K.; Seto, M.; Ma, S.; Novak, Z.; Krout, M. R.; McFadden, R. M.; Roizen, J. L.; Enquist, Jr., J, A.; White, D. E.; Levine, S. R.; Petrova, K. V.; Iwashita, A.; Virgil, S. C.; Stoltz, B. M. *Chem. Eur. J.* **2011**, *17*, 14199–14223. (c) Seto, M.; Roizen, J. L.; Stoltz, B. M. *Angew. Chem., Int. Ed.* **2008**, *47*, 6873–6876.
11. (a) Enders, D.; Reichenbach, L. F. *Synthesis* **2013**, *45*, 959–965. (b) Enders, D.; Barbion, J. *Chem. Eur. J.* **2008**, *14*, 2842–2849. (c) Enders, D.; Breuer, I.; Drosdow, E. *Synthesis* **2005**, 3239–3244.

12. (a) Cagide-Fagín, F.; Nieto-García, O.; Lago-Santomé, H.; Alonso, R. *J. Org. Chem.* **2012**, *77*, 11377–11382. (b) Casey, T. C.; Carlisle, J.; Tizzelli, P.; Male, L.; Spencer, N.; Grainger, R. S. *J. Org. Chem.* **2010**, *75*, 7461–7464.
13. (a) Trajkovic, M.; Balanac, V.; Ferjancic, Z.; Saicic, R. N. *RSC Adv.* **2014**, *4*, 53722–53724. (b) Bergeron-Brlek, M.; Teoh, T.; Britton, R. *Org. Lett.* **2013**, *15*, 3554–3557. (c) Pearson, A. J.; Panda, S.; Bunge, S. D. *J. Org. Chem.* **2013**, *78*, 9921–9928. (d) Marjanovic, J.; Divjakovic, V.; Matovic, R.; Ferjancic, Z.; Saicic, R. N. *Eur. J. Org. Chem.* **2013**, 5555–5560. (e) Sánchez, D.; Bastida, D.; Burés, J.; Isart, C.; Pineda, O.; Vilarrasa, J. *Org. Lett.* **2012**, *14*, 536–539. (f) Niewczas, I.; Majewski, M. *Eur. J. Org. Chem.* **2009**, 33–37. (g) Enders, D.; Narine, A. A. *J. Org. Chem.* **2008**, *73*, 7857–7870. (h) Enders, D.; Terteryan, V.; Palecek, J. *Synthesis* **2008**, 2278–2282. (i) Calderon, F.; Doyaguez, E. G.; Cheong, P. H.-Y.; Fernández-Mayoralas, A.; Houk, K. N. *J. Org. Chem.* **2008**, *73*, 7916–7920. (j) Doyaguez, E. G.; Calderón, F.; Sánchez, F.; Fernández-Mayoralas, A. *J. Org. Chem.* **2007**, *72*, 9353–9356. (k) Enders, D.; Grondal, C. *Angew. Chem., Int. Ed.* **2005**, *44*, 1210–1212.
14. (a) Kim, Y. S.; Lee, J. K.; Pae, A. N.; Cho, Y. S.; Min, S.-J. *Synlett* **2013**, *24*, 2292–2296. (b) Cagide-Fagín, F.; Alonso, R. *Eur. J. Org. Chem.* **2010**, 6741–6747. (c) Kim, H. C.; Kang, S. H. *Angew. Chem., Int. Ed.* **2009**, *48*, 1827–1829. (d) Davis, J. B.; Bailey, J. D.; Sello, J. K. *Org. Lett.* **2009**, *11*, 2984–2987.

Appendix
Chemical Abstracts Nomenclature (Registry Number)

Tris(hydroxymethyl)aminomethane hydrochloride: 1,3-Propanediol, 2-amino-2-(hydroxymethyl)-, hydrochloride (1:1); (1185-53-1)
1,1-Dimethoxycyclohexane: Cyclohexane, 1,1-dimethoxy-; (933-40-4)
para-Toluenesulfonic acid monohydrate: Benzenesulfonic acid, 4-methyl-, hydrate (1:1); (6192-52-5)
Triethylamine: Ethanamine, *N,N*-diethyl-; (121-44-8)
3-Amino-3-hydroxymethyl-1,5-dioxaspiro[5.5]undecane: 1,5-dioxaspiro[5.5]undecane-3-methanol, 3-amino-; (5452-79-9)
Dihydrogen potassium phosphate: Phosphoric acid, potassium salt (1:1); (7778-77-0)
Sodium (meta)periodate: Periodic acid (HIO_4), sodium salt (1:1); (7790-28-5)
1,5-Dioxaspiro[5.5]undecan-3-one; (158632-48-5)

Robert A. Craig II was born in Philadelphia, PA and received his B.S. degree from Davidson College in 2010. He earned his Ph.D. in 2015 from the laboratories of Professor Brian M. Stoltz, at the California Institute of Technology as an NIH predoctoral fellow, focusing on total synthetic efforts toward the polycyclic norcembranoid natural products. In the fall of 2015, he joined the laboratories of Professor Justin Du Bois at Stanford University where he is currently an NIH postdoctoral fellow. His research interests focus on the synthesis and utility of bioactive small molecules to address pertinent biological questions.

Russell C. Smith hails from central IL where he received his B.S. degree in chemistry from Illinois Wesleyan University in 2004 under the guidance of Professor Ram Mohan. His Ph.D. was earned from the University of Illinois Urbana-Champaign in 2010 under the supervision of Professor Scott E. Denmark, investigating the transmetalation step of arylsilanolate cross-coupling reactions. Post-doctoral studies were conducted at the California Institute of Technology with Professor Brian Stoltz on the total synthesis of ineleganolide. Following his educational studies, he joined the staff at Janssen Research and Development working as a medicinal chemist in Immunology.

Beau P. Pritchett received his B.S. degree in Chemistry and B.S.E. degree in Chemical Engineering from Tulane University in 2012. In the fall of 2012, he joined the laboratories of Professor Brian M. Stoltz at Caltech where he has pursued his Ph.D. as an NSF predoctoral fellow. His research interests include chemical synthesis, reaction design, and their applications in natural product synthesis and human medicine.

Benzi I. Estipona moved from Marion, Iowa to join the Class of 2017 at Caltech. He is currently a sophomore majoring in chemistry, and hopes to pursue a career in organometallic chemistry. He currently works as an undergraduate researcher during the summer and academic school year under the mentorship of Professor Brian M. Stoltz. His research is currently focused on the enantioselective total synthesis of the natural product eucomic acid and its glycosylated derivatives.

Brian M. Stoltz was born in Philadelphia, PA in 1970 and obtained his B.S. degree from the Indiana University of Pennsylvania in Indiana, PA. After graduate work at Yale University in the labs of John L. Wood and an NIH postdoctoral fellowship at Harvard in the Corey labs he took a position at the California Institute of Technology. A member of the Caltech faculty since 2000, he currently is a Professor of Chemistry. His research interests lie in the development of new methodology for general applications in synthetic chemistry.

Marcel Weiss was born in Stuttgart, Germany in 1985. He received his Diploma (2010) and his Ph.D. (2015) in Chemistry from the University of Stuttgart where he worked in the group of Professor René Peters on the cooperation of carbophilic Lewis acids in bimetallic catalysts. He then joined the group of Professor Mohammad Movassaghi at the Massachusetts Institute of Technology where his research is currently focused on the total synthesis of complex natural products.

Organic Syntheses

Oxyboration: Synthesis of Borylated Benzofurans

Darius J. Faizi, Nicole A. Nava, Mohammad Al-Amin, and Suzanne A. Blum*[1]

Department of Chemistry, University of California, Irvine, Irvine, California 92617-2025, United States

Checked by Nicholas A. Weires and Neil K. Garg

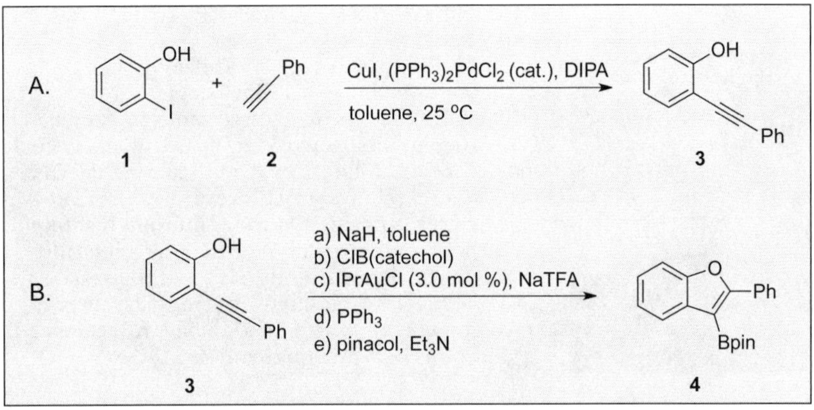

Procedure

Caution! Sodium hydride liberates hydrogen gas, which presents a fire risk. Care should be taken during handling and the chemical should be quenched with caution.

A. *2-(Phenylethynyl)phenol* (**3**). A flame-dried 1-L round-bottomed flask equipped with a stir bar (Note 1) is charged with 2-iodophenol (**1**, 14.0 g, 63.6 mmol, 1.00 equiv), copper iodide (1.09 g, 5.71 mmol, 0.09 equiv), and bis(triphenylphosphine)palladium dichloride (1.33 g, 1.89 mmol, 0.03 equiv) (Notes 2 and 3). The flask is then covered with a rubber septum. A needle is then placed in the septum and the flask is then flushed with nitrogen (Note 4). The flask is then kept under a dynamic nitrogen atmosphere.

DOI: 10.15227/orgsyn.093.0228

Toluene (320 mL) is then added via cannula, followed by diisopropylamine (8.90 mL, 63.6 mmol, 1.00 equiv) using a 24-mL syringe in one portion (Notes 5 and 6). Phenylacetylene (**2**) (10.5 mL, 95.4 mmol, 1.50 equiv) is then added via a 24-mL syringe over five min (Notes 7 and 8, Figure 1). The reaction is stirred for 8.5 h at 25 °C (Note 9).

Figure 1. Step A - Color before and after addition of 2

The septum is removed, and the reaction mixture is filtered through a plug of celite (Note 10). The mixture is then transferred to a 1-L separatory funnel. The filter flask is rinsed with 100 mL of ethyl acetate and added to the separatory funnel. The organic layer is rinsed with saturated aqueous ammonium chloride (1 × 50 mL), saturated aqueous brine (1 × 50 mL), and then transferred to a 1-L Erlenmeyer flask containing anhydrous sodium sulfate (55 g) and allowed to sit for 15 min. The solid sodium sulfate is then removed by filtration through a medium porosity 150-mL filter funnel (200 mmHg applied vacuum), and the resulting solution is collected into a 1-L round-bottomed flask and concentrated by rotary evaporation (35 °C, 10 mmHg) (Note 11). The dark brown oil is then purified by flash chromatography using an eluent mixture of 12% ethyl acetate in hexanes as the eluent (Note 12). The collected fractions are placed in a 250-mL round-bottomed flask and then concentrated by rotary evaporation (35 °C, 10 mmHg) and then placed under high vacuum (25 °C, 0.2 mmHg) for 13 h (Note 13). The brown residue is recrystallized from hexanes to afford 6.34 g

(51.4%) of 2-(phenylethynyl)phenol (**3**) as a brown crystalline solid (Notes 14, 15, 16, and 17).

B. *4,4,5,5-Tetramethyl-2-(2-phenylbenzofuran-3-yl)-1,3,2-dioxaborolane* (**4**). This procedure is carried out in an argon-filled glove box (Note 18). A 500-mL round-bottomed flask is charged with a stir bar, sodium hydride (0.81 g, 31 mmol, 1.0 equiv, 92% wt.), and toluene (192 mL) (Notes 19 and 20). In a separate 20-mL vial, compound **3** (6.1 g, 31 mmol, 1.0 equiv) is dissolved in toluene (15 mL). This solution is drawn into a 24-mL syringe with a long-stem needle. The vial is then rinsed with toluene (3 mL), which is also drawn into the same syringe. This solution is then added to the reaction flask over 30 min (Note 21, Figure 2). The reaction is then stirred for an additional 15 min.

Figure 2. Step B - Part a - Reaction mixture during addition of 3

In a separate 20-mL vial, *B*-chlorocatecholborane (4.8 g, 31 mmol, 1.0 equiv) is dissolved in toluene (15 mL) (Note 22). This solution is then drawn up into another 24-mL syringe equipped with a long stem needle. The vial is rinsed with toluene (3 mL), which is then drawn up into the same syringe. This combined solution is then added to the reaction mixture in the round-bottomed flask over 1 min and then the flask is capped with a glass stopper equipped with a Teflon ring. The reaction mixture is then stirred for 30 min (Note 23). In another 20-mL vial, 1,3-bis(2,6-diisopropylphenyl-imidazol-2-ylidene)gold(I) chloride, also known as IPrAuCl, (0.58 g, 0.94 mmol, 3.0 mol%) is dissolved in toluene (18 mL) (Note 24). This suspension is then transferred via pipette to the reaction mixture over 1 min. The vial is then rinsed with 10 mL toluene and this rinse is added via pipette to the reaction mixture. Sodium trifluoroacetate (1.3 g, 9.4 mmol, 0.30 equiv) is added to a 20-mL vial, and then added in one portion to the reaction mixture (Note 25). This vial is then rinsed with toluene (6 mL), and this toluene is then added to the reaction mixture via pipette. The round-bottomed flask is then capped with a glass stopper equipped with a Teflon ring, and a Keck plastic joint clip. The flask is placed in a pre-heated sand bath (60 °C) and allowed to stir in the glovebox for 21 h (Figure 3)

Figure 3. Step B - Part c - Reaction mixture heated to 60 °C

At this time, TLC indicated that the reaction is complete (Note 26). The mixture is then cooled to room temperature before triphenylphosphine (0.41 g, 1.6 mmol, 5.0 mol%) is added in one portion (Note 27). The reaction flask is then allowed to stir for 3 h. In a separate 100-mL round-bottomed flask, pinacol (11 g, 94 mmol, 3.0 equiv) is dissolved in triethylamine (65 mL, 470 mmol, 15 equiv) (Note 28). This solution is added to the reaction mixture via pipette over 4 min (Note 29). The flask is capped with a glass stopper, and the reaction mixture is stirred for 1 h. The flask is then taken out of the glovebox and the resulting suspension is filtered through celite, using toluene (200 mL) as a rinse (Note 30). The recovered liquid is transferred to a clean 1-L round-bottomed flask and concentrated by rotary evaporation (40 °C, 10 mmHg). The dark brown oil is then dissolved in dichloromethane (20 mL) and transferred to a 100-mL round-bottomed flask and then concentrated by rotary evaporation (30 °C, 10 mmHg). The mixture is then placed under high vacuum (0.2 mmHg) for 16 h (Note 31). Column chromatography using 20% dichloromethane in hexanes yielded 7.26 g (73%) of a light yellow solid after drying under high vacuum for 18 h (25 °C, 0.2 mmHg) (Notes 32, 33, and 34).

Notes

1. A cylindrical-shaped PTFE stir bar with dimensions of 3.7 cm x 0.9 cm x 1.0 cm was used.
2. The submitters sourced 2-iodophenol (98%) from two different suppliers. The submitters and checkers purchased 2-iodophenol (98%) from Acros and used it as received. Another batch was purchased by the submitters from Frontier Scientific, which was contaminated with a large amount of iodine. To remove the iodine, 10 g of the 2-iodophenol was dissolved in 300 mL of ethyl acetate and placed in a 500-mL separatory funnel. The organic layer was then washed with saturated aqueous sodium thiosulfate (3 × 50 mL), brine (1 × 50 mL), and then dried over sodium sulfate (55 g). The solution was then concentrated using rotary evaporation (40 °C, 10 mmHg) and then placed under high vacuum for 24 h (25 °C, 0.2 mmHg). The light yellow oil was then placed in the freezer for 1 h, and the resulting light yellow crystals were used in the reaction.

3. Bis(triphenylphosphine)palladium dichloride (98%) was purchased from Sigma Aldrich and used as received. Copper iodide (98%) was purchased from Strem Chemicals Inc., and used as received.
4. The needle used was a BD PrecisionGlide® disposable needle, 20 gauge, short bevel, purchased from VWR International.
5. Toluene (99.8%, HPLC grade) was purchased from Fisher Scientific and passed through a basic alumina push-still system before use. Diisopropylamine (99.95%) was purchased from Sigma Aldrich and sparged with nitrogen for 15 min before use.
6. After addition of the diisopropylamine, the solution turned from yellow to a red-orange color.
7. Phenylacetylene (98%) was purchased from Alfa Aesar and sparged with nitrogen for 15 min before use.
8. The checkers added phenylacetylene manually using a 20-gauge needle approximately 6 inches in length. The submitters utilized a syringe pump for this addition. After about 1 min, the solution changed from a red-orange to dark brown/black color.
9. The reaction was monitored via thin-layer chromatography on glass-backed silica gel TLC plates with a UV254 indicator, purchased from Merck Millipore International. 2-(Phenylethynyl)phenol, R_f = 0.42 (10% ethyl acetate in hexanes).
10. The filtration was conducted as follows: a medium porosity 350-mL filter funnel with a 9.0 cm I.D. was charged with 64.1 g of celite (Celite 545 was purchased from EMD Millipore and used as received). The solid was then pre-wet with 200 mL of toluene and then connected to a 1-L filter flask. The reaction mixture was then poured over the wet celite, and pressure was applied to pull the solution through the pad (200 mmHg). The celite pad was then rinsed with 200 mL of toluene, and filtered using applied pressure (200 mmHg).
11. During this step, the submitters decided to transfer the solution to a 250-mL round-bottomed flask to aid in the subsequent chromatography step.
12. The column chromatography was run as follows: A flash column with an outer diameter of 11 cm and capacity of 2-L (Pyrex No. 1280) was charged with silica (Davisil 35-70 micron silica purchased from Grace Davidson; used as received) using a wet-pack method (1.0 kg of silica in 1950 mL of hexanes). This gave a silica bed of 9 inches in height. The crude mixture was then dissolved in 50 mL of dichloromethane, and 15.0 g of silica was then added to the 250-mL round-bottomed flask

containing the crude mixture. This mixture was then slowly concentrated using a rotary evaporator (25–35 °C, 50–10 mmHg), yielding a light brown powder. This powder was then dry loaded into the column [Alternatively, the crude mixture could be liquid loaded using a pipette]. Sand was then added to fill 2 inches above the silica. The column was then loaded with the eluent mixture (88:12 hexanes: ethyl acetate). The flow rate was approximately 66 mL/min, and 2500 mL of the eluent were allowed to drain before 300-mL fractions were collected. Fractions 1–13 were then collected as the product fractions; fractions 14-20 were collected as co-elute.

13. The submitters found that trace ethyl acetate would rapidly dissolve the material, resulting in issues during the recrystallization. Placing the compound under high vacuum (25 °C, 0.2 mmHg) for 9 h would remove all of the ethyl acetate and make the recrystallization more effective.
14. The residue obtained from fractions 1–13 were recrystallized as follows: a 100-mL round-bottomed flask was charged with the brown solid, which was then dissolved in boiling hexanes (50 mL), capped with a ground glass stopper, and then allowed to cool to room temperature over 20 h. At this time, the round-bottomed flask was placed in an ice bath for 2 h, and then placed in the freezer (–20 °C) for an additional 2 h. The supernatant is filtered off at reduced pressure (200 mmHg) in a Buchner funnel (9 cm in diameter) with filter paper and the resulting brown solid crystals were washed with chilled hexanes (–20 °C, 1 × 5 mL) and then dried in a 250-mL round-bottomed flask under high vacuum (25 °C, 0.2 mmHg) for 6 h. The first crop of crystals weighed 2.51 g. The filtrate was then concentrated by rotary evaporation (25 °C, 10 mmHg) and then concentrated under high vacuum (25 °C, 0.2 mmHg) for 2 h. The recovered brown solid was then dissolved in boiling hexanes (20 mL), capped with a ground glass stopper, and then allowed to cool to room temperature over 20 h. The cooling and filtration process was repeated as above, and the resulting crystals weighed 1.83 g, with a combined final recovered mass of 4.34 g.
15. The residue obtained from fractions 14–20 (the co-eluted fractions) were recrystallized as follows: a 100-mL round-bottomed flask was charged with the brown solid, which was then dissolved in boiling hexanes (40 mL), capped with a ground glass stopper, and then allowed to cool to room temperature over 20 h. At this time, the round-bottomed flask was placed in an ice bath for 2 h, and then placed in the freezer (–20 °C)

for an additional 2 h. The supernatant is filtered off at reduced pressure (200 mmHg) in a Buchner funnel (9 cm in diameter) with filter paper and the resulting brown solid crystals were washed with chilled hexanes (–20 C, 1 × 5 mL) and then dried in a 250-mL round-bottomed flask under high vacuum (25 °C, 0.2 mmHg) for 6 h. This crop of crystals weighed 2.00 g.

16. The physical properties of 2-(phenylethynyl)phenol follow: R_f = 0.42 (10% ethyl acetate in hexanes, visualized by short-wave UV); ^{1}H NMR ($CDCl_3$, 500 MHz) δ: 5.83 (s, 1H), 6.92 (td, *J* = 7.6, 1.0 Hz, 1H), 6.99 (dd, *J* = 8.2, 0.7 Hz, 1H), 7.25–7.30 (m, 1H), 7.36–7.40 (m, 3H), 7.43 (dd, *J* = 7.7, 1.5 Hz, 1H), 7.52–7.58 (m, 2H). ^{13}C NMR ($CDCl_3$, 125 MHz) δ: 83.1, 96.5, 109.7, 114.9, 120.6, 122.5, 128.7, 129.0, 130.6, 131.76, 131.80, 156.6. IR (film): 3441, 3049, 1596, 1572, 1496, 750, 688 cm^{-1}. Melting Point: 69.5–70.6 °C. HRMS calc for $C_{14}H_{11}O$ $[M+H]^+$: 195.0810. Found 195.0801. The purity of the compound was determined using qNMR: 14.2 mg of the product are dissolved in 0.8 mL of $CDCl_3$. Mesitylene, 11.0 mg (98+%, purchased from Alfa Aesar and used as received), is added. ^{1}H NMR (500 MHz, $CDCl_3$) gave a product purity of 97.8%.
17. A second run on the same scale provided 6.49 g (53%) of the product.
18. This reaction is performed in a nitrogen-filled glove box because the boronic ester intermediate that is formed is highly sensitive to moisture (see Table 1 in the discussion section). The submitters were able to synthesize the desired borylated benzofuran using air-free techniques with septa-capped glassware and dynamic nitrogen at the bench instead of a glovebox, however the yield was substantially lower (50% vs 70% at the 0.86 mmol scale). All glassware used in the glove box was oven-dried overnight before being brought into the glove box.
19. A cylindrical-shaped PTFE stir bar with dimensions of 3.7 cm x 0.9 cm x 1.0 cm was used.
20. Sodium hydride (95 wt%) was purchased from Sigma Aldrich. This material was stored in the nitrogen-filled glovebox, and was titrated to 92% using a spectroscopy-based titration method developed by Hoye.[2] Toluene (99.8%, HPLC grade) was purchased from Fisher Scientific and passed through a basic alumina push-still system, then brought into the glovebox by the checkers using a 500-mL Straus flask. The submitters utilized a push-still system that fed directly into the glovebox. It is important that the toluene is water-free for this reaction.
21. The checkers performed this addition manually. The submitters utilized a syringe pump for this addition. A slight exotherm was noted. Care

should be used during this step, as this reaction liberates hydrogen gas. Careful monitoring and very slow addition are of paramount importance. It is highly recommended to unplug any electronic devices/ionizers that are not in use for this reaction while the highly flammable hydrogen gas is being liberated.

22. The submitters used *B*-chlorocatecholborane (98+%) purchased from TCI America. The checkers used *B*-chlorocatecholborane (97%) purchased from Sigma Aldrich. This solid was stored in the glove box freezer, and subsequently allowed to warm to room temperature before use.
23. It is important to cover the reaction flask at this point, as the newly formed boronic ester is highly moisture sensitive.
24. 1,3-Bis(2,6-diisopropylphenyl-imidazol-2-ylidene)gold(I) chloride (99%) was purchased from Frontier Scientific and used as received.
25. Sodium trifluoroacetate (98%) was purchased from Sigma Aldrich and was dried as follows: A 50-mL round-bottomed flask was charged with ca. 5 g of sodium trifluoroacetate and a stir bar. The solid was dried under high vacuum (0.2 mmHg) at 110 °C for 18 h. The flask was then placed directly into the glovebox while remaining under vacuum.
26. The TLC was conducted using an eluent of 10% methylene chloride in hexanes. 4,4,5,5-Tetramethyl-2-(2-phenylbenzofuran-3-yl)-1,3,2-dioxaborolane: $R_f = 0.29$ (10% DCM in hexanes)
27. Triphenylphosphine (98+%) was purchased from Strem Chemicals and used as received.
28. Pinacol (99%) was purchased from Alfa Aesar and used as received. Triethylamine (99.5%+) was purchased from Sigma Aldrich and passed through a basic alumina push-still system before use. A large excess of triethylamine was used so that it could fully solubilize the pinacol (thus acting as both the base and the solvent)
29. A slight evolution of gas was noted.
30. The filtration was conducted as follows: a medium porosity 350-mL filter funnel with a 9.0 cm I.D. was charged with 64.1 g of celite (Celite 545 was purchased from EMD Millipore and used as received). The solid was then pre-wet with toluene (200 mL) and then connected to a 500-mL filter flask. The reaction mixture was then poured over the wet celite, and pressure was applied to pull the solution through the pad (200 mmHg). The celite pad was then rinsed with toluene (200 mL).
31. Although not necessary, placing the crude reaction mixture under high vacuum (0.2 mmHg) removes the excess pinacol and triethylamine,

which reduces the total amount of material being placed on the column for purification. If the high vacuum step is omitted, it is advised to use a larger chromatographic column than the one described in this text.

32. The column chromatography was run as follows: A flash column with an outer diameter of 7 cm and capacity of 500-mL was charged with silica (Davisil 35-70 micron silica purchased from Grace Davidson; used as received.) using a wet-pack method (575 g of silica in 650 mL of hexanes). This gave a silica bed of 7 inches in length. The crude liquid mixture was then added to the top of the silica bed using a pipette; the reaction flask was rinsed with methylene chloride (3 × 5 mL) and then added to the silica. Sand was then added to fill 1.5 inches above the silica. The column was then loaded with the eluent mixture (80:20 hexanes: methylene chloride). The flow rate was approximately 53 mL/min, and 580 mL were then allowed to drain before fractions were collected in 50-mL test tubes (25 x 150 mm). Fractions 2–23 were collected as the product fractions. Fraction 1 was discarded, as it was contaminated with triphenylphosphine.
33. The physical properties for 4,4,5,5-tetramethyl-2-(2-phenylbenzofuran-3-yl)-1,3,2-dioxaborolane follow: R_f = 0.29 (10% DCM in hexanes, visualized by short-wave UV). ^{1}H NMR ($CDCl_3$, 500 MHz) δ: 1.41 (s, 12H), 7.23-7.31 (m, 2H), 7.38–7.48 (m, 3H), 7.49–7.53 (m, 1H), 7.99–8.03 (m, 1H), 8.14–8.20 (m, 2H); ^{11}B NMR ($CDCl_3$, 96 MHz) δ: 30.8; ^{13}C NMR ($CDCl_3$, 125 MHz) δ: 25.1, 83.8, 110.7, 123.1, 123.2, 124.4, 128.3, 129.2, 131.4, 133.4, 154.8, 163.1. IR (film): 3059, 2977, 1611, 1561, 1389, 1013, 747 cm^{-1}. Melting Point: 89.4–90.8 °C. HRMS calc. for $C_{20}H_{22}BO_3$ $[M+H]^+$: 321.1662. Found 321.1641. The purity of the compound was determined using qNMR: 28.4 mg of the product was dissolved in 0.8 mL of $CDCl_3$. 11.0 mg of mesitylene (98+%, purchased from Alfa Aesar and used as received) was added. ^{1}H NMR (500 MHz, $CDCl_3$) gave a product purity of 97.4%.
34. The checkers also found that when Step B was performed on 15.5 mmol (3.05 g) scale with respect to **3**, the yield of 4,4,5,5-tetramethyl-2-(2-phenylbenzofuran-3-yl)-1,3,2-dioxaborolane was 3.53 g (71%).

Working with Hazardous Chemicals

The procedures in *Organic Syntheses* are intended for use only by persons with proper training in experimental organic chemistry. All hazardous materials should be handled using the standard procedures for work with chemicals described in references such as "Prudent Practices in the Laboratory" (The National Academies Press, Washington, D.C., 2011; the full text can be accessed free of charge at http://www.nap.edu/catalog.php?record_id=12654). All chemical waste should be disposed of in accordance with local regulations. For general guidelines for the management of chemical waste, see Chapter 8 of Prudent Practices.

In some articles in *Organic Syntheses*, chemical-specific hazards are highlighted in red "Caution Notes" within a procedure. It is important to recognize that the absence of a caution note does not imply that no significant hazards are associated with the chemicals involved in that procedure. Prior to performing a reaction, a thorough risk assessment should be carried out that includes a review of the potential hazards associated with each chemical and experimental operation on the scale that is planned for the procedure. Guidelines for carrying out a risk assessment and for analyzing the hazards associated with chemicals can be found in Chapter 4 of Prudent Practices.

The procedures described in *Organic Syntheses* are provided as published and are conducted at one's own risk. *Organic Syntheses, Inc.*, its Editors, and its Board of Directors do not warrant or guarantee the safety of individuals using these procedures and hereby disclaim any liability for any injuries or damages claimed to have resulted from or related in any way to the procedures herein.

Discussion

Boron–element additions are a powerful class of synthetic reactions capable of furnishing boron-containing compounds that are adept at downstream functionalization. First described by Hurd[3] and developed by Brown,[4] hydroboration adds a B–H bond across unsaturated C–C bonds. This method of B–X bond addition has been extended to certain heteroatoms.[5] We recently reported the addition of B–O σ-bonds across alkynes.[6] Prior to this work, B–O σ-bond addition across C–C unsaturated bonds was unknown, likely due to the strength of the B–O σ bond, rendering transition metal-catalyzed oxidative addition, a common technique for borylation, difficult.[7] We hereby describe a robust procedure for the alkoxyboration of alkynes. This alkoxyboration method provides access to bench-stable borylated benzofurans isolated as the potassium organotrifluoroborate salt or the MIDA boronate ester.[8–9]

The outlined procedure results in the formation of 3-borylated benzofuran derivatives (Table 1). Beginning with 2-alkynylphenols (**5**), the requisite boric ester intermediate (**6**) is generated by deprotonation and electrophilic trapping by *B*–chlorocatecholborane. Cyclization is effected by the catalyst IPrAuTFA, generated in situ from IPrAuCl and NaTFA, producing the boronic ester intermediate (**7**). For long-term stability, the catechol boronic ester was then transesterified to the more stable organotrifluoroborate salt or MIDA boronate ester (**8**). A number of functional groups were tolerated by the alkoxyboration reaction, including siloxy (**8a**), amide (**8b**), nitrile (**8c**), and ester (**8e**) groups. The synthesized BF_3K salts[8] and MIDA esters[9] provide potential handles for further downstream functionalization resulting in the rapid synthesis of structurally complex compounds.[6]

Table 1. The Alkoxyboration Reaction and Substrate Scope with Isolated Yields

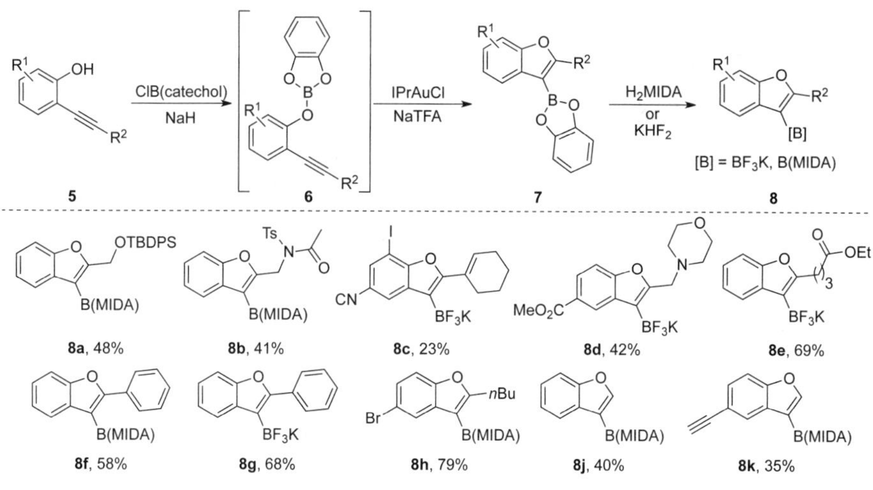

The mechanism of this alkoxyboration reaction is proposed to follow the pathway outlined in Scheme 1. Following the formation of the boronic ester intermediate, the catalyst IPrAuTFA could facilitate the generation of the *O*–heterocyclic scaffold by coordinating to boric ester **6**, yielding species **9**, facilitating a nucleophilic attack that furnishes an organogold intermediate (**12**) and boric ester **11**. These two intermediates recombine to produce borylated alkoxyboration product (**7**) and the catalyst is regenerated. Mechanistic experiments in our group have examined the role of the NaTFA counterion in the alkoxyboration reaction.[13]

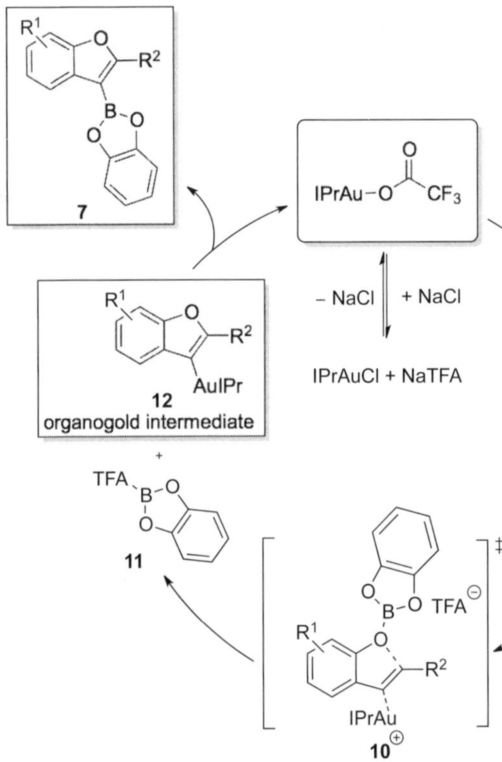

Scheme 1. The Proposed Mechanism for the Oxyboration Reaction

Since our first report of the alkoxyboration reaction, we have optimized several of the reaction parameters to enhance the ease of performing the reaction. Previously, generation of the requisite 2-alkynylphenol starting materials, **5,** required three steps: protection of the phenol group, Sonogashira coupling, and subsequent deprotection. Now using a procedure adapted from Tamaru, we synthesize **3** using a protecting-group free Sonogashira coupling reaction directly from 2-iodophenol (**1**), thereby reducing the total number of synthetic steps to one in order to produce the desired 2-alkynylphenol **3,** which is isolated in a reasonable yield.[10]

In addition to optimizing the synthesis of the starting 2-alkynylphenol substrates **3**, we have examined the possibility of performing the key catalytic alkoxyboration reaction outside the glovebox for operational ease. In order to compare, both reactions were run on the 0.86 mmol scale, with the only variable being in or out of the glovebox. On this smaller scale, a lower yield was obtained for the in-box reaction compared to the large scale reaction reported in this manuscript. Although lower in isolated yield (50% vs. 70%), we have successfully run the oxyboration reaction outside of the glovebox using standard air-free techniques with septa-capped glassware. The lower yield for the out of the glovebox reaction is likely due to hydrolysis of the boric ester intermediate (**6**). For the procedure described in this text, we decided to run the reaction in the glovebox in order to obtain the highest possible yield. Moreover, we have optimized the catalyst quenching (addition of PPh_3) to reduce the total reaction time. Finally, we expanded the transesterification step from MIDA esters and organotrifluoroborate salts to pinacol boronic esters using a procedure adapted from Ingleson.[11] The pinacol boronic esters are valuable because they are well-established cross-coupling partners and Michael addition partners in transition metal chemistry.[12]

In closing, an optimized oxyboration reaction, capable of furnishing 3-borylated benzofurans that are primed for downstream functionalization, is reported. The requisite starting materials are constructed in one step from commercial materials via a protecting-group-free Sonogashira reaction. The reaction conditions are mild and scalable.

References

1. Department of Chemistry, University of California, Irvine, Irvine, CA 92617. blums@uci.edu. We thank the NIH (1R01GM098512-01) for support.
2. Hoye, T. R.; Aspaas, A. W.; Eklov, B. M.; Ryba, T. D. *Org. Lett.* **2005**, *7*, 2205–2208.
3. Hurd, D. T. *J. Am. Chem. Soc.* **1948**, *70*, 2053–2055.
4. Brown, H. C. *Tetrahedron* **1961**, *12*, 117–138.

5. (a) Suginome, M.; Yamamoto, A.; Murakami, M. *J. Am. Chem. Soc.* **2003**, *125*, 6358–6359. (b) Suginome, M. *Chem. Rec.* **2010**, *10*, 348–358. (c) Suginome, M.; Nakamura, H.; Ito, Y. *Chem. Commun.* **1996**, 2777–2778. (d) Ishiyama, T.; Nishijima, K.; Miyaura, N.; Suzuki, A.; *J. Am. Chem. Soc.* **1993**, *115*, 7219–7225. (e) Marder, T. B.; Norman, N. C. *Top. Catal.* **1998**, *5*, 63–73. (f) Lappert, M. F.; Prokai, B. *J. Orgmet. Chem.* **1964**, *1*, 384–400. (g) Hara, S.; Dojo, H.; Takinami, S.; Suzuki, A. *Tetrahedron Lett.* **1983**, *24*, 731–734.
6. Hirner, J. J.; Faizi, D. J.; Blum, S. A. *J. Am. Chem. Soc.* **2014**, *136*, 4740–4745.
7. Sanderson, R. T. *Polar Covalence*; Academic Press: New York, 1983.
8. Molander, G. A.; Ellis, N. *Acc. Chem. Res.* **2007**, *40*, 275–286.
9. Gillis, E. P.; Burke, M. D. *J. Am. Chem. Soc.* **2008**, *130*, 14084–14085.
10. Kimura, M.; Ezoe, A.; Mori, M. Tamaru, Y. *J. Am. Chem. Soc.* **2005**, *127*, 201–209.
11. Del Grosso, A.; Singleton, P.; Muryn, C.; Ingleson, M. *Angew. Chem. Int. Ed.* **2011**, *50*, 2102–2106.
12. Nagaki, A.; Moriwaki, Y.; Yoshida, J. *Chem. Commun.* **2012**, *48*, 11211–11213.
13. Johnson, J. J.; Chong, E.; Tu, K. N.; Blum, S. A. *Organometallics* **2016**, *35*, 655–662.

Appendix
Chemical Abstracts Nomenclature (Registry Number)

2-Iodophenol: Phenol, 2-iodo-; (533-58-4)
Copper iodide: Copper iodide; (1335-23-5)
Bis(triphenylphosphine)palladium dichloride: Palladium, dichlorobis(triphenylphosphine); (13965-03-2)
Toluene: benzene, methyl-; (108-88-3)
Diisopropylamine: ; 2-Propanamine, *N*-(1-methylethyl)-; (108-18-9)
Phenylacetylene: Benzene, ethynyl-; (536-74-3)
Sodium hydride: sodium hydride; (7646-69-7)
B-Chlorocatecholborane: 1,3,2-Benzodioxaborole, 2-chloro- (55718-76-8)

1,3-Bis(2,6-di-isopropylphenylimidazol-2-ylidene)gold(I)chloride: Gold, [1,3-bis[2,6-bis(1-methylethyl)phenyl]-1,3-dihydro-2*H*-imidazol-2-ylidene]chloro-; (852445-83-1)
Sodium trifluoroacetate: Acetic acid, 2,2,2-trifluoro-, sodium salt (1:1) (2923-18-4)
Triphenylphosphine: Phosphine, triphenyl- (603-35-0)
Pinacol: 2,3-Butanediol, 2,3-dimethyl- (76-09-5)
Triethylamine: Ethanamine, *N,N*-diethyl-(121-44-8)

Suzanne A. Blum received her B.S. from the University of Michigan in 2000, where she trained with Profs. Edwin Vedejs and Brian P. Coppola. She earned her Ph.D. degree at the University of California, Berkeley, studying the mechanisms of transition metal reactions, as an NSF Graduate Fellow under the mentorship of Profs. Jonathan A. Ellman and Robert G. Bergman. She completed an NIH Postdoctoral Fellowship with Prof. Christopher T. Walsh at the Harvard Medical School. In 2006, she began her independent career as an Assistant Professor at the University of California, Irvine. She was promoted to Associate Professor in 2012.

Darius J. Faizi received his B.S. in chemistry from the University of California, Los Angeles in 2010, where he studied transition metal catalysis under the guidance of Professor Craig A. Merlic. He is currently pursuing his Ph.D. in organic chemistry under the guidance of Professor Suzanne A. Blum. His graduate research focuses on the transition-metal catalyzed addition of B–O bonds across C–C multiple bonds.

Nicole A. Nava received her B.S. in chemistry from the University of California, Irvine in 2015. While there, she completed undergraduate research under the direction of Professor Suzanne A. Blum. She will begin pursuing a Ph.D. in organic chemistry at Boston College in Fall 2015.

Mohammad Al-Amin was born in Chandpur, Bangladesh. He received his Ph.D. in synthetic organic chemistry in 2010 under the supervision of Professor Masahiro Yoshida and Professor Kozo Shishido at the University of Tokushima as a MEXT scholar. In 2010, he moved to Hokkaido University as a postdoctoral fellow under the direction of Professor Mitsuhiro Arisawa and Professor Satoshi Shuto. He is currently a postdoctoral fellow in the group of Professor Suzanne A. Blum at the University of California, Irvine. His main research interests are the development of new synthetic methods using transition metal catalysts.

Nicholas A. Weires received his B.S. in chemistry from the University of Idaho in 2012, where he studied heterogeneous catalysis under the direction of Professor Jakob Magolan. He is currently pursuing his Ph.D. in organic chemistry in the laboratory of Professor Neil K. Garg at the University of California, Los Angeles. His graduate research has focused on the total synthesis of welwitindolinone alkaloids and the development of new C–C bond forming reactions of amides mediated by nickel catalysis.

Organic Syntheses

Synthesis of 1-Iodopropyne

Samuel G. Bartko, James Deng, and Rick L. Danheiser*

Department of Chemistry, Massachusetts Institute of Technology, Cambridge, MA 02139.

Checked by Joshua Cox and John L. Wood

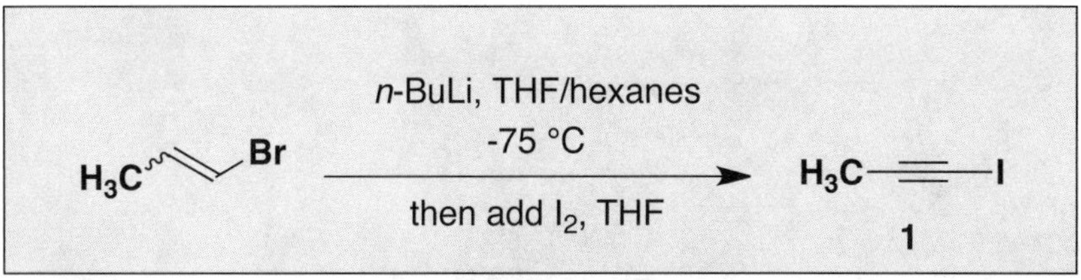

Procedure

1-Iodopropyne (**1**). A 1-L, three-necked, round-bottomed flask is equipped with an overhead mechanical stirrer with a 19 x 60 mm teflon paddle, a 125-mL graduated pressure-equalizing addition funnel topped with a rubber septum with an argon inlet needle, and a rubber septum equipped with a thermocouple probe (Notes 1 and 2) (see Figure 1). The flask is charged with 150 mL of THF (Note 3), cooled in a dry ice/acetone bath, and 1-bromo-1-propene (15 mL, 21 g, 174 mmol, 1.5 equiv) (Note 4) is added in one portion by syringe. The resulting solution is cooled to –73 °C (internal temperature) and *n*-BuLi solution (114 mL, 293 mmol, 2.2 equiv) (Note 5) is added dropwise via the graduated addition funnel at such a rate so as to maintain the internal temperature between –58 and –63 °C (Note 6). The walls of the addition funnel are rinsed with two 5-mL portions of THF delivered by syringe. The resulting pale yellow reaction solution is stirred further for 2 h at –73 °C (internal temperature), and a solution of iodine (29.7 g, 117 mmol, 1.0 equiv) in 80 mL of THF (Note 7) is then added via cannula directly into the flask via the septum at such a rate so as to maintain the internal temperature between –58 and –63 °C (Note 8).

DOI 10.15227/orgsyn.093.0245

Figure 1. Reaction Assembly

The resulting turbid brown solution is stirred at –73 °C (internal temperature) for 20 min, the dry ice/acetone bath is replaced with a room temperature water bath, and the reaction mixture is then allowed to warm to 0 °C (internal temperature) over 5 min. At this point a solution of 5.5 g of NaH_2PO_4 in 125 mL of deionized water is added to the reaction mixture via the addition funnel over 2 min (Note 9).

The contents of the reaction flask are transferred to a 1-L separatory funnel (the flask is rinsed with ca. 20 mL of pentane and 20 mL of water) (Note 10) and the milky aqueous layer is separated and extracted with pentane (3 x 50 mL). The combined organic layers are washed with 100 mL of saturated NaCl solution, dried over 10 g of $MgSO_4$ for ca. 10 min, and gravity filtered through filter paper with the aid of 25 mL of pentane into a 1-L round-bottomed flask containing a 32 x 15 mm, Teflon-coated, oval magnetic stir bar. The filtrate is concentrated to a volume of approximately 25 mL by distillation at atmospheric pressure under argon (oil bath temperature gradually raised from 50 to 95 °C) through a 32-cm Vigreux column (Notes 11 and 12) into a 1-L flask cooled at 0 °C (see Figure 2).

Figure 2. Initial Concentration of Reaction Product Solution

The residual yellow solution (ca. 25 mL) in the distillation pot is transferred to a 50-mL, round-bottomed flask (containing a 13 x 8 mm, Teflon-coated, oval magnetic stir bar) using 10 mL of pentane to rinse the 1-L flask. To concentrate the resulting solution, the flask is equipped with a 26-cm Vigreux column (Note 11) topped with a short path distillation head with a 50-mL pear-shaped receiving flask (Note 12). The receiving flask is cooled at 0 °C with an ice water bath (Figure 3). Distillation is carried out at atmospheric pressure, gradually raising the temperature of the oil bath until the vapor temperature at the distillation head reaches 85 °C. In this fashion, the pentane and most of the remaining THF and hexane are removed by distillation, concentrating the product solution to a volume of ca. 20 mL. (Notes 13 and 14).

Figure 3. Further Concentration of Reaction Product Solution

The reaction product is then distilled using the setup shown in Figure 4 (Note 12). The 50-mL flask containing the product is equipped with a 6-cm vacuum-jacketed micro Vigreux column (Figure 4) with a short-path distillation head attached to a 50-mL, three-necked, pear-shaped flask fitted with two glass stoppers secured with Teflon tape. The receiving flask is cooled in an ice water bath. Distillation of the product is carried out at atmospheric pressure under argon by raising the oil bath temperature from 25 to 130 °C over 30 min and then gradually from 130 to 150 °C over 30 min. The product is collected in a single fraction as 15.8–16.4 g (81–84%) of colorless liquid, 95% purity by QNMR (Note 15). Although this material is sufficiently pure for most purposes, product with greater than 98% purity can be obtained by low-temperature recrystallization according to the following procedure.

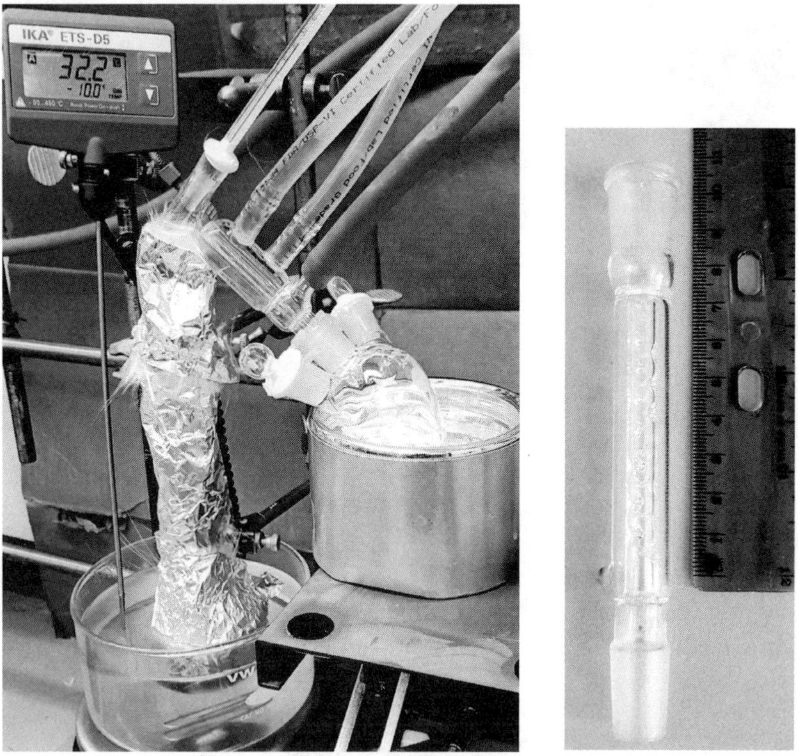

Figure 4. Distillation of Iodopropyne

Low-temperature recrystallization of the distilled product is carried out using the setup shown in Figure 5. The three-neck receiving flask from the distillation is disconnected from the distillation head and the center neck is equipped with a rubber septum pierced with a metal spatula (Figure 6) and the left neck is equipped with a pressure-equalized, jacketed 50-mL addition funnel fitted with a rubber septum and argon inlet needle. The right neck is fitted with a rubber septum pierced with a 16-gauge cannula with filter paper secured onto the blunt end with Teflon tape (Figure 7) and connected to a 50-mL two-neck pear-shaped flask. Initially the end of the cannula filter is maintained above the level of the contents of the flask.

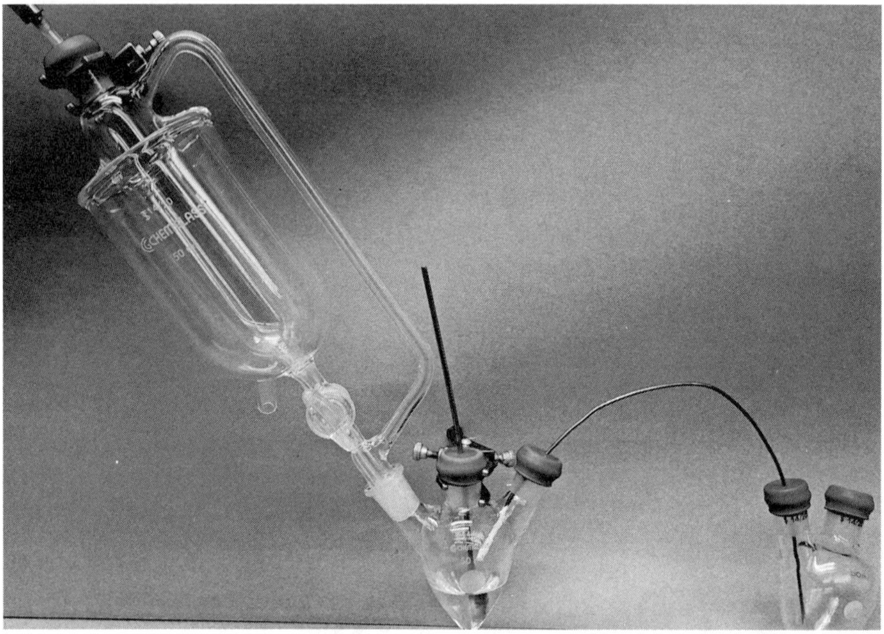

Figure 5. Set-up for Low-Temperature Recrystallization of Iodopropyne

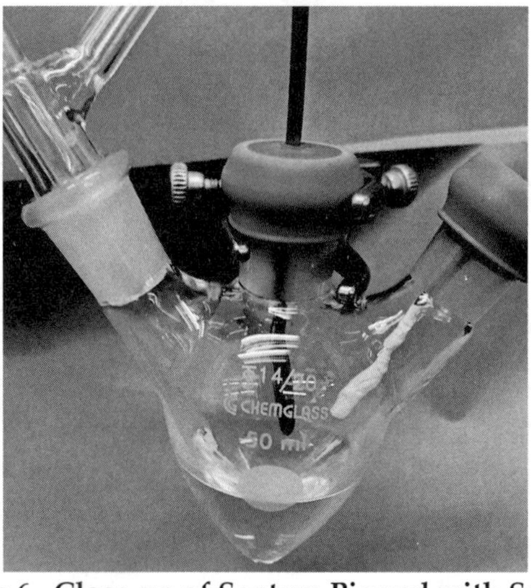

Figure 6. Close-up of Septum Pierced with Spatula

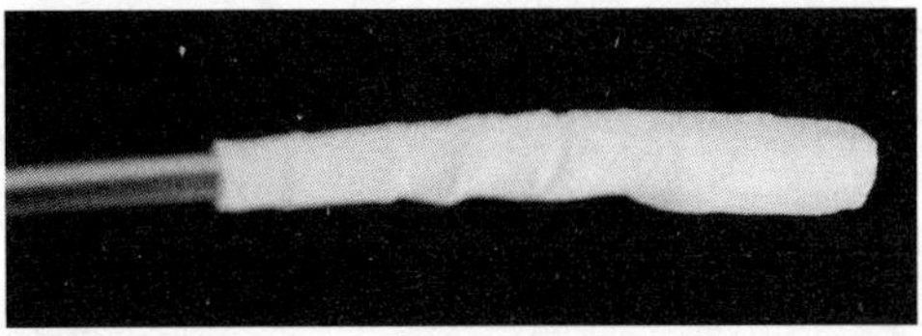

Figure 7. Close-up View of Cannula Filter

Pentane (4.4 mL, Note 16) is added to the three-necked flask by syringe and a homogeneous solution is obtained by stirring with the spatula. The flask is placed in an acetone bath, and small pieces of dry ice are gradually added to the bath to slowly lower the bath temperature to –78 °C over a period of 90 min. When crystal formation begins, the metal spatula is used to periodically scrape crystals from the sides of the flask and to break up aggregates of crystals. The mixture is maintained at -78 °C for an additional 90 min, using the metal spatula to stir and scrape the crystals as needed to prevent a single large aggregate mass from forming (Figure 8).

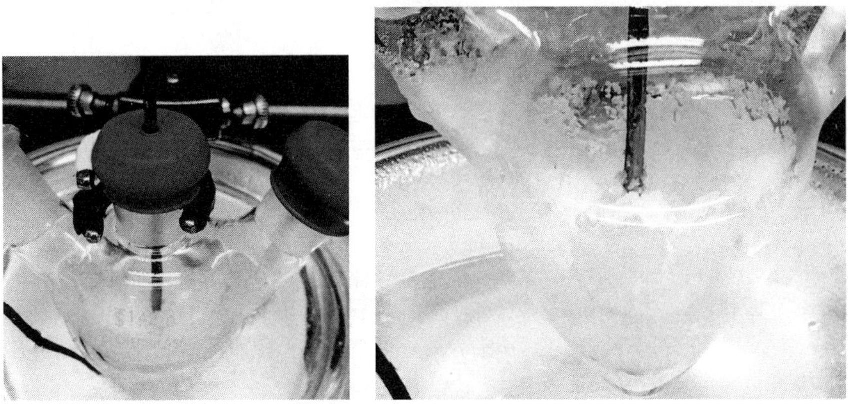

Figure 8. Appearance of Iodopropyne Crystals

The end of the cannula filter is then lowered to the bottom of the flask and the supernatant pentane is removed through the cannula by increasing the argon pressure (Note 17). The addition funnel is charged with 18 mL of pentane which is cooled to -78 °C by filling the jacket with dry ice/acetone. A portion of cold pentane (ca. 3 mL) is added to the crystals from the addition funnel, the cannula filter is raised above the level of the mixture, and the crystals are stirred with the metal spatula. The cannula filter is then lowered, a short outlet needle is inserted through the septum of the two-necked receiving flask, and the liquid is removed through the cannula by increasing the argon pressure. This washing process is repeated with five additional 3-mL portions of cold pentane. After removal of the last portion of pentane, the metal spatula is used to distribute the crystals along the walls of the flask so as to expose as much surface area as possible for subsequent drying.

With the flask still immersed in the dry ice/acetone bath, the addition funnel is rapidly replaced with an inlet adapter connected to a double manifold. The septa on the other necks (with metal spatula and cannula filter) are quickly replaced with glass stoppers, which are secured to the flask with Teflon tape (Note 18). Residual pentane is removed in vacuo (0.5-1.0 mmHg, 1 h) at –78 °C. The flask is back filled with argon, and the dry ice/acetone bath is replaced with a room temperature water bath. Upon warming to room temperature, the colorless crystals melt to afford 13.2 g (68%, >98% pure by QNMR) (Note 19) of 1-iodopropyne as a colorless liquid (Notes 20 and 21).

Notes

1. The reaction flask (equipped with the addition funnel) was dried in an oven (160 °C) overnight, assembled while hot, flushed with argon, and then maintained under an atmosphere of argon during the course of the reaction.
2. Mechanical stirring is preferred due to the formation of a precipitate in some runs at the stage of iodine addition (vide infra). In these cases difficulties were encountered in maintaining stirring when magnetic stirring was employed.
3. Tetrahydrofuran (ultra low water) was purchased from J. T. Baker and purified by pressure filtration through activated alumina prior to use. The Checkers used non-stabilized THF, purchased from Fisher Scientific and passed through two packed columns of neutral alumina.
4. 1-Bromo-1-propene (mixture of E and Z isomers, 98%) was purchased from Sigma-Aldrich and distilled at atmospheric pressure under argon (bp 58–60 °C) prior to use. NMR analysis of the distilled 1-bromo-1-propene revealed a 3:1 ratio of *Z*/*E* isomers.

5. *n*-Butyllithium (2.58 M in hexanes) was purchased from Sigma-Aldrich and titrated prior to use.[2] The butyllithium was transferred to the addition funnel via cannula.
6. The addition of *n*-BuLi solution required 30 min on this scale.
7. Iodine (99.8%) was purchased from Alfa Aesar and used as received. The iodine was weighed into a dry 100-mL two-necked, pear-shaped flask equipped with a magnetic stirbar and dissolved in 70 mL of THF by stirring under argon for ca. 30 min. The iodine solution was transferred to the reaction flask by cannula with the aid of two 5-mL portions of THF to rinse the flask.
8. Initially the addition of the brown iodine solution temporarily darkens the reaction mixture, which then returns to a pale yellow color. In some (but not all) runs a white precipitate was observed upon addition of the first drops of iodine solution. Toward the end of the iodine addition, the reaction mixture appears as a turbid brown solution. Addition of the iodine solution required 15 min on this scale.
9. The addition of aqueous NaH_2PO_4 solution leads to separation of the reaction mixture into a yellow top layer and a cloudy white bottom layer.
10. Pentane (reagent grade, 98%) was purchased from Sigma-Aldrich and used as received.
11. The column length refers to the effective length, measured from the lowest to highest indent in the Vigreux column. The column is insulated with a wrapping of glass wool held in place with a covering of aluminum foil. Concentration through a 30-cm Vigreux column was found necessary to obtain clean separation of the solvents from iodopropyne.
12. The Vigreux columns, distillation heads, and receiving flasks were all dried in an oven (160 °C) overnight and allowed to cool in a dessicator box before use.
13. The bath temperature initially is raised to 90 °C and held there for ca. 45 min while the THF and pentane distill (distillation head temperature 65-68 °C). The last of the solvent distills as the bath temperature is raised to 120 °C, and the distillation head temperature then falls to 25 °C. When the bath temperature is further raised to 130-135 °C, the iodopropyne begins to distill (distillation head temperature increases from 25 to 85 °C) at which point the distillation is halted.
14. The Submitters report that quantitative ^1H NMR spectroscopy (dimethyl terephthalate as internal standard) indicated that the residual

orange liquid after concentration consists of 91% 1-iodopropyne, 2.1% 1-bromobutane, 1.8% (*E*)-1-iodopropene (see Discussion Section), and trace amounts of hexane and THF (amounts not determined due to overlap with bromobutane resonances). It is assumed that several percent of polymeric material is present.

15. The material collected has bp 105-107 °C. The Submitters report that quantitative ^{1}H NMR spectroscopy (dimethyl terephthalate as internal standard) indicated that the distilled product consists of 95% 1-iodopropyne, 2% 1-bromobutane, 1% (*E*)-1-iodopropene, and trace amounts of hexane and THF (amounts not determined due to overlap with bromobutane resonances).
16. The Submitters found that best results were obtained in the recrystallization by using 0.25 mL of pentane per 1 g of crude product.
17. It is important that all of the supernatant liquid is removed prior to adding the next wash of cold pentane. The Submitters note that this was facilitated by stirring the crystals periodically with the spatula during the removal of liquid.
18. Shaking the cannula filter and spatula before removal from the flask may be necessary to dislodge any crystals that adhere to them.
19. A second run of the reaction at full scale provided 14.1 g (73%, >98% pure by QNMR) of the product.
20. At –78 °C the desired product is a colorless solid that melts at ca. –30 °C. Iodopropyne is stable to storage under argon in the freezer (–20 °C) but turns pink when exposed to light.
21. 1-Iodopropyne (**1**) has the following physical and spectroscopic properties: ^{1}H NMR ($CDCl_3$, 400 MHz) δ: 2.02 (s); ^{13}C NMR ($CDCl_3$, 150 MHz) δ –8.2, 5.8, 89.9; IR (thin film, ATR) 2959, 2915, 2848, 2197, 1438, 1012 cm^{-1}; HRMS (ESI) *m/z*: Calcd. for C_3H_3I $[M+H]^+$: 166.9352, found: 166.9349. The purity was determined to be 98.4% by quantitative NMR spectroscopy with methyl salicylate as internal standard.

Working with Hazardous Chemicals

The procedures in *Organic Syntheses* are intended for use only by persons with proper training in experimental organic chemistry. All hazardous materials should be handled using the standard procedures for work with chemicals described in references such as "Prudent Practices in the Laboratory" (The National Academies Press, Washington, D.C., 2011; the full text can be accessed free of charge at http://www.nap.edu/catalog.php?record_id=12654). All chemical waste should be disposed of in accordance with local regulations. For general guidelines for the management of chemical waste, see Chapter 8 of Prudent Practices.

In some articles in *Organic Syntheses*, chemical-specific hazards are highlighted in red "Caution Notes" within a procedure. It is important to recognize that the absence of a caution note does not imply that no significant hazards are associated with the chemicals involved in that procedure. Prior to performing a reaction, a thorough risk assessment should be carried out that includes a review of the potential hazards associated with each chemical and experimental operation on the scale that is planned for the procedure. Guidelines for carrying out a risk assessment and for analyzing the hazards associated with chemicals can be found in Chapter 4 of Prudent Practices.

The procedures described in *Organic Syntheses* are provided as published and are conducted at one's own risk. *Organic Syntheses, Inc.*, its Editors, and its Board of Directors do not warrant or guarantee the safety of individuals using these procedures and hereby disclaim any liability for any injuries or damages claimed to have resulted from or related in any way to the procedures herein.

Discussion

Halo alkynes serve as versatile building blocks for the construction of organic molecules and have found wide application in the synthesis of substituted alkynes including acetylenic natural products.[3] Nucleophilic addition and substitution reactions of halo alkynes have been known for many years,[3a] and recently halo alkynes have been shown to function as reaction partners in [2 + 2] cycloadditions as well.[4] Most importantly, however, halo alkynes also participate in a variety of transition-metal catalyzed carbon-carbon bond forming reactions, including Cadiot-Chodkiewicz reactions leading to 1,3-diynes, "inverse Sonogashira couplings" for the synthesis of aryl acetylenes, syntheses of 1,3-enynes and aryl and hetaryl acetylenes via cross-coupling reactions with organoboron, -tin, and –zinc compounds, and NHK reactions with aldehydes. In addition, research in our laboratory and that of Hsung has led to the development of practical and efficient methods for the synthesis of ynamides via the N-alkynylation of various amine derivatives using bromo and iodo alkynes.[5]

Halo alkynes are most often prepared by dehydrohalogenation of 1,1-dihalo alkenes, via halogenation of terminal alkynes, and by the synthetic elaboration of metal halo acetylides (i.e., X-C≡C-Met).[2] In connection with a recent natural product total synthesis project in our laboratory, we required a practical method for accessing multigram quantities of 1-bromo- or 1-iodo-1-propyne. While these simple acetylenic halides have numerous applications as synthetic building blocks (vide infra), only the iodide is commercially available, and as of this writing iodopropyne typically sells for $350-750 per gram.

Previously, Brandsma has reported the synthesis of 1-bromopropyne by the Straus reaction of propyne gas with potassium hypobromite in a two-phase mixture of water and pet ether.[6] We found this procedure less than ideal for several reasons. In the first place, propyne gas is expensive and rather inconvenient to handle. In addition, the product, 1-bromopropyne, is quite volatile (bp 64 °C) and has been reported to be dangerously reactive with air.[6] Iodopropyne, by contrast, is less volatile (lit. bp 109-110 °C) and is reported to be much more stable to air than the bromo derivative.[7]

Several laboratories have previously described the preparation of 1-iodopropyne. In 1943, Cleveland and Murray reported the generation of 1-iodopropyne by reaction of the Grignard derivative of propyne with molecular iodine.[7] However, no yield was reported for this reaction which has the disadvantage of requiring the use of propyne gas. Diner and Lown have reported the preparation of iodopropyne from propyne in 45% yield by reaction with an iodonium nitrate reagent.[8] Brandsma has described a more attractive route to iodopropyne via reaction of 1,2-dibromopropane with lithium amide in liquid ammonia followed by addition of molecular iodine (Scheme 1).[9] While this procedure produces iodopropyne in "excellent yield", it does require the handling of liquid ammonia and care must also be taken to avoid the formation of explosive nitrogen triiodide.

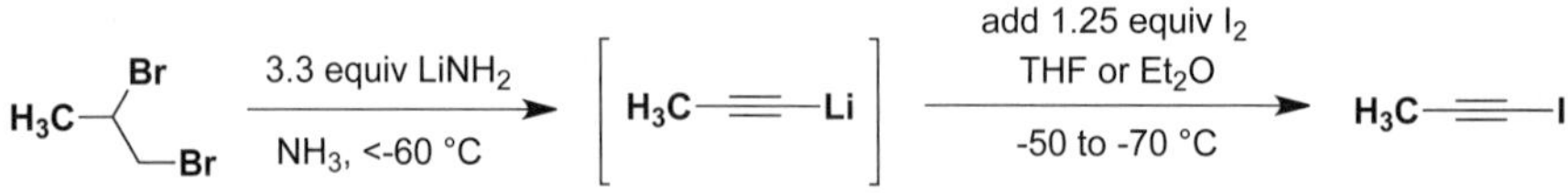

Scheme 1. Synthesis of Iodopropyne According to Brandsma

Robertson reported a more convenient route to iodopropyne in 2012[10] involving a variant of the Hofmeister halo alkyne synthesis introduced by Isobe and coworkers.[11] As shown in Scheme 2, reaction of 1-(trimethylsilyl)propyne with *N*-iodosuccinimide in the presence of $AgNO_3$ provided 1-iodopropyne in modest yield. While we regarded the Robertson approach to be the most attractive synthesis of iodopropyne reported to date, we undertook the development of an alternative route that would not require the use of relatively expensive 1-(trimethylsilyl)propyne.

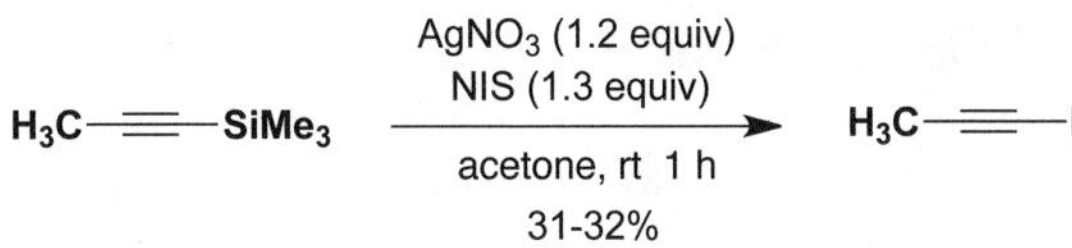

Scheme 2. Synthesis of Iodopropyne According to Robertson

Suffert has reported a convenient method for the generation of propynyllithium by the reaction of relatively inexpensive (*Z*/*E*)-1-bromopropene with *n*-butyllithium.[12] Propynyllithium generated in this fashion can be trapped with carbonyl electrophiles (aldehydes, ketones, acyl chlorides, and Weinreb amides)[12] or protonated to form solutions of propyne that can then be employed in Sonogashira coupling reactions.[13] Suffert has suggested that this transformation proceeds via a mechanism related to the Fritsch-Buttenberg-Wiechell reaction, i.e., via deprotonation at the C-1 carbon of 1-bromopropene, followed by α-elimination and rearrangement to generate propyne that then reacts with a second equivalent of butyllithium. The use of *n*-butyllithium is crucial to the success of this method; with *s*-butyllithium or *t*-butyllithium halogen-metal exchange was reported to be a significant side reaction producing 1-propenyllithium that then reacts with the electrophilic trapping agent to afford substituted alkenes.

We anticipated that reaction of propynyllithium generated via the Suffert method with molecular iodine could provide an efficient, economical, and convenient route to 1-iodo-1-propyne. In the event, application of the Suffert chemistry for the synthesis of iodopropyne turned out to present certain practical challenges not encountered in reactions with the carbonyl electrophiles studied previously.

Scheme 3 outlines the optimal reaction conditions employed in the procedure described in this article. Not unexpectedly, the stoichiometry employed in the reaction proved to be important. For example, employing a larger excess of butyllithium led to the formation of 1-iodobutane and *n*-octane, byproducts that proved difficult to separate from the desired 1-iodopropyne. The use of THF as reaction solvent is also necessary, as reaction in diethyl ether proceeds very slowly as reported by Suffert.[12a]

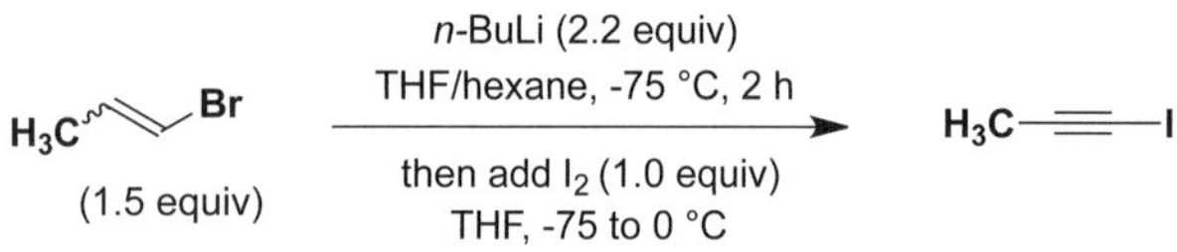

Scheme 3. Optimized Reaction Conditions

Development of an effective protocol for the isolation and separation of iodopropyne from trace byproducts required considerable experimentation. Careful distillation at atmospheric pressure was effective in separating the product (bp 105-107 °C) from THF and hexane, and provided material determined to be 95% pure by quantitative NMR analysis. The principal impurities were found to be 1-bromobutane and (*E*)-1-iodopropene, byproducts formed by halogen metal exchange of bromopropene with butyllithium and iodination of the resulting propenyllithium.

Separation of these byproducts from iodopropyne was not trivial as the boiling points of 1-bromobutane (102 °C) and 1-iodopropene (90 °C) are very similar to that of the desired product. Although distillation provides 1-iodopropyne with purity (95%) sufficient for many applications, we investigated the development of a procedure for separation of the iodopropyne from the trace byproducts. As described in this article, further purification of iodopropyne can be achieved by low-temperature crystallization from pentane, furnishing material in very good yield and with >99% purity.

Synthetic Applications of 1-Iodopropyne

A number of reports describe the application of 1-iodopropyne as a useful synthetic building block and as an intermediate in total syntheses of acetylenic natural products. For example, Mukai and coworkers employed a Cadiot-Chodkiewicz coupling of iodopropyne in their syntheses of the endiyne natural product (-)-Al-2 and related compounds.[14] Similar coupling reactions were featured in Robertson's approach to the same class of natural products.[10]

Fürstner and Dierkes employed the coupling of an organocopper compound with iodopropyne in their total synthesis of (*S,S*)-(+)-dehydrohomoancepsenolide.[15] Mound et al. demonstrated that iodopropyne can be coupled to aryl boronates under Suzuki conditions[16] and Vermeer et al. have reported the coupling of iodopropyne with allenylcopper compounds to furnish allenynes.[17] Oxidative carbonylation of iodopropyne has been reported by Bruk and Tempkin.[18] Finally, we have employed iodopropyne in N-alkynylation reactions for the synthesis of ynamide benzannulation reaction partners.[19]

References

1. Department of Chemistry, Massachusetts Institute of Technology, Cambridge, MA 02139. Email: danheisr@mit.edu. We thank the National Science Foundation (CHE-1464799) for generous financial support.
2. The Checkers titrated butyllithium with diphenylacetic acid according to the method of Kofron, W. G.; Baclawski, L. M. *J. Org. Chem.* **1976**, *41* 1879-1880. The Submitters titrated the reagent with menthol in THF at 0 °C with 1,10-phenanthroline as indicator using the general method of Watson, S. C.; Eastham, J. F. *J. Organomet. Chem.* **1967**, *9*, 165-167.
3. For reviews on the synthesis and chemistry of halo alkynes, see (a) Delavrenne, S. Y.; Viehe, H. G. In *Chemistry of Acetylenes*; Viehe, H. G., Ed.; Marcel Dekker: New York, 1969; pp 651–750. (b) Hopf, H.; Witulski, B. In *Modern Acetylene Chemistry*; Stang, P. J., Diederich, F., Eds.; VCH: Weinheim, 1995; pp 33–66. (c) Brandsma, L. *Synthesis of Acetylenes, Allenes, and Cumulenes, Methods and Techniques*; Elsevier, Ltd.: Oxford, 2004; pp 191–202. (d) Witulski, B.; Alayrac, C. In *Science of Synthesis;* de Meijere, A., Ed.; Thieme: Stuttgart, 2005; Vol. 24, pp 905-932. (e) Wu, W.; Jiang H. *Acc. Chem. Res.* **2014,** *47*, 2483–2504.
4. For example, see Wang, Y.-P.; Danheiser, R. L. *Tetrahedron Lett.* **2011,** *52*, 2111–2114.
5. For examples with detailed procedures published in this journal, see (a) Kohnen, A. L.; Dunetz, J. R.; Danheiser, R. L. *Org. Synth.* **2007**, *84*, 88–101. (b) Sagamanova, I. K.; Kurtz, K. C. M.; Hsung, R. P. *Org. Synth.*

2007, *84*, 359–367. (c) Read, J. M.; Wang, Y.-P.; Danheiser, R. L. *Org. Synth.* **2015**, *92*, 156–170.

6. Brandsma, L.; Verkruijsse, H. D. *Synthesis* **1990**, 984–985.
7. Cleveland, F. F.; Murray, M. J. *J. Chem. Phys.* **1943**, *11*, 450–454. With regard to 1-bromo-1-propyne, these authors report " . . . the pure compound ignites quickly when it comes in contact with air and gives dense clouds of black smoke having an irritating odor".
8. Diner, U. E.; Lown, J. W. *Can. J. Chem.* **1971**, *49*, 403–415.
9. Brandsma, L. *Synthesis of Acetylenes, Allenes, and Cumulenes, Method and Techniques;* Elsevier: Amsterdam, 2004, pp 199–201.
10. Wu, B.; Feast, G. C.; Thompson, A. L.; Robertson, J. *J. Org. Chem.* **2012,** *77*, 10623–10630.
11. (a) Nishikawa, T.; Shibuya, S.; Hosokawa, S.; Isobe, M. *Synlett* **1994**, 485–486. (b) Hofmeister H.; Annen K.; Laurent H.; Wiechert R.; *Angew Chem, Int Ed.* **1984**, *23*, 727–729.
12. (a) Suffert, J.; Toussaint, D. *J. Org. Chem.* **1995,** *60*, 3550–3553. (b) Toussaint, D.; Suffert, J. *Org. Synth.* **1999,** *76*, 214–220.
13. Abraham, E.; Suffert, J. *Synlett* **2002,** *2*, 328–330.
14. Miyakoshi, N.; Aburano, D.; Mukai, C. *J. Org. Chem.* **2005,** *70*, 6045–6052.
15. Fürstner, A.; Dierkes, T. *Org. Lett.* **2000,** *2*, 2463–2465.
16. Mound, W. R.; Scutt, J. N.; Slater, M.; Willetts, N. J. (Syngenta Ltd.); US Patent WO2014096289, June 26, 2014.
17. Ruitenberg, K.; Meijer, J.; Bullee, R. J.; Vermeer, P. *J. Organomet. Chem.* **1981**, *217*, 267–271.
18. Bruk, L. G.; Tempkin, O. N. *Inorg. Chim. Acta* **1998,** *280*, 202–210.
19. Lam, T. Y.; Wang, Y.; Danheiser, R. L. *J. Org. Chem.* **2013,** *78*, 9396–9414.

Appendix
Chemical Abstracts Nomenclature (Registry Number)

1-Iodopropyne: 1-Propyne, 1-iodo; (624-66-8)
1-Bromo-1-propene: 1-Propene, 1-bromo-; (590-14-7)
n-BuLi: *n*-Butyllithium; (109-72-8)
Iodine: Iodine; (7553-56-2)

Rick L. Danheiser received his undergraduate education at Columbia where he carried out research in the laboratory of Professor Gilbert Stork. He received his Ph.D. at Harvard in 1978 working under the direction of E. J. Corey on the total synthesis of gibberellic acid. Dr. Danheiser is the A. C. Cope Professor of Chemistry at MIT where his research focuses on the design and invention of new annulation and cycloaddition reactions, and their application in the total synthesis of biologically active compounds.

Samuel G. Bartko was born in Littleton, Colorado and received a S.B. degree in Chemistry at the University of Wyoming in 2014. He is currently pursuing a Ph.D. degree in the Department of Chemistry at the Massachusetts Institute of Technology under the direction of Professor Rick Danheiser. The main focus of Sam's current research is the development of new strategies for the efficient synthesis of highly substituted pyridines based on cycloadditions of highly conjugated molecules.

James Deng was born in New Haven, Connecticut and is currently an undergraduate chemistry major at the Massachusetts Institute of Technology and is expecting to graduate with a B.S. degree in chemistry in 2017. James joined the laboratory of Professor Danheiser as an undergraduate researcher in January 2014 with his main project focusing on the investigation of synthetic routes to bioactive polyacetylene natural products.

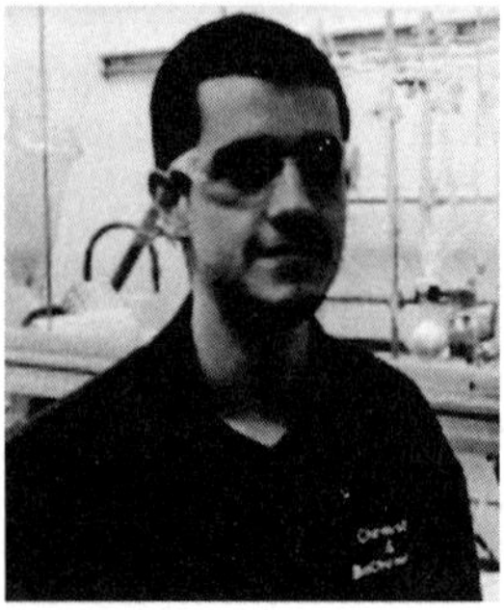

Joshua Cox received his B.S. degree from Texas A&M in chemistry in 2014. He is currently pursuing a Ph.D. degree in the Department of Chemistry at Baylor University under the direction of John L. Wood.

Organic Syntheses

Potassium *tert*-Butoxide–Catalyzed Dehydrogenative Cross-Coupling of Heteroarenes with Hydrosilanes

Anton A. Toutov, Wen-Bo Liu, Brian M. Stoltz,*[1] and Robert H. Grubbs*

Division of Chemistry and Chemical Engineering, California Institute of Technology, Mail Code 101-20, Pasadena, CA 91125

Checked by Jose M. Medina and Neil K. Garg

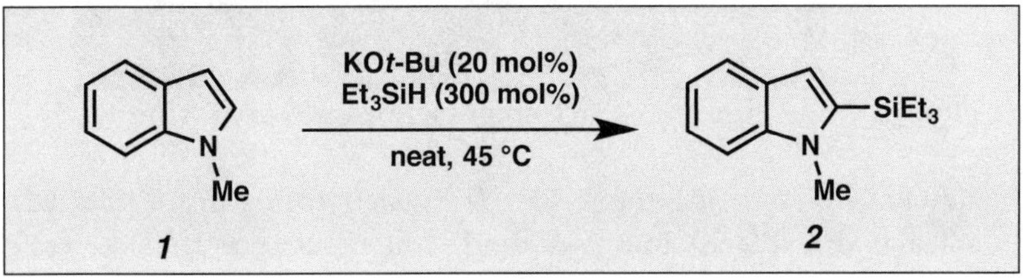

Procedure

A. *1-Methyl-2-(triethylsilyl)-1H-indole* (**2**). An oven-dried 100-mL round-bottom Schlenk flask is equipped with a Teflon-coated magnetic stir bar (25 x 8 mm) and capped with a rubber septum (Note 1). The side arm of the flask is connected to Schlenk line and the flask is cooled to room temperature under vacuum and back-filled with nitrogen. Potassium *tert*-butoxide (1.34 g, 12.0 mmol, 0.2 equiv) is added to the flask under positive nitrogen flow (Notes 2 and 3), and then the flask is evacuated and back-filled with nitrogen three times. *N*-Methylindole (7.48 mL, 7.86 g, 60.0 mmol, 1.0 equiv) and triethylsilane (28.6 mL, 20.82 g, 179.5 mmol, 3.0 equiv) are added sequentially via syringe through the septum, resulting in a yellow heterogeneous mixture (Figure 1). After the septum is replaced with a glass stopper (Note 4), the reaction mixture is degassed (Note 5) and stirred at 45 °C for 76 h (Note 6), resulting in a dark purple solution. The heating bath is removed and the reaction mixture is allowed to cool to ambient temperature (~25 °C), then anhydrous diethyl ether (30 mL) is added

Figure 1. Reaction vessel prior to degassing

slowly while stirring (Note 7). The reaction mixture is filtered through a silica pad (Note 8) into a 500-mL round-bottom flask as a clear yellow solution. The reaction flask is rinsed with diethyl ether (3 x 10 mL) and transferred to the top of a silica pad. The silica pad is washed with diethyl ether (3 x 40 mL) and the combined filtrates are concentrated *in vacuo* using a rotary evaporator to give an orange liquid (Note 9). This liquid is transferred to a 50-mL round-bottom flask, and diethyl ether (3 x 5 mL) is used to assistant the transfer. The resulting solvents are concentrated *in vacuo* using a rotary evaporator (Note 9) and put under high vacuum (ca. 0.03 mmHg) for 30 min. An egg-shaped magnetic stir bar (16 x 6 mm) is added to this flask

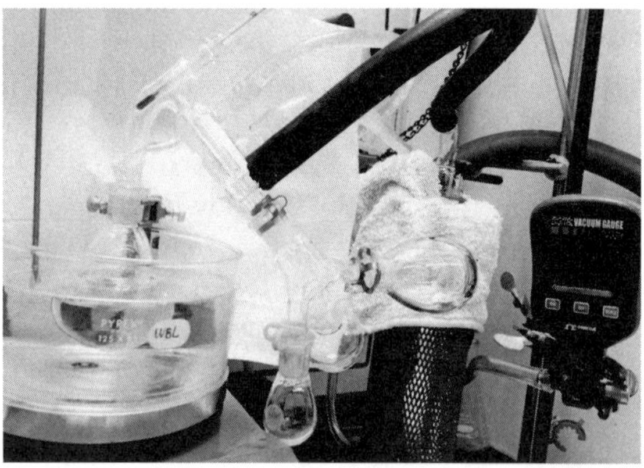

Figure 2. Vacuum distillation apparatus

and a distillation apparatus with a thermometer installed at the distillation head is attached (Figure 2). The product 1-methyl-2-(triethylsilyl)-1*H*-indole **2** (10.34–10.55 g, 70–72% yield) is isolated by vacuum distillation as a pale yellow oil (Figure 3) (Notes 10 and 11).

Figure 3. Product (2), a pale yellow oil

Notes

1. The flask and stir bar are dried in a 120 °C oven for 2 days before use. To limit the possibility of contamination from adventitious transition metal species, a brand-new stir bar is used for each reaction.
2. The submitters purchased the following reagents from commercial sources and used them without further purification: KO*t*-Bu (sublimed grade, 99.99% trace metals basis, Aldrich), *N*-methylindole (>96%, TCI) and triethylsilane (99%, in Sure/Seal™, Aldrich). The checkers purchased and used without further purification KO*t*-Bu (98%, Strem Chemicals) instead.
3. Potassium *tert*-butoxide (KO*t*-Bu) is hygroscopic and kept in a desiccator or a glove box. It should be manipulated as quickly as possible to avoid hydrolysis.
4. The Schlenk flask is uncapped and quickly capped with a lightly greased glass stopper (oven-dried and cooled down in a desiccator) under a positive stream of nitrogen.

5. Freeze-pump-thaw cycles (x 4 times) are used for degassing: 1) close th sidearm stopcock and leave the attached hose of the Schlenk line on vacuum 2) put this flask into liquid nitrogen until the mixture is frozen, open th stopcock to vacuum and pump off atmosphere for 10 min; 3) close th stopcock and thaw the mixture in an *i*-PrOH bath until it just melts; 4) repea steps (2) – (3) for 4 cycles, and 5) fill the flask with nitrogen gas.
6. An oil bath is used for heating. The mixture is stirred at 800 rpm throughou the reaction.
7. *Caution*: Gas bubbles through the liquid vigorously.
8. Silica pad (diameter: 4 cm, height: 3 cm) is packed with silica gel (10 g). Th silica gel is purchase from Silicycle (SilaFlash P60, 60 Å, 40–63 µm).
9. Rotary evaporation is performed at 25 mmHg and 25 °C.
10. Vigorous stirring (700 rpm) and full vacuum are maintained throughout th distillation. The condenser is cooled with ice-water (using a circulating ice water pump), and a liquid nitrogen cold trap is connected between th distillation apparatus and high vacuum pump (see Figure 2 for setup). Th checkers used chilled water (from the fume-hood faucet) to cool th condenser. Distillation procedure: 1) The oil bath is heated to 85 °C slowly When the vacuum stabilizes around 0.24 mmHg (submitters pressur reading = 0.1 mmHg), a forerun comes off as a colorless liquid and th thermometer reads 23–28 °C.; 2) When the drip rate decreases, the heatin bath is increased to 140 °C and more undesired liquid comes over a 0.24 mmHg (submitters pressure reading = 0.08 mmHg) when th thermometer reads ~70 °C; 3) Again, as the drip rate decreases, the heatin bath is increased to 165 °C, and when the vacuum is lowered to 0.23 mmH (submitters pressure reading = 0.07 mmHg), desired 1-methyl-2 (triethylsilyl)-1*H*-indole (**2**) product is isolated in a separate receiving flas (100–110 °C) as a pale yellow oil (Figure 3).
11. The product has been characterized as follows: R_f = 0.5 (10% CH_2Cl_2 i hexanes, EMD gel 60 F254 pre-coated plates (0.5 mm), visualized by shor wave UV); ^{1}H NMR (500 MHz, $CDCl_3$) δ: 0.88 – 0.95 (m, 6H), 0.98 – 1.03 (m 9H), 3.85 (s, 3H), 6.73 (d, *J* = 1.0 Hz, 1H), 7.09 (ddd, *J* = 7.9, 7.0, 1.0 Hz, 1H 7.23 (ddd, *J* = 8.3, 7.0, 1.3 Hz, 1H), 7.34 (dq, *J* = 8.3, 1.0 Hz, 1H), 7.63 (dt, *J* = 7.9, 1.1 Hz, 1H); ^{13}C NMR (125 MHz, $CDCl_3$) δ: 4.1, 7.7, 33.1, 109.1, 113.0 119.1, 120.7, 122.0, 128.7, 138.4, 140.3. IR (Neat film) 3058, 2953, 2909, 2874 1608, 1492, 1464, 1414, 1372, 1356, 1325, 1299, 1233, 1167, 1102, 1069, 1005, 973 797 cm^{-1}; HRMS (ESI+) calcd for $C_{15}H_{24}NSi$ [M+H]$^+$: 246.16725, found 246.16679; Anal. calcd for $C_{15}H_{23}NSi$: C, 73.40; H, 9.45: N, 5.71; found: C, 73.2 H, 9.64; N, 5.7. The checkers determined the purity of the compound usin

qNMR: 14.7 mg of the product are dissolved in 0.9 mL of $CDCl_3$. Hexamethylbenzene, 9.8 mg (99+%, purchased from Alfa Aesar and used as received), is added. ^1H NMR (500 MHz, $CDCl_3$) gave a product purity of 98.5%.

Working with Hazardous Chemicals

The procedures in *Organic Syntheses* are intended for use only by persons with proper training in experimental organic chemistry. All hazardous materials should be handled using the standard procedures for work with chemicals described in references such as "Prudent Practices in the Laboratory" (The National Academies Press, Washington, D.C., 2011; the full text can be accessed free of charge at http://www.nap.edu/catalog.php?record_id=12654). All chemical waste should be disposed of in accordance with local regulations. For general guidelines for the management of chemical waste, see Chapter 8 of Prudent Practices.

In some articles in *Organic Syntheses*, chemical-specific hazards are highlighted in red "Caution Notes" within a procedure. It is important to recognize that the absence of a caution note does not imply that no significant hazards are associated with the chemicals involved in that procedure. Prior to performing a reaction, a thorough risk assessment should be carried out that includes a review of the potential hazards associated with each chemical and experimental operation on the scale that is planned for the procedure. Guidelines for carrying out a risk assessment and for analyzing the hazards associated with chemicals can be found in Chapter 4 of Prudent Practices.

The procedures described in *Organic Syntheses* are provided as published and are conducted at one's own risk. *Organic Syntheses, Inc.*, its Editors, and its Board of Directors do not warrant or guarantee the safety of individuals using these procedures and hereby disclaim any liability for any injuries or damages claimed to have resulted from or related in any way to the procedures herein.

Discussion

Heteroarylsilanes are valuable synthetic intermediates which are widely used in complex molecules synthesis,[2,3] medicinal chemistry and drug discovery,[4,5] and advanced materials and polymers synthesis.[6,7] The traditional methods for the synthesis of these motifs rely on deprotonation of heteroarenes by employin pyrophoric organometallic reagents (i.e., BuLi), followed by substitutio reactions with silicon electrophiles. Direct dehydrogenative cross-coupling of C-H bonds of heteroarenes with Si–H bonds of hydrosilanes is an effectiv alternative method for the construction of C–Si bonds.[8] Despite significan advances in C–H activation recently, there are only a handful of methods tha enable such a transformation. However, these examples are all limited to the us of precious metal catalysts (Ru,[9] Ir,[10,11] and Rh[12,13]). Moreover, stoichiometri sacrificial hydrogen acceptors are necessary for these transformations.

We have established a method for such a dehydrogenative cross-couplin reaction using potassium *tert*-butoxide as a catalyst that provides a unique wa to access a wide array of heteroarylsilanes.[14,15] As described above, the desire silylation product **2** was obtained in good yield (70–72%). This procedur features readily available and inexpensive starting materials, mild reactio conditions, and good functional group tolerance.

The procedure reported here is applicable to a range of heteroarenes an hydrosilanes, giving access to synthetically valuable heteroarylsilanes on multi gram scales.[14,15]

References

1. A.A.T. and W.-B.L. contributed equally to this work. Division of Chemistr and Chemical Engineering, California Institute of Technology, Pasaden California, 91125; E-mail: rhg@caltech.edu, stoltz@caltech.edu. Thi publication is based on work supported by the NSF under the CCI Center fo Selective C–H Functionalization (CHE-1205646) and under CHE-1212767. W. B.L. wishes to thank the Shanghai Institute of Organic Chemistry (SIOC) an Prof. S.-L. You for a postdoctoral fellowship. A.A.T. is grateful to the Resnic Sustainability Institute at Caltech and to Dow Chemical for a predoctora fellowship, and to NSERC for a PGS D fellowship. We also thank Dr. Sco Virgil and the Caltech Center for Catalysis and Chemical Synthesis for acces to analytical equipment.
2. Ball, L. T.; Lloyd-Jones, G. C.; Russell, C. A. *Science* **2012**, *337*, 1644–1648.
3. Denmark, S. E.; Baird, J. D. *Chem. –Eur. J.* **2006**, *12*, 4954–4963.
4. Showell, G. A.; Mills, J. S. *Drug Discov. Today* **2003**, *8*, 551–556.
5. Franz, A. K.; Wilson, S. O. *J. Med. Chem.* **2013**, *56*, 388–405.
6. Zhang, F.; Wu, D.; Xu, Y.; Feng, X. *J. Mater. Chem.* **2011**, *21*, 17590–17600.

7. Tamao, K.; Uchida, M.; Izumizawa, T.; Furukawa, K.; Yamaguchi, S. *J. Am. Chem. Soc.* **1996**, *118*, 11974–11975.
8. Cheng, C.; Hartwig, J. F. *Chem. Rev.* **2015**, *115*, 8946–8975.
9. Lu, B.; Falck, J. R. *Angew. Chem. Int. Ed.* **2008**, *47*, 7508–7510.
10. Klare, H. F.; Oestreich, M.; Ito, J.-i.; Nishiyama, H.; Ohki, Y.; Tatsumi, K. *J. Am. Chem. Soc.* **2011**, *133*, 3312–3315.
11. Cheng, C.; Hartwig, J. F. *J. Am. Chem. Soc.* **2015**, *137*, 592–595.
12. Cheng, C.; Hartwig, J. F. *Science* **2014**, *343*, 853–857.
13. Cheng, C.; Hartwig, J. F. *J. Am. Chem. Soc.* **2014**, *136*, 12064–12072.
14. Toutov, A. A.; Liu, W.-B.; Betz, K. N.; Fedorov, A.; Stoltz, B. M.; Grubbs, R. H. *Nature* **2015**, *518*, 80–84.
15. Toutov, A. A.; Liu, W.-B.; Betz, K. N.; Stoltz, B. M.; Grubbs, R. H. *Nat. Protoc.* **2015**, *10*, 1897.

Appendix
Chemical Abstracts Nomenclature (Registry Number)

Potassium *tert*-butoxide: 2-Propanol, 2-methyl-, potassium salt (1:1); (865-47-4)
N-methylindole: 1-*H*-Indole, 1-methyl-; (603-76-9)
1-Ethyl-2-(triethylsilyl)-1*H*-indole (**2**): 1-*H*-Indole, 1-methyl-2-(triethylsilyl)-; (1085787-53-6)
Triethylsilane: Silane, triethyl-; (617-86-7)

Anton A. Toutov was born in Kiev, Ukraine. He obtained his B.S. degree in chemistry in 2011 from Queen's University in Kingston, Canada, where he developed new heterocyclic lithiation chemistries under the supervision of Victor Snieckus. He then moved to Caltech where he is currently a 5th year Ph.D. candidate in the laboratory of Robert H. Grubbs. His research focuses on using Earth-abundant elements in the development of efficient and sustainable chemical reactions.

Wen-Bo Liu was born in China and he received his Bachelor's Degree in Chemistry from the Nankai University in 2006. He obtained his Ph.D. in organic chemistry (2011) from the Shanghai Institute of Organic Chemistry (SIOC) under the supervision of Professor Li-Xin Dai and Professor Shu-Li You. Then he joined the laboratory of Professor Brian M. Stoltz at Caltech as a postdoctoral scholar, working on asymmetric catalysis and sustainable chemistry.

Brian M. Stoltz was born in Philadelphia, PA in 1970 and obtained his B.S. degree from the Indiana University of Pennsylvania in Indiana, PA. After graduate work at Yale University in the labs of John L. Wood and an NIH postdoctoral fellowship at Harvard with E. J. Corey, he took a position at the California Institute of Technology. A member of the Caltech faculty since 2000, he is currently Professor of Chemistry. His research interests lie in the development of new methodology for general applications in synthetic chemistry.

Robert H. Grubbs obtained his B.S. degree from the University of Florida. After receiving his Ph.D. from Columbia University under the supervision of Professor Ronald Breslow, he joined the lab of James P. Collman as an NIH Postdoctoral Fellow at Stanford University. Then he started his academic career as an assistant professor in chemistry at Michigan State University in 1969. He moved to California Institute of Technology in 1978 and currently, he is the Victor & Elizabeth Atkins Professor of Chemistry.

Jose M. Medina was born in Peru. He received his Bachelor's Degree in Chemistry from CSU Channel Islands in 2012. He is currently pursuing his Ph.D. in organic chemistry in the laboratory of Professor Neil K. Garg at the University of California, Los Angeles. His graduate research has focused on harnessing cyclic alkynes for the synthesis of heterocyclic compounds and the development of nickel-catalyzed C–C bond forming reactions from amides.

Organic Syntheses

Synthesis of α-Carboline

Linli He, Shawn P. Allwein, Benjamin J. Dugan, Kyle W. Knouse, Gregory R. Ott, and Craig A. Zificsak*[1]

Discovery and Product Development, Teva Branded Pharmaceutical Products R&D, Inc., Malvern and West Chester, PA, USA

Checked by Kevin Kou and Richmond Sarpong

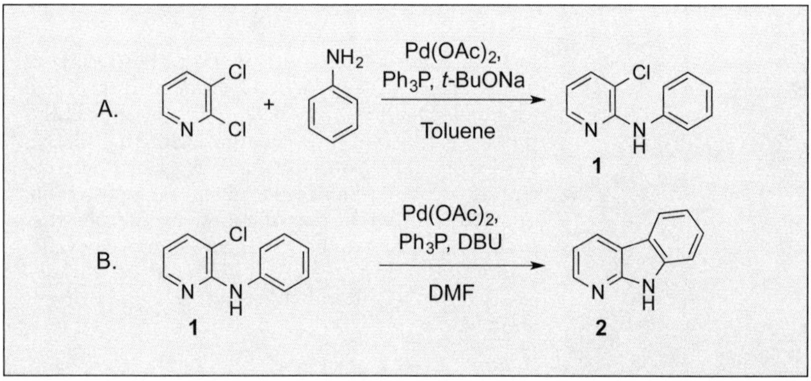

Procedure

A. *3-Chloro-N-phenyl-pyridin-2-amine* (**1**). An air-dried, 1-L, three-necked, round-bottomed flask is equipped with a Teflon-coated, egg-shaped, magnetic stirring bar (1/2"x 5/8"), rubber septum, air-jacketed condenser with nitrogen inlet and thermometer (Figure 1). The flask is evacuated (20 mmHg), backfilled with nitrogen, then charged with 2,3-dichloropyridine (30.0 g, 203 mmol, 1.0 equiv) (Note 1), palladium(II) acetate (1.14 g, 5.08 mmol, 0.025 equiv), triphenylphosphine (3.99 g,

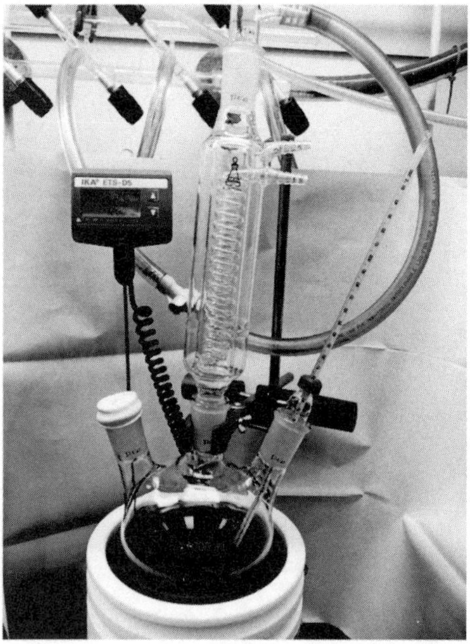

Figure 1. Reaction Setup: Steps A and B

15.2 mmol, 0.075 equiv) and sodium *tert*-butoxide (29.9 g, 311 mmol, 1.53 equiv). The septum is replaced and toluene (210 mL) is added via syringe, the resulting light brown solution is stirred for 10 min before aniline (20.0 mL, 219 mmol, 1.05 equiv) is added via syringe in one portion. The resulting dark brown mixture is heated to 100 °C in a heating mantle over 20 min and stirred for 16 h (Note 2). The reaction is cooled to 0 °C in an ice bath and 2 M HCl (300 mL) is added. The ice bath is removed and the mixture is stirred for 10 min at room temperature before filtering through a pad of Celite 545 (15 g) in a 350-mL, medium-porosity, fritted funnel. The filtrate is poured into a 1-L separatory funnel, the layers are separated, and the organic phase is further extracted with 2 M HCl (2 x 50 mL). The combined acidic extracts are cooled to 0 °C in an ice bath and dichloromethane (100 mL) is added. The biphasic solution is then treated with 19 M NaOH (50 mL) and transferred to a 1-L separatory funnel. The phases are separated and the aqueous phase is further extracted with dichloromethane (3 x 50 mL). The combined organic extracts are dried over sodium sulfate (10 g) for 10 min, diluted with hexane (250 mL), then filtered through a pad of silica gel (90 g) (Note 3), rinsing with dichloromethane:hexane (1:1, 500 mL) (Figure 2). The filtrate is collected into two 1-L flasks. The first 500 mL fraction is colorless and the second 500 mL fraction has a subtle tint of yellow (Figure 3). Both fractions were concentrated to give a yellow liquid. ^{1}H NMR analyses of both fractions were identical; therefore the two fractions were combined to give 34.5 g (82%) of 3-chloro-*N*-phenyl-pyridin-2-amine as a yellow liquid, which formed beige crystals when cooled in a −20 °C freezer (Notes 4 and 5).

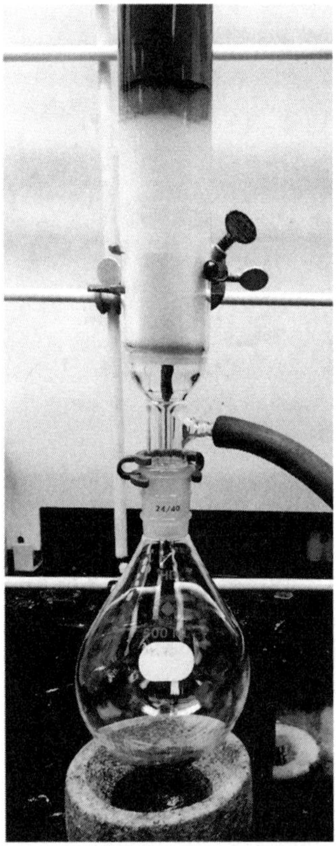

Figure 2. Filtration through silica gel

Figure 3. Fractions 1 and 2 following filtration through silica gel

B. *α-Carboline* (**2**). An air-dried, 1-L, three-necked, round-bottomed flask is equipped with a Teflon-coated, egg-shaped, magnetic stirring bar (1/2″x 5/8″), rubber septum, air condenser with nitrogen inlet and thermometer (Figure 1). The flask is evacuated with high vacuum, backfilled with nitrogen then charged with 3-chloro-*N*-phenyl-pyridin-2-amine (35.0 g, 171 mmol, 1.0 equiv), palladium(II) acetate (0.96 g, 4.3 mmol, 0.025 equiv), and triphenylphosphine (3.36 g, 12.8 mmol, 0.075 equiv). The septum is replaced and *N,N*-dimethylformamide (150 mL) is added via syringe. The resulting solution is stirred for 10 min before 1,8-diazabicyclo[5.4.0]undec-7-ene (52.0 mL, 52.9 g, 348 mmol, 2.0 equiv) is added via syringe (Note 6). The resulting dark amber solution is heated to 130 °C in a heating mantle over 25 min and stirred for 48 h (Note 7). The reaction is cooled to room temperature and filtered through a sintered glass funnel into a 2-L round-bottomed flask, equipped with a Teflon-coated, egg-shaped, magnetic stirring bar (1/2″x5/8″). The sintered glass funnel is replaced with an addition funnel and water (850 mL) is added over 25 min

Figure 4. Golden yellow suspension after precipitation with water

with stirring (exothermic). The golden yellow suspension (Figure 4) is stirred for 30 min before the solids are collected in a Büchner funnel. The flask and solids are washed with *N,N*-dimethylformamide/water (1:3, 100 mL) followed by water (600 mL). The yellow solids are dried in a vacuum desiccator for 1 h, transferred to a 1-L Erlenmeyer flask equipped with a Teflon-coated, egg shaped, magnetic stirring bar (1/2″x 5/8″). Acetonitrile (300 mL) is added to the flask and the red suspension is stirred for 15 min at room temperature before the solids are collected in a Büchner funnel. The flask and solids are washed with cold acetonitrile until the filtrate is clear (120 mL). The solids are transferred to a 500-mL round-bottomed flask and dried under vacuum for 16 h to afford α-carboline as a light green solid (24.9 g, 87% yield) (Note 8). The isolated solids (24.9 g) are heated in toluene (700 mL) to an internal temperature of 100 °C in a 1-L Erlenmeyer flask (Note 9) for 20 min with stirring (Note 10). The heater and stirrer are turned off and the mixture is gradually cooled to room temperature over 90 min (Note 11) then chilled in an ice bath for 20 min (Note 12). The solids are collected by filtration, washed with toluene (100 mL) and dried on a Büchner funnel and then in vacuo to afford **2** as beige or gray crystals (23.4 g, 81%) (Figure 5) (Notes 13 and 14).

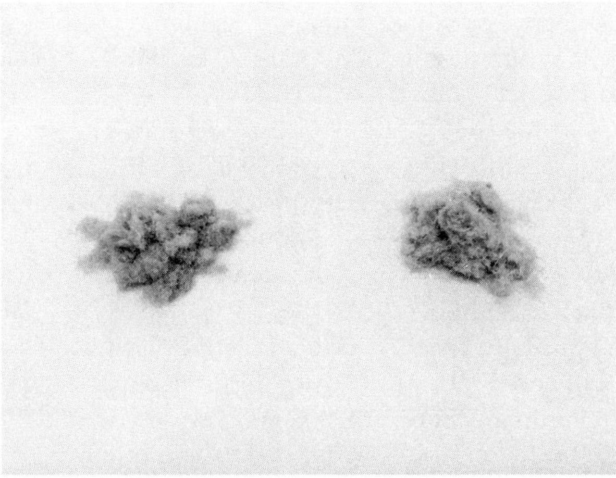

Figure 5. Checker submitted: Left: α-Carboline obtained as fluffy, gray needles after recrystallization (run 1); Right: Fluffy, beige needles after recrystallization (run 2)

Notes

1. 2,3-Dichloropyridine (98%, Oakwood Chemical) and palladium(II) acetate (98%, Strem Chemicals) were used as received. Triphenylphosphine (99%), sodium *tert*-butoxide (98%), toluene (99.8%, Extra Dry, AcroSeal®) and aniline (99.8%) were purchased from Acros Organics and used as received.
2. The reaction is monitored by UPLC-MS (Waters Acquity Ultra Performance LC with a Waters Aquity UPLC BEH C18 1.7 μm 2.1x50 mm column paired with a Micromass-ZQ 2000 quadrupole mass spectrometer with electrospray ionization. Solvent A consisted of 0.1% formic acid in 19:1 water:acetonitrile; solvent B consisted of 0.1% formic acid in acetonitrile. The target column temperature was 45 °C, with a run time of two minutes, a flow rate of 0.60 mL/min, and a solvent mixture of 5%A:95%B ramping to 100%B at 1.7 min, holding for 0.6 min. Retention time of 1.38 min for **1**.
3. Silica gel (Sorbent Technologies, 40-75 μm particle size, 60Å pore size) was placed into a fritted glass funnel (50 mm diameter) and house vacuum was used to pull the solution through the silica gel.

4. 3-Chloro-*N*-phenyl-pyridin-2-amine exhibits the following spectroscopic properties: ^{1}H NMR (500 MHz, DMSO-d_6) δ: 6.82 (dd, *J* = 7.7, 4.8 Hz, 1H), 6.98 (app. t, *J* = 7.3 Hz, 1H), 7.24–7.32 (m, 2H), 7.65–7.70 (m, 2H), 7.78 (dd, *J* = 7.7, 1.5 Hz, 1H), 8.09 (dd, *J* = 4.8, 1.5 Hz, 1H), 8.34 (s, 1H); ^{13}C NMR (125 MHz, DMSO-d_6) δ: 115.5, 115.6, 120.5, 122.0, 128.3, 137.4, 140.4, 145.6, 151.3; IR (neat, ATR): 3400, 3060 (weak), 3031 (weak), 3010 (weak), 1597, 1578, 1515, 1495, 1439, 1402, 1127, 1035, 1028, 779, 754, 741, 690, 545, 504 cm^{-1}. Anal calcd for $C_{11}H_9N_2Cl$: C, 64.56; H, 4.43; N, 13.69. Found: C, 64.49; H, 4.38; N, 13.84. The Checkers determined purity to be 99.7% by quantitative NMR versus an authentic sample of caffeine (99.9% purity, Alfa Aesar) using 24.0 mg of **1** and 24.7 mg of caffeine in DMSO-d_6 (ca. 1.5 mL). mp 49–50 °C. R_f = 0.75 (2:1 hexanes/EtOAc).
5. A second run afforded 36.1 g (87% yield) of **1** as an off-white solid. In this run, the Celite pad was rinsed with 2 M HCl (2 × 50 mL) and these rinses used to extract the organic layer.
6. *N,N*-Dimethylformamide (99.8%, Extra Dry, AcroSeal®) and 1,8-diazabicyclo[5.4.0]undec-7-ene (98%) were purchased from Acros Organics and used as received.
7. The reaction is monitored by UPLC-MS as specified in Note 2. Retention time of 0.86 min for **2**.
8. Purity was determined to be 98.7% by quantitative NMR versus an authentic sample of caffeine (99.9% purity, Alfa Aesar) using 26.6 mg of **2** and 25.8 mg of caffeine in DMSO-d_6 (ca. 1.5 mL). mp 218–223 °C. A second run afforded 25.2 g (87% yield) as a pale green solid.
9. The flask was wrapped in aluminum foil.
10. The bulk of the solids fully dissolve, but small particles do remain.
11. The Erlenmeyer flask was not removed from the heat plate, but the flask was allowed to gradually cool as the heat plate slowly cooled.
12. An internal temperature of 0 °C is reached.
13. α-Carboline exhibits the following spectroscopic properties: ^{1}H NMR (500 MHz, DMSO-d_6) δ: 7.17–7.25 (m, 2H), 7.42–7.48 (m, 1H), 7.51 (d, *J* = 8.0 Hz, 1H), 8.15 (d, *J* = 7.8 Hz, 1H), 8.41 (dd, *J* = 4.8, 1.6 Hz, 1H), 8.49 (dd, *J* = 7.7, 1.6 Hz, 1H), 11.80 (s, 1H); ^{13}C NMR (125 MHz, DMSO-d_6) δ: 111.3, 115.0, 115.2, 119.4, 120.4, 121.2, 126.6, 128.4, 138.8, 146.1, 151.9; IR (film, ATR): 3400, 3129 (weak), 3048 (weak), 2984 (weak), 1597, 1577, 1515, 1494, 1456, 1416, 1403, 1275, 1231, 750, 738, 726, 690, 543 cm^{-1}. Purity was determined to be 99.9% by quantitative NMR versus an authentic sample of caffeine (99.9% purity, Alfa Aesar.) using 26.6 mg of

2 and 25.8 mg of caffeine in DMSO-d_6 (ca. 1.5 mL). Anal calcd for $C_{11}H_8N_2$: C, 78.55; H, 4.79; N, 16.66. Found: C, 78.36; H, 4.61; N, 16.56. mp 217–221 °C. R_f = 0.24 (2:1 hexanes/EtOAc).

14. A second run starting from 35.5 g of 3-chloro-*N*-phenyl-pyridin-2-amine yielded a pale green solid (25.2 g, 87% yield), which was recrystallized in toluene to give beige needles (24.1 g, 83% yield).

Working with Hazardous Chemicals

The procedures in *Organic Syntheses* are intended for use only by persons with proper training in experimental organic chemistry. All hazardous materials should be handled using the standard procedures for work with chemicals described in references such as "Prudent Practices in the Laboratory" (The National Academies Press, Washington, D.C., 2011; the full text can be accessed free of charge at http://www.nap.edu/catalog.php?record_id=12654). All chemical waste should be disposed of in accordance with local regulations. For general guidelines for the management of chemical waste, see Chapter 8 of Prudent Practices.

In some articles in *Organic Syntheses*, chemical-specific hazards are highlighted in red "Caution Notes" within a procedure. It is important to recognize that the absence of a caution note does not imply that no significant hazards are associated with the chemicals involved in that procedure. Prior to performing a reaction, a thorough risk assessment should be carried out that includes a review of the potential hazards associated with each chemical and experimental operation on the scale that is planned for the procedure. Guidelines for carrying out a risk assessment and for analyzing the hazards associated with chemicals can be found in Chapter 4 of Prudent Practices.

The procedures described in *Organic Syntheses* are provided as published and are conducted at one's own risk. *Organic Syntheses, Inc.*, its Editors, and its Board of Directors do not warrant or guarantee the safety of individuals using these procedures and hereby disclaim any liability for any injuries or damages claimed to have resulted from or related in any way to the procedures herein.

Discussion

Since the first preparation of α-carboline (**2**),[2] this heterocycle has found utility as a building block in natural product and medicinal chemistry[3] as well as in optoelectronic materials.[4] Literature procedures for the preparation of **2** fall into two general catagories: thermolysis of a pyridylbenzotriazole,[2] with extrusion of nitrogen leading to bond formation (a modified Graebe-Ullman reaction) or formation of the C4a-C4b bond from an aryl halide–aryl-H intramolecular coupling.[5,6]

Reviewing the known syntheses, two deficiencies with the preparation of **2** are of particular note: reactions are limited to small scale, producing less than 1 g of material, and only moderate to low yields are obtained after silica gel chromatography. Our focus was to find a method to prepare multiple grams of **2** with minimal or no chromatography.

Using the palladium-mediated synthesis of **2** from Cuny et al.[6] as a starting point, we chose to run discrete reaction steps in order to allow for purification of the Buchwald-Hartwig coupling product. Use of toluene as solvent gave an improvement in reaction rate and permitted an acceptable concentration (1M) of substrate. The use of an air jacketed condenser is needed in order to condense the *tert*-butanol generated from the base. Extraction of the product into aqueous HCl allowed for removal of the triphenylphosphine, triphenylphosphine oxide, and metal-ligand complex. Following pH adjustment and extraction, filtration of the organic extract through a plug of silica gel, and resubjecting the second fraction to another filtration, removes residual palladium complex and excess aniline. Isolation of **1** in >99% purity allowed for minimal purification in the second stage. The biaryl coupling was found to proceed readily under conditions similar to those used by Cuny,[6] substituting DMF for DMA and the less expensive triphenylphosphine ligand in place of tricyclohexylphosphine tetrafluoroborate salt. Addition of water to the reaction mixture allowed for precipitation of the bulk of the product, with minimal losses to the filtrate. These initial solids did contain significant amounts of ligand and palladium complex, which was removed in a single trituration in acetonitrile to furnish material at 98% purity. Higher purity product was readily obtained with good recovery by a recrystallization in toluene.

Lastly, using the optimized conditions used for the formation of **2**, we tested the generality of the two steps with a selection of aniline coupling partners. As shown in Table 1, all of the anilines coupled to 2,3-

dichloropyridine (3.0 g scale) in good yield and were isolated by silica gel chromatography following workup. The second stage coupling, when carried out in a discrete step, furnished low to good yields of the substituted carbolines. Products were purified by chromatography due to inconsistent precipitate formation from DMF/water. Notably, the yield of analog **2g** was low, but in the one-pot-two-step protocol of Cuny, no product had formed.[6] Analog **2e**, with an *o*-fluoro substituent, suffered from lower yield than the literature.

Table 1: Functionalized α-carboline analogs prepared by steps A and B

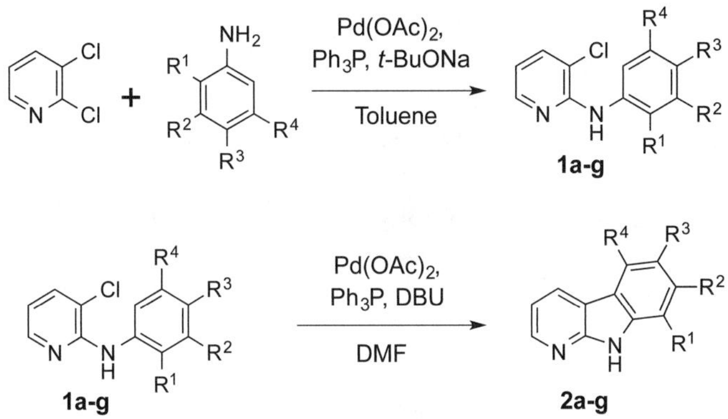

Entry	R¹	R²	R³	R⁴	#	Yield (%)	#	Yield (%)
1	H	H	H	H	**1a**	85	**2a**	84
2	H	H	OMe	H	**1b**	91	**2b**	77
3	H	H	F	H	**1c**	83	**2c**	84
4	OMe	H	H	H	**1d**	82	**2d**	89
5	F	H	H	H	**1e**	82	**2e**	35
6	Me	H	H	Me	**1f**	81	**2f**	52
7	H	OMe	H	OMe	**1g**	93	**2g**	39

References

1. L.H., S.P.A.: Chemical Process Research and Development, Discovery & Product Development, Teva Branded Pharmaceutical Products R&D, Inc., Malvern, PA; B.J.D., K.W.K., G.R.O. and C.A.Z.: Medicinal Chemistry, Discovery & Product Development, Teva Branded Pharmaceutical Products R&D, Inc., West Chester, PA, USA. Email: Craig.Zificsak@Tevapharm.com
2. Lawson, W.; Perkin, W. H.; Robinson, R. *J. Chem. Soc., Trans.* **1924**, *125*, 626–657.
3. For a current review, see: Wadsworth, A. D.; Naysmith, B. J.; Brimble, M. A. *Eur. J. Med. Chem.* **2015**, *97*, 816–829.
4. Byeon, S. Y.; Jeon, S. K.; Hwang, S.-H.; Lee, J. Y. *Dyes Pigments* **2015**, *120*, 258–264.
5. Iwaki, T.; Yasuhara, A.; Sakamoto, T. *J. Chem. Soc., Perkin Trans. 1* **1999**, 1505–1510.
6. Laha, J. K.; Petrou, P.; Cuny, G. D. *J. Org. Chem.* **2009**, *74*, 3152–3155.

Appendix
Chemical Abstracts Nomenclature (Registry Number)

2,3-Dichloropyridine: Pyridine, 2,3-dichloro-; (2402-77-9)
Palladium(II) acetate: Acetic acid, palladium(2+) salt; (3375-31-3)
Triphenylphosphine: Phosphine, triphenyl-; (603-35-0)
Aniline: Benzenamine; (62-53-3)
Sodium *tert*-butoxide: 2-Propanol, 2-methyl-, sodium salt; (865-48-5)
3-Chloro-*N*-phenyl-pyridin-2-amine: 2-Pyridinamine, 3-chloro-*N*-phenyl-; (6604-94-0)
1,8-Diazabicyclo[5.4.0]undec-7-ene: Pyrimido[1,2-*a*]azepine, 2,3,4,6,7,8,9,10-octahydro-; (6674-22-2)
α-Carboline: 9*H*-Pyrido[2,3-*b*]indole; (244-76-8)

Dr. Linli He was born in P.R. China, where he obtained his B.S. degree in chemistry from ZheJiang University. He moved to US in 1994 and received his Ph.D. degree in organic chemistry in 2000 from The City University of New York under the guidance of Prof. Robert Bittman. He then joined the Process Research and Development team of Rhodia Pharma Solutions. He accepted a position at Teva Global R&D (formerly Cephalon, Inc.) in 2005. His research interest is mainly involved in organic process research and development of active pharmaceutical ingredients.

Dr. Shawn P. Allwein was born and raised in Pennsylvania. He received a B.S. degree in chemistry from Shippensburg University then his Ph.D. from the University of Arizona, working in the laboratories of Prof. Jon Rainier. Following an NIH post-doctoral fellowship with Prof. Michael Crimmins at the University of North Carolina-Chapel Hill, he started his industrial career with the Department of Process Research at Merck and Co., Inc. In 2006. He then joined the department of Chemical Process Research and Development at Cephalon/Teva Pharmaceuticals where he is currently an Associate Director.

Benjamin J. Dugan received a B.S. degree in Chemistry from the University of Delaware in 1993 under the tutelage of the late Dr. Cynthia McClure. He began his career at FMC Corporation in the agricultural products division. In 2006, he moved to Cephalon, Inc., acquired by Teva Pharmaceutical Industries Ltd. in 2011, and engaged in oncology research focused on small molecule, ATP competitive, kinase inhibitors culminating with the discovery of CEP-33779. He is currently a Research Scientist focused on the development of novel, bioactive small molecules for treatment of central nervous system disorders.

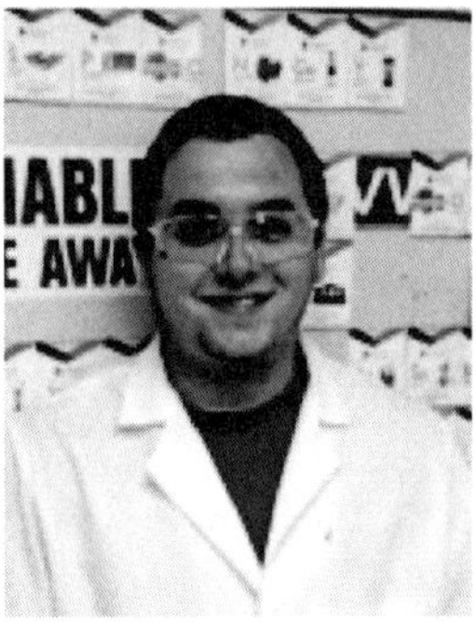

Kyle W. Knouse was born in Pennsylvania in 1993. He completed his bachelor's degree in Chemistry at Temple University in 2016. At Temple, he performed undergraduate research under the guidance of Professor William Wuest. During the summer of 2015 he undertook an internship at Teva Pharmaceutical Industries Ltd. under the supervision of Dr. Craig Zificsak. He recently began studies towards his Ph.D. in Organic Chemistry at the Scripps Research Institute.

Dr. Gregory R. Ott received his B.A. in Chemistry from The Johns Hopkins University in 1992. After obtaining his Ph.D. in Organic Chemistry at The University of Pennsylvania under the direction of Prof. Amos B. Smith, III, Gregory served as a post-doctoral fellow in the laboratories of Prof. Clayton Heathcock at UC, Berkeley. His industrial career began at Dupont Pharmaceuticals in 1999 and he transitioned to Bristol-Myers Squibb in 2001. In 2005, Greg joined Cephalon, Inc. which was acquired by Teva Pharmaceutical Industries Ltd. in 2011. Greg is currently the Director of Medicinal Chemistry at Teva.

Dr. Craig A. Zificsak completed his undergraduate education at the University of Wisconsin–Stevens Point and received his Ph.D. at the University of Minnesota in 2002, working under the direction of Prof. Richard P. Hsung on the total synthesis of indole alkaloids. Following post-doctoral research at Johnson & Johnson, he began his career at Cephalon, Inc. in 2004, which was acquired by Teva Pharmaceutical Industries Ltd. in 2011. He is currently a Research Scientist focused on the design and synthesis of biologically active compounds for oncology and diseases of the central nervous system.

Dr. Kevin Kou was born and raised in Victoria, B.C. (Canada). As part of the B.Sc./Co-op program at the University of Victoria, he acquired industrial experience working at CANTEST, Neuromed Technologies, Targanta Therapeutics, and TRIUMF (radiochemistry), and worked with Prof. Jeremy Wulff for his honor's thesis. He obtained his Ph.D in transition metal catalysis from the University of Toronto with Prof. Vy Dong and is currently a post-doctoral fellow with Prof. Richmond Sarpong at UC Berkeley.

Organic Syntheses

Synthesis of Methyl 1-Formylcyclopropanecarboxylate utilizing Ti-Claisen Condensation

Yuichiro Ashida, Satomi Kajimoto, Hidefumi Nakatsuji,* and Yoo Tanabe[1]*

Department of Chemistry, School of Science and Technology, Kwansei Gakuin University, 2-1 Gakuen, Sanda, Hyogo 669-1337, Japan

Checked by Bryon K. Anderson, Yu-Wen Huang, and John L. Wood

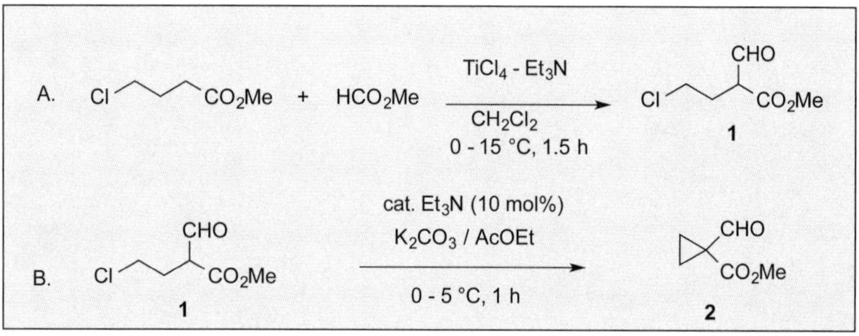

Procedure

A. Methyl 4-chloro-2-formylbutanoate (**1**). An oven-dried 500-mL, three-necked (24/40), round-bottomed flask equipped with a Teflon-coated magnetic stirring bar (egg-shaped, 32 mm length x 15 mm diameter), an internal thermometer, a 50-mL pressure-equalizing addition funnel fitted with a nitrogen inlet (central neck), and a second 60-mL pressure-equalizing addition funnel is charged with methyl 4-chlorobutanoate (12 mL, 13.7 g, 100 mmol, 1 equiv) (Notes 1 and 2), HCO_2Me (18 mL, 18 g, 300 mmol, 3 equiv) (Note 3), and CH_2Cl_2 (100 mL) (Notes 4 and 5) (Figure 1). The stirred solution is immersed in an ice bath, cooling the internal temperature to 0 °C, and $TiCl_4$ (24 mL, 41.7 g, 220 mmol, 2.2 equiv) is added dropwise through a 60-mL dropping funnel (Figure 1, right side) (Notes 6 and 7) over a period of 20 min, while maintaining the internal temperature at 5–10 °C (Note 8).

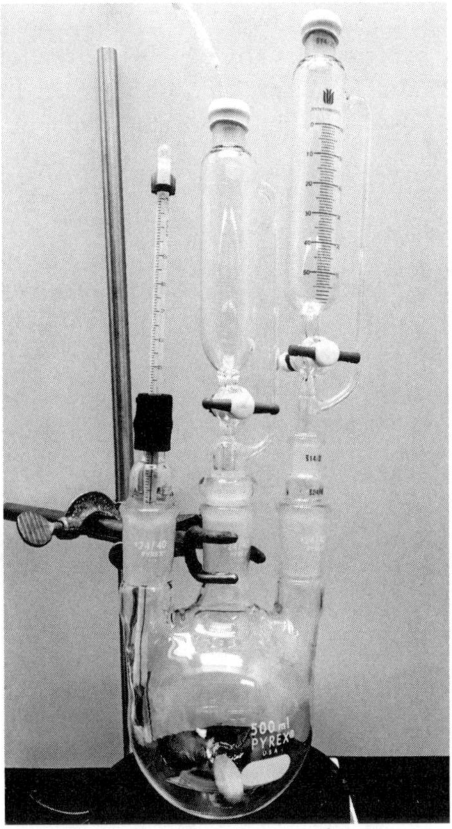

Figure 1. Reaction Set-up for Step A

Triethylamine (36 mL, 26.3 g, 260 mmol, 2.6 equiv) (Note 9) is then added dropwise to the vigorously stirred yellow reaction mixture over a period of 30 min using the 50-mL addition funnel in the center neck of the flask, while maintaining the internal temperature at 15 °C or lower (Note 10) (Figure 2). After complete addition, the dark orange reaction is stirred (500 rpm) at 0 °C for 1 h (Note 11), then quenched dropwise with water (100 mL) over a period of 10 min to maintain the internal temperature at 10 °C or lower (Note 12). The biphasic mixture is then transferred to a 500-mL round-bottomed flask and the initial reaction flask is rinsed with EtOAc (2 x 10 mL). The solution is concentrated using a rotary evaporator (22 °C, 46 mmHg). The mixture is then transferred to a 500-mL separatory funnel with EtOAc (50 mL), and the aqueous phase is separated and re-extracted with EtOAc (50 mL). The combined organic phase is washed with water

(100 mL) and brine (50 mL), dried over Na_2SO_4 (20 g), filtered through a 150-mL medium porosity sintered glass funnel and concentrated using a rotary evaporator (45 °C, 25 mmHg) to furnish α-formyl ester **1** as a yellow liquid (16.39 g), which is used for the next step without any purification (Notes 13 and 14).

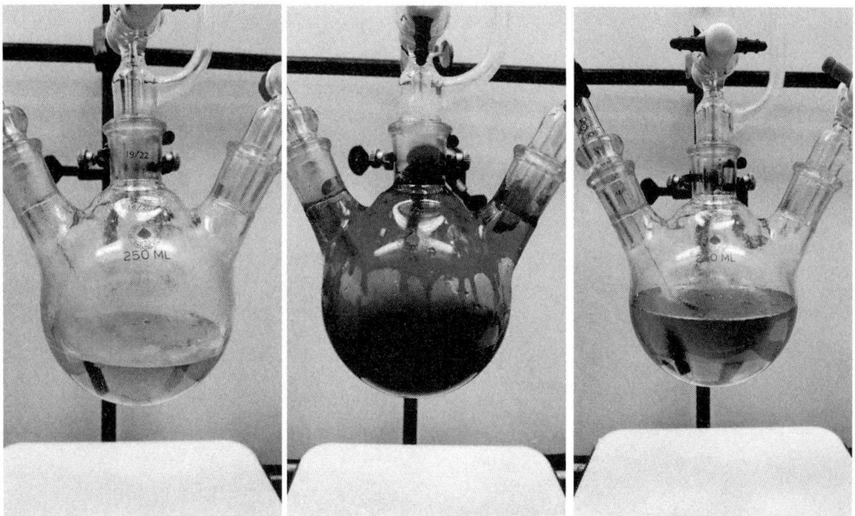

Figure 2. Color Transitions Observed in Step A

after $TiCl_4$ addition after Et_3N addition after quench with H_2O

*B. Methyl 1-formylcyclopropanecarboxylate [**2**] (Note 15).* An oven-dried 250-mL, three-necked (24/40), round-bottomed flask equipped with a Teflon-coated magnetic stirring bar (egg-shaped, 26 mm length x 13 mm diameter), an internal thermometer, a glass stopper (central neck), and a Dryrite drying tube (Note 16) (Figure 3) is charged with crude α-formyl ester **1** (16.39 g) in AcOEt (100 mL). The light orange solution is stirred and immersed in an ice bath, cooling the internal temperature to 0 °C, and then potassium carbonate (K_2CO_3) (13.9 g, 100 mmol, 1 equiv) (Note 17) is added portionwise (split into five equal parts) over 10 min after temporarily removing the glass stopper (Note 18). Immediately after the addition is complete, triethylamine (1.4 mL, 1.00 g, 10.0 mmol, 0.1 equiv) is added in one portion.

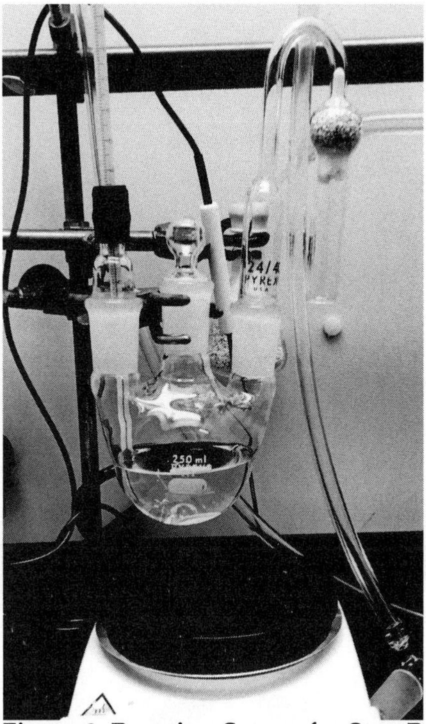

Figure 3. Reaction Set-up for Step B

After stirring (600 rpm) the suspension at 0 °C for 1 h, the reaction is quenched with water (100 mL) and transferred to a 500-mL separatory funnel. The initial reaction flask is rinsed with EtOAc (2 x 5 mL) and H_2O (2 x 5 mL), which are added to the separatory funnel. The organic phase is separated and the aqueous phase is re-extracted with EtOAc (20 mL). The combined organic phase is washed with water (20 mL) and brine (20 mL), dried over Na_2SO_4 (20 g), filtered through a 150-mL medium porosity sintered glass funnel, then concentrated under reduced pressure using a rotary evaporator (45 °C, 25 mmHg) to furnish an amber-colored liquid. The obtained crude product (ca. 12 g) is moved into a 25-mL round-bottomed flask with a Teflon-coated magnetic stir bar (Note 19) (Figure 4). Distillation while immersed in a temperature-controlled oil bath under reduced pressure (84–86 °C, 19–25 mmHg) provides the desired product **2** (8.68 g, 69% overall yield) as a colorless liquid (Notes 20, 21, and 22).

Figure 4. Distillation Set-up and Pure Final Product

Notes

1. The methyl 4-chlorobutanoate, methyl formate, and dichloromethane must be added by temporary removal of one of the addition funnels followed by purging of the system with nitrogen.
2. The checkers used methyl 4-chlorobutyrate (98+%) from Acros Organics. The submitters used methyl 4-chlorobutanoate (GC purity >98%) purchased from Tokyo Chemical Industry Co., Ltd. and used as received.
3. The checkers used methyl formate (97%, pure) from Acros Organics. The submitters used methyl formate (HCO_2Me) (GC purity >95%) purchased from Tokyo Chemical Industry Co., Ltd. and used as received.

4. The checkers used non-stabilized dichloromethane (20-L drum, ACS Reagent) from J. T. Baker, which was then passed through two packed columns of neutral alumina in a solvent purification system manufactured by SG Water U.S.A., LLC. The submitters used dichloromethane (CH_2Cl_2) (purity 99.5%) was purchased from Wako Pure Chemical Industries, Ltd. and used as received without any purification.
5. The submitters studied the use of CH_2Cl_2 and toluene as reaction solvents and noted the reaction to be homogeneous with the former, whereas the use toluene results in formation of yellow precipitate and a viscous reaction mixture. The checkers employed only CH_2Cl_2.
6. The checkers used titanium (IV) chloride (sure-sealed 200 g bottle, ReagentPlus, 99.9% trace metal basis) obtained from Sigma-Aldrich. The submitters used titanium tetrachloride ($TiCl_4$) (99.0%, 500 g bottle) purchased from Wako Pure Chemical Industries, Ltd. and used as received.
7. The checkers charged the 60 mL addition funnel with $TiCl_4$ (from a sure-sealed bottle) using a syringe. The submitters report delivering the $TiCl_4$ using a 10 mL pipet, wherein the operation should be rapidly and carefully conducted to take care of white smoke evolution.
8. The submitters note this step to be slightly exothermic when using addition rates of 24 mL of $TiCl_4$ over 5–10 min. However, the checkers observed a steady temperature at 5-10 °C when adding 24 mL $TiCl_4$ dropwise over 20 min. A feature not noted by the submitters is the formation of yellow crystals around the tip of the addition funnel. These crystals tend to fall off with time and slowly dissolve in the dichloromethane.
9. The checkers purchased triethylamine (≥99.5%) from Sigma-Aldrich and distilled it from CaH_2 immediately prior to use. The submitters used triethylamine (Et_3N) (purity 99%) purchased from Wako Pure Chemical Industries, Ltd. and used it as received.
10. The submitters note the reaction to be considerably exothermic. The checkers found that by adding the Et_3N dropwise over 30 min (approx. 1-1.2 mL per minute) it was possible to maintain an internal temperature at 10 °C or lower without affecting the formation of the expected dark orange reaction mixture.
11. The checkers note that the reaction progress can be monitored by ^1H NMR.

12. The submitters caution that this quench is exothermic. The checkers found that adding the water dropwise over 10 min was sufficient to maintain an internal temperature at 10 °C or lower, while the submitters' addition over 5 min was sufficient to maintain temperatures below 20 °C.
13. The checkers performed two half-scale and two full-scale reaction. The crude yields were 97% (7.94 g), 99% (8.14 g), 98% (16.13 g), and 99% (16.39 g) respectively. Analysis of the crude reaction mixture by ^{1}H NMR and ^{13}C NMR revealed that the mixture consists of three tautomers **1a**, (*Z*)-**1b**, (*E*)-**1b** and a very small amount of by-product tentatively assigned as (*E*)-**1x**, in an approximate ratio of 36:44:16:4.
14. Although not necessary for step B the submitters and checkers established in parallel studies that this reaction mixture could be purified via flash chromatography through SiO_2. In the checker's hands 5 g of the crude product was purified using 25 g of silica (Silicycle Silica, Flash P60, 40-63 μm, 230-400 mesh) loaded into a 30 mm diameter column. The column was slurry packed, the sample loaded with hexane and then eluted with a gradient of EtOAc/hexanes (5% increasing by approximately 2% every 10-12 fractions). Fractions were collected (6.7 mL) at a flow rate of 0.8 mL/sec. Overall, 72 fractions were collected and product was observed at fractions 10-32. These fractions were concentrated in vacuo to furnish 2.51 g of a colorless oil, which was determined by ^{1}H NMR to be (Z)-**1b** and trace **1a**. The submitters note the composition of the chromatographed material (5 g) to be a mixture of **1** (4.45 g, 85%) with cyclopropane **2** (ca. 6% based on ^{1}H NMR). The checkers did not observe cyclopropane **2** at this stage. Compound **1** has been found to slightly decompose on silica and distillation results in decomposition. Compound **1** gradually solidified at ambient temperature, and over a week undergoes slow tautomerization to enols (Z)-**1b** and (*E*)-**1b**. The checkers note the purified sample of (Z)-**1b** was observed to undergo slow crystallization, which upon collection and trituration (hexanes) of the resulting white solid revealed them to be (*E*)-**1b** by ^{1}H NMR.

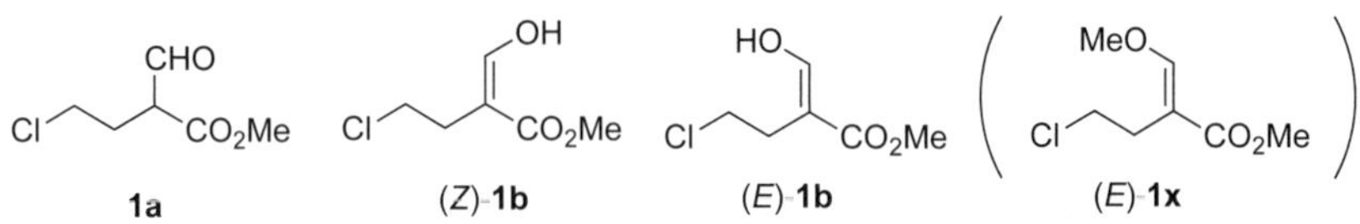

Physical and spectroscopic properties of (*Z*)-**1b**: colorless oil; ^{1}H NMR (400 MHz, $CDCl_3$) δ: 2.52 (td, *J* = 7.0, 0.7 Hz, 2H), 3.55 (t, *J* = 6.9 Hz, 2H), 3.80 (s, 3H), 7.11 (d, *J* = 12.8 Hz, 1H), 11.49 (d, *J* = 12.7 Hz, 1H); ^{13}C NMR (101 MHz, $CDCl_3$) δ: 31.2, 44.05, 51.8, 101.1, 163.3, 172.1; IR (neat) 2957, 1721, 1672, 1612, 1446, 1397, 1350, 1328, 1281, 1222, 1189, 1166, 1124, 989, 956, 811, 740, 653, 574, 452 cm^{-1}; HRMS (+ESI) *m/z* $[M + H]^+$ calcd for $C_6H_9ClO_3$ 165.0313, found 165.0313; Anal. Calcd for $C_6H_9ClO_3$: C, 43.79; H, 5.51; Cl, 21.54. Found; C, 43.56; H, 5.48; Cl, 21.28.

Those of (*E*)-**1b**: colorless crystals; mp 72–80 °C; ^{1}H NMR (400 MHz, $CDCl_3$) δ: 2.77 (t, *J* = 7.0 Hz, 2H), 3.63 (t, *J* = 6.9 Hz, 2H), 3.73 (s, 3H), 6.17 (br s, 1H), 7.77 (d, *J* = 9.0 Hz, 1H); ^{13}C NMR (101 MHz, $CDCl_3$) δ: 27.4, 43.6, 51.7, 107.3, 155.3, 168.8; IR (neat) 3208, 1667, 1634, 1444, 1397, 1331, 1308, 1283, 1203, 1169, 1105, 746, 731 cm^{-1}.

15. Step B should be carried out within a week due to the sensitivity of the starting material.
16. The half-scale reaction utilized an oven-dried 100-mL, three-necked, round-bottomed flask equipped with a Teflon-coated magnetic stirring bar (egg-shaped, 20 mm length x 10 mm diameter), an internal thermometer, a glass stopper (central neck) and a $CaCl_2$ drying tube.
17. The checkers used potassium carbonate (anhydrous, 99%) from Alfa Aesar.
18. The submitters note this to be a slightly exothermic reaction. The checkers did not observe a rise in temperature when adding the potassium carbonate in five equal portions over 10 min. However, the quench with water after reaction completion is observed to be exothermic, but the temperature can be maintained at 5–10 °C if the water is added over 5 min. Reaction progress monitoring in step B is possible by TLC using $KMnO_4$ staining [10% AcOEt in hexanes; R_f = 0.23 (**1**) and 0.33 (**2**)] or by ^{1}H NMR comparison of aliquots.
19. Full-scale utilized a 25-mL round-bottomed flask equipped with a magnetic stirring bar (egg-shaped, 18 mm length x 10 mm diameter) with a short path distillation apparatus (110 mm height x 110 mm width). Half scale employed a 10-mL round-bottomed flask and a magnetic stirring bar (rod shaped, 10 mm length x 3 mm diameter). The receiving flask is cooled to 0 °C by immersion in an ice-water bath.
20. The submitters note: 1st fraction: 42–62 °C / 20 mmHg (bath temp. 82–87 °C), 0.24 g. 2nd fraction: 62–64 °C / 17 mmHg (bath temp. 87–111 °C), 8.82 g (overall yield, 69% in 2 steps). 3rd fraction: 64–52 (fade out) °C / 17 mmHg (bath temp. 111–123 °C), 0.17 g. The

submitters also noted the bp to be 59–63 °C / 16 mmHg and the purity based on quantitative ^{1}H NMR analysis was 97-99%. The checkers performed a fractional distillation collecting distillate boiling at 84–86 °C / 25 mmHg (bath temp. 104–124 °C). The checkers note: 1st fraction (135 mg, boiling temp. 84–86 °C / 25 mmHg, bath temp. 100–104 °C). 2nd fraction (8.68 g, boiling temp. 84–86 °C / 25 mmHg, bath temp. 104–124 °C). The collection of the 2nd fraction was not stopped until the internal temperature dropped.

21. The checkers performed two half-scale and two full-scale reactions. The yields after distillation were 73% (4.70 g), 68% (4.33 g), 77% (9.82 g), and 69% (8.68 g) respectively.
22. Physical and spectroscopic properties of **2**: colorless liquid; ^{1}H NMR (400 MHz, $CDCl_3$) δ: 1.58–1.62 (m, 2H), 1.64–1.68 (m, 2H), 3.80 (s, 3H), 10.37 (s, 1H); ^{13}C NMR (101 MHz, $CDCl_3$) δ: 22.5, 33.4, 52.3, 171.4, 198.6; IR (neat) 2958, 2868, 1700, 1440, 1319, 1285, 1196, 1147, 1085, 1047, 1002, 959, 888, 810, 781, 742, 698, 474 cm^{-1}; HRMS (+ESI) *m/z* $[M + H]^+$ calcd for $C_6H_9O_3$ 129.0546, found 129.0543; quantitative ^{1}H NMR analysis was performed with ethylene carbonate (purchased from TCI, purity >99.0%) as the internal standard and obtained in 97.6% purity.

Working with Hazardous Chemicals

The procedures in *Organic Syntheses* are intended for use only by persons with proper training in experimental organic chemistry. All hazardous materials should be handled using the standard procedures for work with chemicals described in references such as "Prudent Practices in the Laboratory" (The National Academies Press, Washington, D.C., 2011; the full text can be accessed free of charge at http://www.nap.edu/catalog.php?record_id=12654). All chemical waste should be disposed of in accordance with local regulations. For general guidelines for the management of chemical waste, see Chapter 8 of Prudent Practices.

In some articles in Organic Syntheses, chemical-specific hazards are highlighted in red "Caution Notes" within a procedure. It is important to recognize that the absence of a caution note does not imply that no significant hazards are associated with the chemicals involved in that procedure. Prior to performing a reaction, a thorough risk assessment should be carried out that includes a review of the potential hazards associated with each chemical and experimental operation on the scale that is planned for the procedure. Guidelines for carrying out a risk assessment and for analyzing the hazards associated with chemicals can be found in Chapter 4 of Prudent Practices.

The procedures described in Organic Syntheses are provided as published and are conducted at one's own risk. Organic Syntheses, Inc., its Editors, and its Board of Directors do not warrant or guarantee the safety of individuals using these procedures and hereby disclaim any liability for any injuries or damages claimed to have resulted from or related in any way to the procedures herein.

Discussion

Methyl or ethyl 1-formylcyclopropanecarboxylate (**2**) or (**2′**) is a unique bifunctional compound with both aldehyde and ester functionalities at the same C-1 position in a simple cyclopropane molecule. Cyclopropane **2** or **2′**, therefore, serves as a useful synthetic building block, especially for medicinal and process chemistry, and natural product synthesis. As illustrated in Figure 5, characteristic cyclopropane segments are installed in various pharmaceuticals utilizing **2** or **2′**. The key feature is the chemoselective condensation of the aldehyde group in preference to the ester group.

(i) A traditional barbituric acid analog **I** containing a 5-spirocyclopropane moiety for a dihydroorotate dehydrogenase inhibitor,[2] (ii) arylpyrazole compound **II** containing cyclopropanecarboxamide for a parasiticidal agent,[3] (iii) arylsulfonylpiperazine **III** containingcyclopropanecarboxamide for a 11β-hydroxysteroid dehydrogenase inhibitor,[4] (iv) 5,7,8,9-tetrahydropyrimido[4,5-b][1,4]diazepin-6-ones compound **IV** containing 3-spirocyclopropane moiety for a protein kinase inhibitor,[5] (v) oxo-substituted aza-heterocylic compound **V** containing cyclopropane-carboxylic acid for the treatment and/or prevention of cardiovascular conditions,[6] (vi) oxazolo[5,4-b]pyridine-5-yl compound **VI** containing cyclopropanecarboxylate

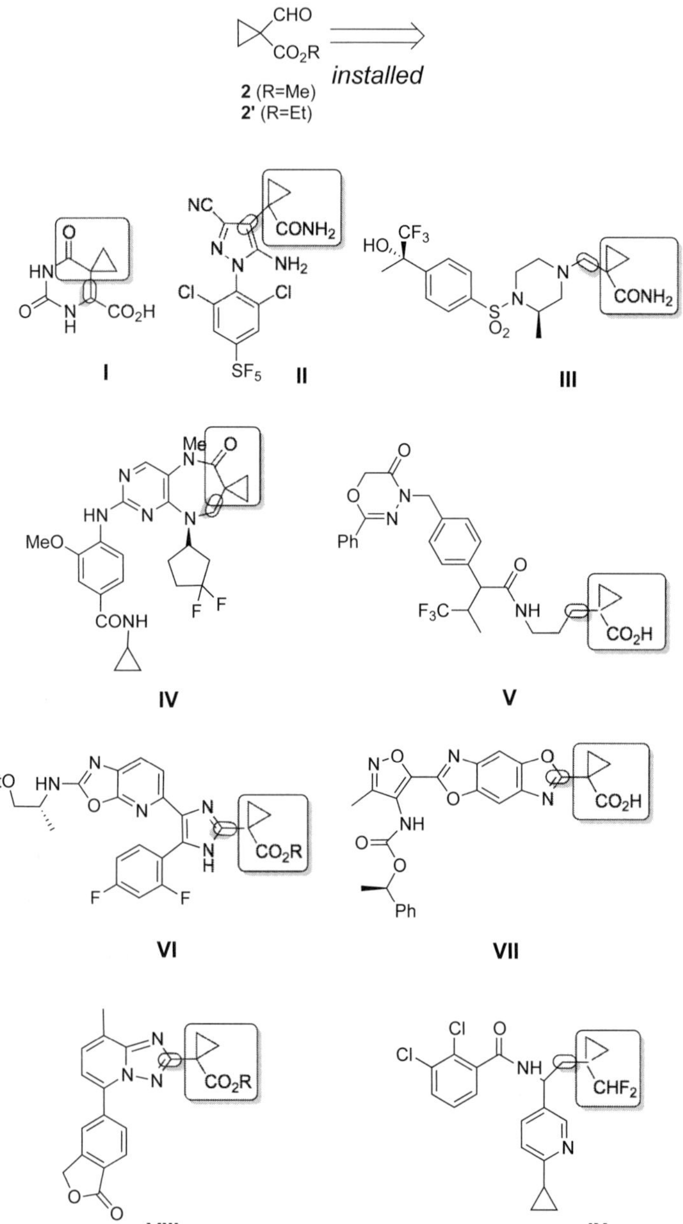

Figure 5. Pharmaceuticals incorporating cyclopropanecarboxylic acid or ester segments utilizing methyl or ethyl 1-formylcyclopropanecarboxylate (2) or (2′)

for the treatment of cancer,[7] (vii) 2,6-disubstituted benzobisoxazole compound **VII** containing cyclopropanecarboxylic acid for lysophosphatidic acid receptor antagonists,[8] (viii) [1,2,4]triazolopyridine compound **VIII** containing cyclopropanecarboxylate for phosphodiestererase inhibitors,[9] and (ix) 3-pyridyl-substituted benzamide compound **IX** containing 1-(difluoromethyl)cyclopropane for purinergic $2X_7$ ($P2X_7$) receptor inhibitors.[10]

As described above, **2** or **2′** has a significant role in the structural scaffolds of a variety of pharmaceuticals possessing cyclopropanecarboxylic acid derivatives. Noteworthy is that application of this manipulation has increased as a screening technique to discover new pharmaceuticals, likely because cyclopropanes are requisite isosteres for the corresponding dimethyl compounds.

On the other hand, cyclopropane **2** contributed as the starting compound to a formal synthesis of aspidospermine, a distinctive aspidosperma alkaloid,[11,12] in that a notable acid-catalyzed thermal rearrangement of cyclopropyl imine intermediate to 2-pyrroline is the key starting step (Scheme 1).[13]

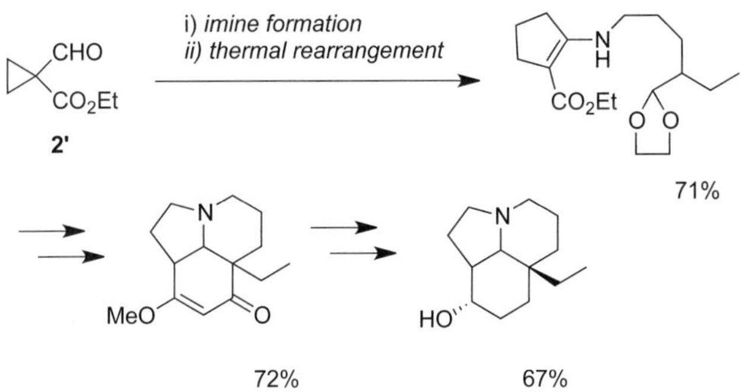

Scheme 1. Formal synthesis of aspidospermine alkaloid starting from ethyl 1-formylcyclopropanecarboxylate (2′)

On the whole, the reported synthetic methods for **2** or **2′** are categorized into four approaches.

(i) As illustrated in Scheme 2, Ayers' half reduction protocol of methyl and ethyl cyclopropanedicarboxylates (**3** and **3′**) is the most representative.[14] Commercially available **3** and **3′** (ca. twice as expensive as methyl 4-

chlorobutanoate) were converted by the treatment with more than 2.0 equiv of Li(*t*BuO)$_3$AlH, not to the desired aldehyde **2** and **2′** directly, but to alcohols **4** and **4′** in 88% and 79% yield, respectively.[4,6,9] Dess-Martin (DM) oxidation of **4** or **4′** using ca. 2 equiv of DM periodinane successfully afforded **2** (24%) or **2′** (76%). The DIBAL reduction method with **3** was also applied, but required harsh conditions such as –78 °C and 7 h.[5]

Although this approach is likely the most accessible, Li(*t*BuO)$_3$AlH is quite expensive (ca. $ 150 / 100 mL, 1.0 M) among commercially available hydride reagents and is not hydride atom-economical.

(ii) As an alternative method to (i),[7] **3** was converted by a half-hydrolysis reaction to monocarboxylic acid **5**, which was transformed to **2** through mixed anhydride formation and successive $NaBH_4$ reduction to give common intermediate **4**. TEMPO oxidation of **4** with trichloroisocyanuric acid afforded the desired product **2**, although an accurate yield was not described. This approach, however, is not straightforward and requires tedious procedures.

(iii) As depicted in Scheme 3, this approach utilizes the notable protocol of A. I. Meyer's group.[15,16] Ethyl cyanoacetate was converted to ethyl 1-cyanocyclopropanecarboxylate **6** (commercially available in 5-g scale, but extremely expensive), which is transformed to masked aldehyde **8** through 1,3-dioxadine formation and successive $NaBH_4$ reduction. Finally, acid hydrolysis of **8** gave the desired compound **2′**. This method also requires four steps with high (80 °C) and low (□40 °C) temperature reactions, the use of large amounts of conc. H_2SO_4, and steam distillation purification.

(iv) Scheme 4 depicts a method starting from γ–butyrolactone developed by Kuraray's group,[17] which is the most relevant for our strategy. γ–Butyrolactone was α-formylated using HCO_2Me/NaH and protected with an ethoxycarbonyl group to give **9**. Conventional ring opening with chlorination using $SOCl_2$ and $ZnCl_2$ in EtOH gave precursor **10**. Finally, cyclopropanation concomitant with deprotection was performed to afford **2′**. This approach required a protective and deprotective sequence and afforded a moderate total yield (26%).

Due to the utility of **2** or **2′**, 5-100 g scale production methods have been disclosed in recent medicinal chemistry patents. The reported synthetic methods for **2** or **2′**, however, require column chromatographic purification despite the high volatility, or crude product is used in the next condensation step without purification. Our concise and straightforward method involves purification by simple distillation (the boiling point was documented for the first time) without the use of column chromatography, and is performed within short reaction and purification periods.

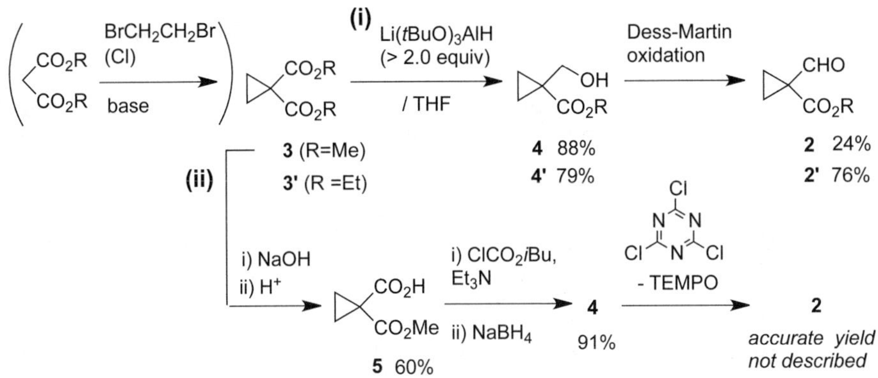

Scheme 2. Half-reduction method of cyclopropane precursor 3 or 3′ derived from dimethyl malonate

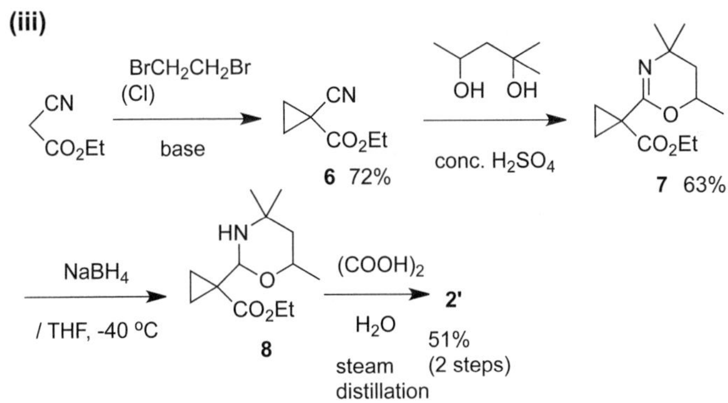

Scheme 3. A. I. Meyers' transformation method starting from ethyl cyanoacetate

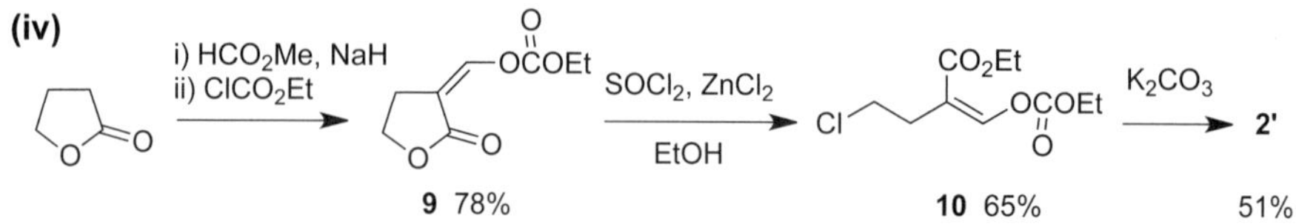

Scheme 4. Kuraray group's method starting from γ-butyrolactone

Among the various carbon homologation methods, α-formylation of simple esters with HCO_2Me is a well-recognized useful reaction. A literature survey (SciFinder®) revealed reports of ca. 100 examples utilizing base reagent (e.g. NaOR, NaH, LDA, and LiHMDS)-mediated methods and 5 examples using $TiCl_4$/amine-mediated methods. In general, a major conventional reaction using bases (e.g. NaOR, NaH) requires long reaction periods and results in moderate yield in almost all cases. LDA- and LiHMDS-promoted methods are superb with regard to yield but require rigorous procedures (reaction time schedule and accurate reagent equivalents) and low temperature (–78 °C).

α-Formylation of simple esters utilizing $TiCl_4$/amine-mediated (Ti-Claisen) condensation[18,19] for the synthesis of α-formylated esters **11** is depicted in Table 1 (13 examples). Titanium (Ti) (*or* Zr)-self-Claisen condensations between two of the same esters,[20-22] Ti-crossed-Claisen condensations between esters or acids with acid chlorides,[23,24] and Ti-Dieckmann (intramolecular Claisen) condensations[25-27] have several advantages, including: (i) powerful C-C bond forming reactivity; (ii) highly available reagents with robust reactions; (iii) accessible temperature (0 °C to ambient); (iv) compatibility with base-labile functional groups such as γ-halogeno, γ–ketone carbonyl, etc., despite the high reactivity. On the other hand, a mild variant Ti-Claisen condensation method using ketene silyl acetals with acid chlorides also satisfies the four listed features [(i)-(iv)].[28]

The present α-formylation reaction of methyl 4-chlorobutanoate (**1**) is a distinctive example of the compatibility with a base-sensitive γ-chloro group. Synthesis of **2** is not possible using the base-mediated α-formylation method due to undesirable and predominant cyclopropane formation leading to methyl cyclopropanecarboxylate.

In conclusion, the present straightforward and robust synthetic strategy for the unique, but useful, building block **2** utilizing Ti-Claisen condensation provides a new promising avenue, especially for pharmaceutical syntheses.

Table 1. α-Formylation of esters utilizing Ti-Claisen condensation

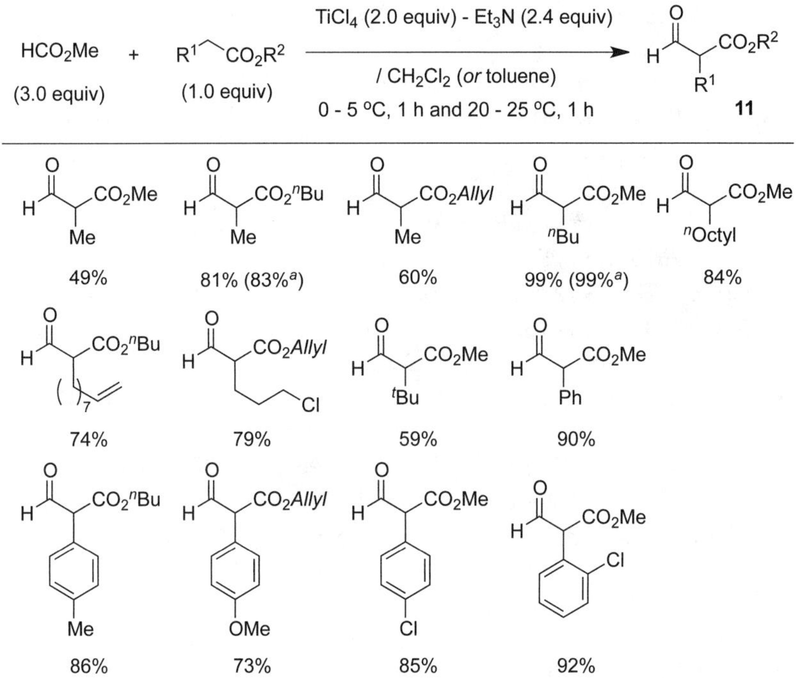

[a] Use of toluene solvent

References

1. Department of Chemistry, School of Science and Technology, Kwansei Gakuin University, 2-1 Gakuen, Sanda, Hyogo, 669-1337, Japan. Email: tanabe@kwansei.ac.jp. This research was partially supported by Grant-in-Aids for Scientific Research on Basic Areas (B) "18350056", Priority Areas (A) "17035087" and "18037068", and Exploratory Research "17655045" from the Ministry of Education, Department of Chemistry, School of Science and Technology, Kwansei Gakuin University, 2-1 Gakuen, Sanda, Hyogo, 669-1337, Japan
2. Husbands, S.; Fraser, W.; Suckling, C. J.; Wood, H. C. S. *Tetrahedron* **1995**, *51*, 865–870.
3. Billen, D.; Boyle, J.; Critcher, D. J.; Gethin, D. M.; Hall, K. T.; Kyne, G. M. WO 2006/134468 A1, p. 54.

4. Sun, D.; Wang, Z.; Cardozo, M.; Choi, R.; DeGraffenreid, M.; Di, Y.; He, X.; Jaen, J. C.; Labelle, M.; Liu, J.; Ma, J.; Miao, S.; Sudom, A.; Tang, L.; Tu, H.; Ursu, S.; Walker, N.; Yan, X.; Ye, Q.; Powers, J. P. *Bioorg. Med. Chem. Lett.* **2009**, *19*, 1522–1527.
5. Pierard, F.; Charrier, J-D. WO 2009/023269.
6. Lampe, T.; Hahn, M.; Stasch, J-P.; Schlemmer, K-H.; Wunder, F.; Heitmeier, S.; Griebenow, N.; el Sheikh, S.; Li, V. M-J.; Becker, E-M.; Stoll, F.; Knorr, A. WO 2010/102717 A1, p. 111.
7. Coates, D. A.; Gilmour, R.; Martin, J. A.; Martin de la Nava, E. M. WO 2012/074761 A1, p. 25.
8. Buckman, B.; Nicholas, J. B.; Emayan, K.; Seiwert, S. D. WO 2013/025733 A1, p. 413.
9. Nielsen, S. F.; Larsen, J. C. H. WO 2013/092739 A1, p. 24.
10. Kilburn, J. P.; Rasmussen, L. K.; Jessing, M.; Eldemenky, E. M.; Chen, B.; Jiang, Y.; Hopper, A. T. WO 2014/057078 A1, p. 93.
11. Stork, G.; Dolfini, J. E. *J. Am. Chem. Soc.* **1963**, *85*, 2872–2873.
12. Ban, Y.; Sato, Y.; Inoue, I.; Nagai, M.; Oishi, T.; Terashima, M.; Yonemitsu, O.; Kanaoka, Y. *Tetrahedron, Lett.* **1965**, 2261–2268.
13. (a) Stevens, R. V.; Fitzpatrick, J. M.; Kaplan, M.; Zimmerman, R. L. *J. Chem. Soc. D,* **1971**, 857–858. (b) Stevens, R. V.; DuPree, L. E. *J. Chem. Soc. D,* **1970**, 1585–1586.
14. Ayers, T. A. *Tetrahedron Lett.* **1999**, *40*, 5467–5470.
15. (a) Meyers, A. I.; Nabeya, A.; Adickes, H. W.; Politzer, I. R.; Malone, G. R.; Kovelesky, A. C.; Nolen, R. L.; Portnoy, R. C. *J. Org. Chem.* **1973**, *38*, 36–56. An additional relevant route can include selective reduction of the analogous nitrile carboxylic acid: (b) Singh, R. K.; Danishefsky, S. *J. Org. Chem.* **1975**, *40*, 2969–2970. (c) Nicolaou, K. C.; Petasis, N. A.; Li, W. S.; Ladduwahetty, T.; Randall, J. L.; Webber, S. E.; Hernandez, P. E. *J. Org. Chem.* **1983**, *48*, 5400–5403.
16. Fry, J. L.; Ott, R. A. *J. Org. Chem.* **1981**, *46*, 602–607.
17. Ujita, K.; Kanehira, K. JP 2002/105029.
18. Nakatsuji, H.; Nishikado, H.; Ueno, K.; Tanabe, Y. *Org. Lett.* **2009**, *11*, 4258–4261.
19. Tanabe, Y. *Bull. Chem. Soc. Jpn.* **1989**, *62*, 1917–1924.
20. Yoshida, Y.; Hayashi, R.; Sumihara, H.; Tanabe, Y. *Tetrahedron Lett.* **1997**, *38*, 8727–8730.
21. Tanabe, Y.; Hamasaki, R.; Funakoshi, S. *Chem. Commun.* **2001**, 1674–1675.

22. Nakatsuji, H.; Ashida, Y.; Hori, H.; Sato, Y.; Honda, A.; Taira, M.; Tanabe, Y. *Org. Biomol. Chem.* **2015**, *13*, 8205–8210; See its references 10 and 17.
23. Misaki, T.; Nagase, R.; Matsumoto, K.; Tanabe, Y. *J. Am. Chem. Soc.* **2005**, *127*, 2854–2855.
24. Nagase, R.; Oguni, Y.; Ureshino, S.; Mura, H.; Misaki, T.; Tanabe, Y. *Chem. Commun.* **2013**, *49*, 7001–7003.
25. Crane, S. N.; Corey, E. J. *Org. Lett.* **2001**, *3*, 1395–1397.
26. Tanabe, Y.; Makita, A.; Funakoshi, S.; Hamasaki, R.; Kawakusu, T. *Adv. Synth. Catal.* **2002**, *344*, 507–510.
27. Tanabe, Y.; Manta, N.; Nagase, R.; Misaki, T.; Nishii, Y.; Sunagawa, M.; Sasaki, A. *Adv. Synth. Catal.* **2003**, *345*, 967–970.
28. Iida, A.; Nakazawa, S.; Okabayashi, T.; Horii, A.; Misaki, T.; Tanabe, Y. *Org. Lett.* **2006**, *8*, 5215–5218.

Appendix
Chemical Abstracts Nomenclature (Registry Number)

Methyl 4-chlorobutanoate (3153-37-5)
Methyl formate (HCO_2Me) (107-31-3)
Titanium tetrachloride ($TiCl_4$) (7550-45-0)
Triethylamine (Et_3N) (121-44-8)
Potassium carbonate (K_2CO_3) (584-08-7)

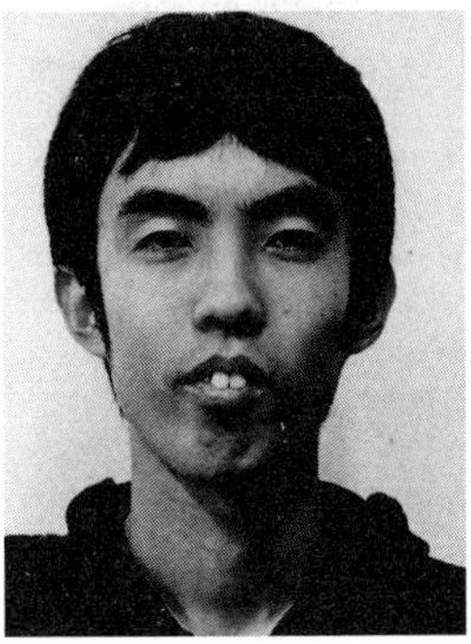

Yuichiro Ashida was born in Fukuchiyama, Kyoto, Japan, in 1989. He received his B. S. degree (2012), and M. S. degree (2014) from Kwansei Gakuin University under the direction of Professor Yoo Tanabe. Presently, he is engaged in his doctoral studies on the development of (*E*)-, (*Z*)-stereocomplementary parallel synthesis of multi-substituted α,β-unsaturated esters utilizing (*E*)-, (*Z*)-stereodefined enol tosylates and phosphonates, which is directed for process chemistry.

Satomi Kajimoto was born in Hyogo, Japan, in 1991. She received her B. S. degree (2014) from Kwansei Gakuin University under the supervision of Prof. Yoo Tanabe. Her graduate research focuses on the practical preparation of α–formyl esters from simple esters and methyl formate utilizing Ti-Claisen condensation and its utilization towards useful synthetic building blocks.

Hidefumi Nakatsuji received his B. S. degree (2005) and his Ph. D. degree (2010) from Kwansei Gakuin University under the direction of Professor Yoo Tanabe. Immediately, Dr. Nakatsuji moved to Nagoya University (Professor Kazuaki Ishihara's group) and studied as JSPS Postdoctoral Fellowship and CREST project researcher until 2014. Next, he was promoted to Associate Professor of Yoo Tanabe's group. His research interests are the development of chiral phosphine and phosphine oxide organocatalysts for a MCR type cyclization and of condensation reactions for cost-effective reactions directed for process chemistry.

Yoo Tanabe received his bachelor's degree at Tokyo in the laboratory of Professor Kenji Mori. He received his Ph.D. at Tokyo Inst. Technology under the direction of Professor Teruaki Mukaiyama on the development of practical acylation reactions. After leaving Sumitomo Chemical Co. Ltd, Dr. Tanabe moved to Kwansei Gakuin University in 1991 as Associate Professor and promoted to full Professor in 1997. His research focuses on the exploitation of useful synthetic reactions directed for process chemistry: concise synthesis of useful fine chemicals and of total synthesis of biologically active natural products.

Bryon K. Anderson obtained his B.S. in Chemistry at Montana State University, Bozeman in 2010. He subsequently obtained his Ph.D. at Montana State University in Organic Chemistry under the direction of Professor Tom Livinghouse in 2015 on the synthesis of the enantiopure tetracyclic cores of asparagamine A and stemofoline alkaloids. He is currently a post-doctoral research associate in the laboratory of Professor John L. Wood at Baylor University.

Yu-Wen Huang was born in Hsinchu, Taiwan (R.O.C.) in 1982. He received his bachelor's degree from National Cheng Kung University in 2005. He then joined M.S. program at National Tsing Hua University, whereas he obtained his master degree under the supervision of Professor Shang-Cheng Hung in carbohydrate synthesis. In 2016, he received his Ph.D. degree at University of Rochester advised by Professor Alison J. Frontier on 1,6-conjugate addition initiated Nazarov reaction and a sequential 1,5-hydride transfer chemistry. He is currently a post-doctoral fellow in CPRIT lab (Baylor University) with Professor Ke Kong and John L. Wood.

Organic Syntheses

Nickel-Catalyzed Suzuki–Miyaura Coupling in *t*-Amyl Alcohol for the Preparation of 5-(Furan-3-yl)pyrimidine

Liana Hie and Neil K. Garg*[1]

Department of Chemistry and Biochemistry, University of California, Los Angeles

Checked by Kristen Flynn, Alyssa Antropow, and Mohammad Movassaghi

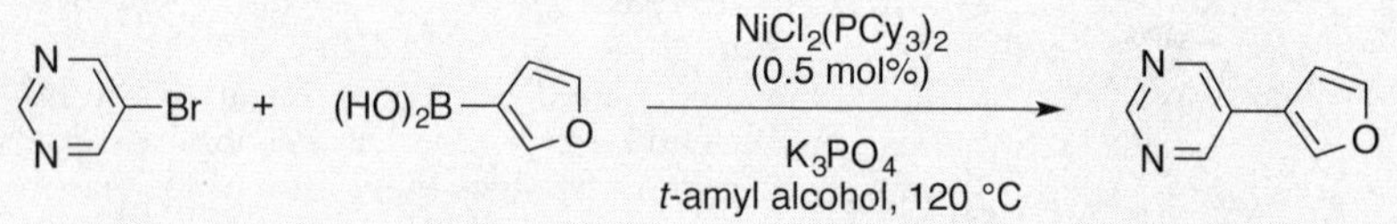

Procedure

5-(Furan-3-yl)pyrimidine. A 1000-mL round-bottomed flask is charged with a Teflon-coated, octagon-shaped magnetic stir bar (4.0 x 0.5 cm) and K_3PO_4 (60.0 g, 283 mmol, 4.50 equiv), and then attached to a gas inlet adapter (90°). The flask and its contents are flame-dried (Note 1) under oil pump vacuum (1.0 mmHg) for ca. 5 min (Note 2). After cooling to 23 °C under nitrogen, the gas inlet adapter is removed and the flask is charged with 5-bromopyrimidine (10.0 g, 63.0 mmol, 1 equiv), 3-furanylboronic acid (17.6 g, 158 mmol, 2.50 equiv), and $NiCl_2(PCy_3)_2$ (0.220 g, 0.315 mmol, 0.500 mol%) (Note 3). The flask is equipped with a 24/40 oven-dried reflux condenser and a septum with a needle, and then evacuated and backfilled with nitrogen (3x). *tert*-Amyl alcohol (210 mL, 0.3 M) is added via cannula from an oven-dried, argon purged graduated cylinder through the septum under a flow of argon. The resulting reddish heterogeneous mixture is stirred at 23 °C for 1 h under positive nitrogen pressure (Figure 1) (Note 4).

Published on the Web 10/19/16

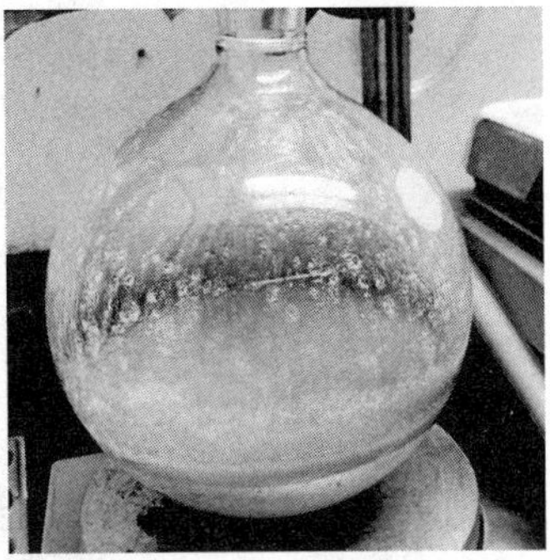

Figure 1. Reddish heterogeneous mixture formed after mixing the reagents (Photo provided by Submitters)

The reaction vessel containing a brown heterogeneous mixture and equipped with a reflux condenser is placed into a pre-heated oil bath set at a temperature of 120 °C (Figure 2) (Note 5).

Figure 2. Brown heterogeneous mixture resulted after stirring at 120 °C for 1 h (Photo provided by Submitters)

After stirring for 1 h, the reaction vessel containing a brown greenish heterogeneous mixture is removed from the oil bath and cooled to 23 °C. Thin layer chromatography is performed to check the progress of the reaction (hexanes:EtOAC = 3:1, R_f = 0.31) (Figure 3). Into the flask is added deionized water (100 mL) and Et_2O (200 mL), and the resulting mixture is stirred for 15 min (Figure 4).

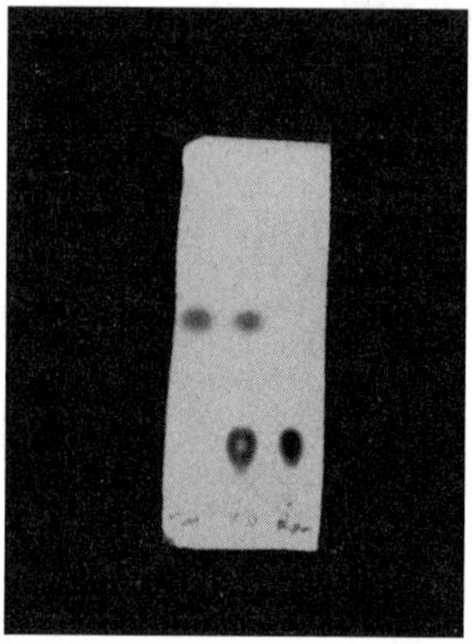

Figure 3. Thin layer chromatography of the crude reaction mixture (left lane = 5-bromopyrimidine, middle lane = cross spots of 5-bromopyrimidine and the reaction mixture, right lane = reaction mixture) (Photo provided by Submitters)

Figure 4. Brown homogenous mixture was formed upon stirring with water and diethyl ether (Photo provided by Submitters)

The combined mixture is then poured into a 1000-mL separatory funnel. The reaction flask is rinsed once with H_2O (100 mL) and Et_2O (100 mL), and both washes are transferred to the separatory funnel (Figure 5a). The organic and aqueous layers are separated. The organic layer is washed sequentially with 1.0 M NaOH (2 x 200 mL) (Figure 5b) and saturated sodium chloride solution (200 mL) (Figure 5c).

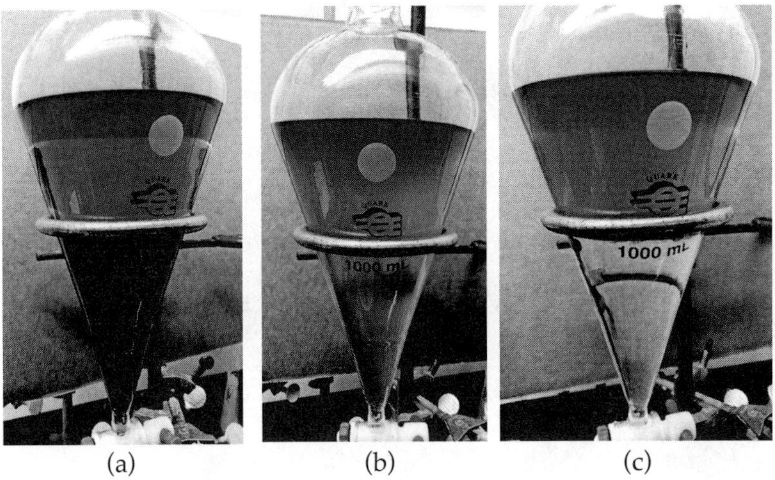

Figure 5. (a) The reaction mixture with water and diethyl ether in a separatory funnel. (b) The organic layer and 1.0 M NaOH. (c) The organic layer and saturated sodium chloride (Photo provided by Submitters)

The organic layer is then dried over Na_2SO_4 (100 g) for 15 min and filtered over a 150-mL glass-fritted filter packed with celite (30 g). The drying agent is rinsed with Et_2O (100 mL) and the combined organics are concentrated by rotary evaporation (275 mmHg, 30 °C to remove Et_2O and 30 mmHg, 50 °C to remove *tert*-amyl alcohol, The temperature was controlled by a water bath). In a 500-mL round-bottomed flask fitted with a 24/40 reflux condenser, the tan solid is dissolved in 200 mL of hexanes, which is brought to 80 °C by heating in an oil bath (Notes 6 and 7). Upon complete dissolution, the filtrate is allowed to cool to room temperature and then cooled at –20 °C for 12 h. The light tan crystals that form are collected by filtration on a 150 mL fritted funnel and washed with 100 mL of ice-cold hexane (0 °C). The resulting crystals are transferred to a 100 mL round-bottomed flask and then dried under oil pump vacuum (1.0 mmHg, 23 °C) for 8 h. The mother liquor is concentrated and a second recrystallization is performed to afford additional 5-(furan-3-yl)pyrimidine (total yield of both crops is 7.60 g, 83% yield) as a light tan crystalline solid (Notes 8 and 9).

Figure 6. Product collected as light tan crystalline solid (Photo provided by Submitters)

Notes

1. Heat resistant gloves should be worn to prevent burns.
2. Submitters observed bubbling upon flame-drying likely due to water content. Checkers used anhydrous base and did not observe bubbling.
3. Potassium phosphate tribasic (anhydrous) was purchased from Alfa-Aesar. 5-Bromopyrimidine (98%) and 3-furanylboronic acid (96%) were purchased from Combi-Blocks, Inc. and used as received. Bis(tricyclohexylphosphine)nickel(II) chloride (97%) and *tert*-amyl alcohol (ReagentPlus, 99%) were purchased from Sigma-Aldrich and used as received.
4. The reaction mixture was stirred at 900 rpm.
5. Temperature was monitored by a thermocouple placed directly in the oil bath.
6. Submitters recommend an extra quantity of EtOAc can be added to aid dissolution, if necessary. Checkers did not use any EtOAc during recrystallizations.
7. Hot filtration using a 150 mL fritted funnel can be performed to remove any brown residue, if needed. Checkers performed recrystallization with and without hot filtration, and no difference in purity was observed. Procedure for hot filtration: Glassware had been heated in the

oven at 200 °C for ca. 30 min and was used within five min of removal from the oven. The hot hexanes solution (80 °C) is filtered through a glass-fritted funnel into an Erlenmeyer flask. If room-temperature glassware is used for this procedure, the product precipitates.

8. A second run was performed at half scale, which provided 3.75 g (82%) of the product.
9. This compound exhibits the following physiochemical properties: mp = 80–81 °C; TLC (hexanes:EtOAc = 3:1), R_f = 0.31; ^{1}H NMR (400 MHz, $CDCl_3$) δ: 6.73 (dd, *J* = 1.9, 0.9 Hz, 1H), 7.57 (t, *J* = 1.7 Hz, 1H), 7.84 (dd, *J* = 1.5, 0.9 Hz, 1H), 8.85 (s, 2H), 9.12 (s, 1H); ^{13}C NMR (125 MHz, $CDCl_3$) δ: 108.1, 120.0, 126.7, 139.5, 144.7, 153.7, 157.2; IR (film): 3136, 3112, 3041, 1969, 1957, 1880, 1607, 1575, 1549, 1513, 1438, 1414, 1340, 1187, 1159, 1091, 1061, 1012, 924, 871, 790, 732, 721, 645, 634; HRMS-ESI (*m*/*z*) cacld for $C_8H_7N_2O$ $[M+H]^+$: 147.0553, found: 147.0547; Anal. Calcd for $C_8H_6N_2O$: C 65.75, H 4.14, N 19.17, found: C 65.65, H 4.25, N 19.01.

Working with Hazardous Chemicals

The procedures in *Organic Syntheses* are intended for use only by persons with proper training in experimental organic chemistry. All hazardous materials should be handled using the standard procedures for work with chemicals described in references such as "Prudent Practices in the Laboratory" (The National Academies Press, Washington, D.C., 2011; the full text can be accessed free of charge at http://www.nap.edu/catalog.php?record_id=12654). All chemical waste should be disposed of in accordance with local regulations. For general guidelines for the management of chemical waste, see Chapter 8 of Prudent Practices.

In some articles in *Organic Syntheses*, chemical-specific hazards are highlighted in red "Caution Notes" within a procedure. It is important to recognize that the absence of a caution note does not imply that no significant hazards are associated with the chemicals involved in that procedure. Prior to performing a reaction, a thorough risk assessment should be carried out that includes a review of the potential hazards associated with each chemical and experimental operation on the scale that is planned for the procedure. Guidelines for carrying out a risk assessment

and for analyzing the hazards associated with chemicals can be found in Chapter 4 of Prudent Practices.

The procedures described in *Organic Syntheses* are provided as published and are conducted at one's own risk. *Organic Syntheses, Inc.*, its Editors, and its Board of Directors do not warrant or guarantee the safety of individuals using these procedures and hereby disclaim any liability for any injuries or damages claimed to have resulted from or related in any way to the procedures herein.

Discussion

Transition metal-catalyzed cross-couplings are versatile methods for carbon–carbon (C–C) and carbon–heteroatom bond formation.[2] Traditionally, palladium catalysts have been most commonly utilized for this type of transformation; however, there has been a significant interest in the development of cross-coupling reactions that rely on non-precious metal catalysts. In particular, nickel has emerged as an alternative to palladium due to several factors. For example, nickel is significantly more abundant, much less expensive, and also possesses a much lower CO_2 footprint.[3] Moreover, nickel catalysts possess the unique ability to activate a wide range of electrophilic coupling partners, well beyond the scope of traditional cross-couplings that use palladium catalysis.[3]

Our research groups and many others have developed new protocols for aryl C–C and C–X bond formations using nickel catalysts.[4,5,6,7] In our efforts to render our methodologies more industrially friendly, we have developed the nickel-catalyzed Suzuki–Miyaura coupling in 'green' solvents,[8,9] such as 2-Me-THF and *tert*-amyl alcohol.[10] The methodology proceeds under mild conditions and has a broad scope with respect to the boronic acids, the electrophiles (halides and phenolic derivatives), and heterocycle constituents.[4a]

A variety of heterocyclic halides (i.e., chlorides, bromides, iodides) and phenol-derived electrophiles (i.e., carbamates, sulfamates, mesylates, and pivalates) were coupled with aryl boronic acids to give coupled products (**1–7**, Table 1). These results also demonstrate that heterocycles such as pyridine, pyrimidine, pyrazole, quinolone, isoquinoline, and carbazole are tolerated under these coupling conditions.

Table 1. Coupling of heterocyclic substrates with aryl boronic acids.[a,b]

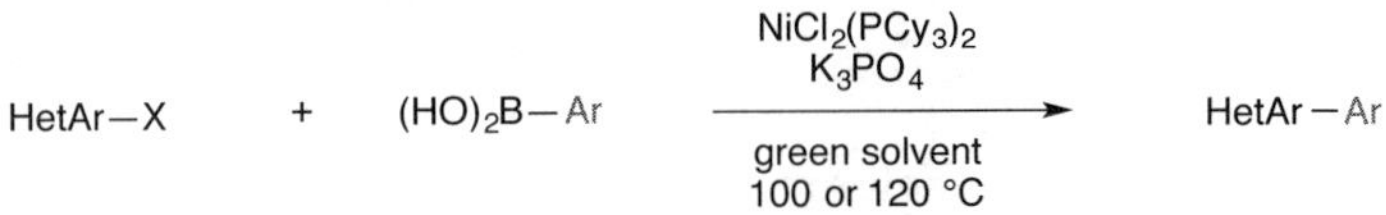

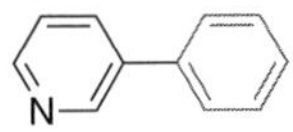

1
X = Cl
100% yield (t-amyl alcohol)
100% yield (2-Me-THF)

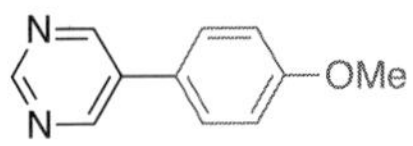

2
X = Br
94% yield (t-amyl alcohol)[c]
80% yield (2-Me-THF)[c]

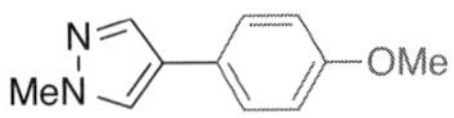

3
X = I
35% yield (t-amyl alcohol)[c]
49% yield (2-Me-THF)[c]

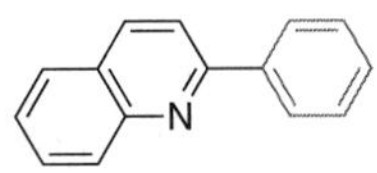

4
X = OCONEt$_2$
63% yield (t-amyl alcohol)[c]
48% yield (2-Me-THF)[c]

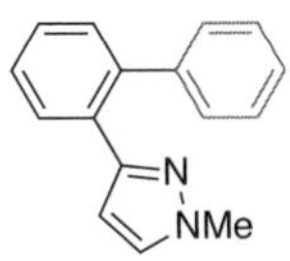

5
X = OSO$_2$NMe$_2$
94% yield (t-amyl alcohol)[c]
99% yield (2-Me-THF)[c]

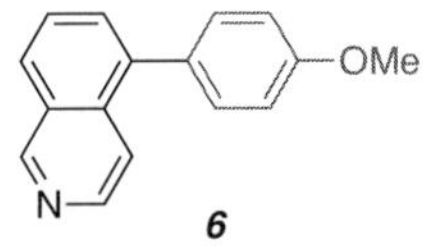

6
X = OMs
96% yield (t-amyl alcohol)
95% yield (2-Me-THF)

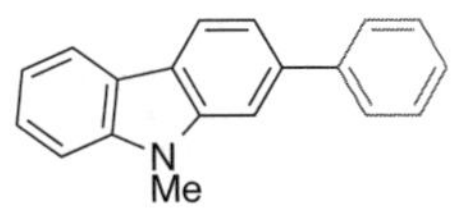

7
X = OPiv
92% yield (t-amyl alcohol)[c]
89% yield (2-Me-THF)[c]

[a] Conditions: $NiCl_2(PCy_3)_2$ complex (5 mol %), halide substrate (1.00 equiv), aryl boronic acid (2.50 equiv), K_3PO_4 (4.50 equiv), hexamethylbenzene (0.10 equiv), 100 or 120 °C, 12 h. [b] Yield of product determined by ^{1}H NMR analysis of the crude reaction mixtures using hexamethylbenzene as an internal standard. [c] Conditions: $NiCl_2(PCy_3)_2$ complex (10 mol %), halide substrate (1.00 equiv), aryl boronic acid (4.00 equiv), K_3PO_4 (7.20 equiv), hexamethylbenzene (0.10 equiv), 100 or 120 °C, 12 h.

The methodology could also be applied to access bis(heterocyclic) frameworks, which are prevalent motifs in numerous bioactive molecules, but are sometimes difficult to access via palladium-catalyzed methods.[11] Pyridine- and pyrimidine-containing substrates were coupled with heterocyclic boronic acids to give bis(heterocycles) **8** and **9** in good to excellent yields with 1 mol% of the Ni precatalyst (Table 2). Isoquinoline- and dihydrobenzofuran-containing substrates were coupled with heterocyclic boronic acids to forge bis(heterocycles) **10** and **11** in excellent yields using 5 mol% Ni precatalyst. It should be noted that the couplings shown in Tables 1 and 2 could be achieved in the presence of 2-Me-THF or *tert*-amyl alcohol as the reaction medium.

Tableb 2. Couplings of heterocyclic boronic acids with heterocyclic substrates. [a,b]

$$\text{HetAr–X} \quad + \quad (\text{HO})_2\text{B–Ar} \xrightarrow[\text{green solvent, 100 or 120 °C}]{\text{NiCl}_2(\text{PCy}_3)_2,\ \text{K}_3\text{PO}_4} \text{HetAr–Ar}$$

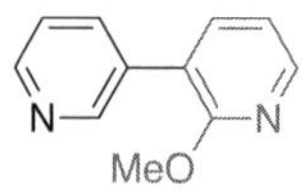

8
X = Cl
98% yield (*t*-amyl alcohol)
91% yield (2-Me-THF)

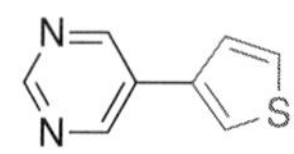

9
X = Br
86% yield (*t*-amyl alcohol)
93% yield (2-Me-THF)

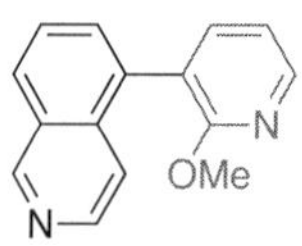

10
X = OMs
100% yield (*t*-amyl alcohol)[c]
96% yield (2-Me-THF)[c]

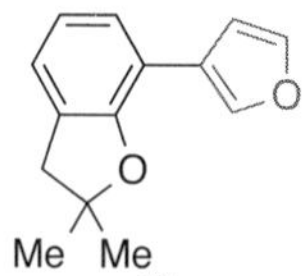

11
X = OSO_2NMe_2
100% yield (*t*-amyl alcohol)[d]
100% yield (2-Me-THF)[d]

[a] Conditions: $NiCl_2(PCy_3)_2$ complex (1 mol %), halide substrate (1.00 equiv), aryl boronic acid (2.50 equiv), K_3PO_4 (4.50 equiv), hexamethylbenzene (0.10 equiv), 100 or 120 °C, 12 h. [b] Yield of product determined by ^{1}H NMR analysis of the crude reaction mixtures using hexamethylbenzene as an internal standard. [c] Conditions: $NiCl_2(PCy_3)_2$ complex (5 mol %), halide substrate (1.00 equiv), aryl boronic acid (4.00 equiv), K_3PO_4 (7.20 equiv), hexamethylbenzene (0.10 equiv), 100 or 120 °C, 12 h. [d] Conditions: $NiCl_2(PCy_3)_2$ complex (10 mol %), halide substrate (1.00 equiv), aryl boronic acid (4.00 equiv), K_3PO_4 (7.20 equiv), hexamethylbenzene (0.10 equiv), 100 or 120 °C, 12 h.

References

1. Department of Chemistry and Biochemistry, University of California, Los Angeles. E-mail: neilgarg@chem.ucla.edu. The authors are grateful to Boehringer Ingelheim, the ACS Green Chemistry Institute's

Pharmaceutical Roundtable, Bristol-Myers Squibb, the Camille and Henry Dreyfus Foundation, the A. P. Sloan Foundation, the UCLA Gold Shield Alumnae, and the University of California, Los Angeles, for financial support. We are grateful to Bristol-Myers Squibb, the ACS Division of Organic Chemistry, the Foote and Majeti-Alapati families for fellowship support (L.H.). These studies were also supported by shared instrumentation grants from the NSF (CHE-1048804) and the National Center for Research Resources (S10RR025631).

2. (a) Hassan, J.; Sevignon, M.; Gozzi, C.; Schulz, E.; Lemaire, M. *Chem. Rev.* **2002**, *102*, 1359–1470. (b) Muci, A. R.; Buchwald, S. L. In *Topics in Current Chemistry*; Miyaura, N., Ed.; Springer Verlag: New York, 2002; Vol. *219*, pp 131–209. (c) Jiang, L.; Buchwald, S. L. In *Metal-Catalyzed Cross-Coupling Reactions*, 2nd ed.; Meijere, A., Diederich, F., Eds.; Wiley-VCH: Weinheim, 2004; pp 699–760. (d) Corbet, J. P.; Mignani, G. *Chem. Rev.* **2006**, *106*, 2651–2710. (e) Negishi, E. Bull. *Chem. Soc. Jpn*. **2007**, *80*, 233–257. (f) Shen, H. C. In *Application of Transition Metal Catalysis in Drug Discovery and Development: An Industrial Perspective*; Crawley, M. L., Trost, B. M., Eds.; John Wiley & Sons, Inc.: Hoboken, NJ, 2012; pp 25–96.

3. For selected recent reviews in nickel catalysis, see: (a) Tasker, S. Z.; Standley, E. A.; Jamison, T. F. *Nature* **2014**, *509*, 299–309. (b) Mesganaw, T.; Garg, N. K. *Org. Process Res. Dev.* **2013**, *17*, 29–39. (c) Yu, D.-G.; Li, B.-J.; Shi, Z.-J. *Acc. Chem. Res.* **2010**, *43*, 1486–1495. (d) Rosen, B. M.; Quasdorf, K. W.; Wilson, D. A.; Zhang, N.; Resmerita, A.-M.; Garg, N. K.; Percec, V. *Chem. Rev.* **2011**, *111*, 1346–1416. (e) Li, B.-J.; Yu, D.-G.; Sun, C.-L.; Shi, Z.-J. *Chem.–Eur. J.* **2011**, *17*, 1728–1759. (f) Cornella, J.; Zarate, C.; Martin, R. *Chem. Soc. Rev.* **2014**, *43*, 8081–8097. (g) Yamaguchi, J.; Muto, K.; Itami, K. *Eur. J. Org. Chem.* **2013**, 19–30.

4. (a) Ramgren, S. D.; Hie, L.; Ye, Y.; Garg, N. K. *Org. Lett.* **2013**, *15*, 3950–3953. (b) Hie, L.; Chang, J. J.; Garg, N. K. *J. Chem. Educ.* **2015**, *92*, 571–574. (c) Hie, L.; Fine Nathel, N. F.; Hong, X.; Yang, Y.-F.; Houk, K. N.; Garg, N. K. *Angew. Chem. Int. Ed.* **2016**, *55*, 2810–2814. (d) Weires, N. A.; Baker, E. L.; Garg, N. K. *Nat. Chem.* **2016**, *8*, 75–79. (e) Hie, L.; Fine Nathel, N. F.; Shah, T. K.; Baker, E. L.; Hong, A.; Fang, Y.-F.; Liu, P.; Houk, K. N.; Garg, N. K. *Nature* **2015**, *524*, 79–83. (f) Fine Nathel, N. F.; Kim, J.; Hie, L.; Jiang, X.; Garg, N. K. *ACS Catal.* **2014**, *4*, 3289–3293. (g) Ramgren, S. D.; Hie, L.; Ye, Y.; Garg, N. K. Org. Lett. 2014, 15, 3950–3953. (h) Hie, L.; Ramgren, S. D.; Mesganaw, T.; Garg, N. K. *Org. Lett.* **2012**, *14*, 4182–4185.

5. For selected examples of aryl C–C bond couplings using Ni catalysis, see: (a) Guan, B.-T.; Wang, Y.; Li, B.-J.; Yu, D.; Shi, Z.-J. *J. Am. Chem. Soc.* **2008**, *130*, 14468–14470. (b) Quasdorf, K. W.; Tian, X.; Garg, N. K. *J. Am. Chem. Soc.* **2008**, *130*, 14422–14423. (c) Tobisu, M.; Shimasaki, T.; Chatani, N. *Angew. Chem. Int. Ed.* **2008**, *47*, 4866– 4869. (d) Quasdorf, K. W.; Riener, M.; Petrova, K. V.; Garg, N. K. *J. Am. Chem. Soc.* **2009**, *131*, 17748–17749. (e) Antoft-Finch, A.; Blackburn, T.; Snieckus, V. *J. Am. Chem. Soc.* **2009**, *131*, 17750– 17752. (f) Xi, L.; Li, B.-J.; Wu, Z.-H.; Lu, X.-Y.; Guan, B.-T.; Wang, B.- Q.; Zhao, K.-Q.; Shi, Z.-J. *Org. Lett.* **2010**, *12*, 884–887. (g) Quasdorf, K. W.; Antoft-Finch, A.; Liu, P.; Silberstein, A. L.; Komaromi, A.; Blackburn, T.; Ramgren, S. D.; Houk, K. N.; Snieckus, V.; Garg, N. K. *J. Am. Chem. Soc.* **2011**, *133*, 6352–6363. (h) Zhang, N.; Hoffman, D. J.; Gutsche, N.; Gupta, J.; Percec, V. *J. Org. Chem.* **2012**, *77*, 5956– 5964. (i) Ge, S.; Hartwig, J. F. *Angew. Chem. Int. Ed.* **2012**, *51*, 12837– 12841.
6. For selected examples of aryl C–N bond couplings using Ni catalysis, see: (a) Ogata, T.; Hartwig, J. F. *J. Am. Chem. Soc.* **2008**, *130*, 13848–13849. (b) Gao, C.-Y.; Yang, L.-M. *J. Org. Chem.* **2008**, *73*, 1624–1627. (c) Fors, B. P.; Watson, D. A.; Biscoe, M. R.; Buchwald, S. L. *J. Am. Chem. Soc.* **2008**, *130*, 13552–13554. (d) Tobisu, M.; Shimasaki, T.; Chatani, N. *Chem. Lett.* **2009**, *38*, 710–711. (e) Shimasaki, T.; Tobisu, M.; Chatani, N. *Angew. Chem. Int. Ed.* **2010**, *49*, 2929–2932. (f) Mesganaw, T.; Silberstein, A. L.; Ramgren, S. D.; Fine Nathel, N. F.; Hong, X.; Liu, P.; Garg, N. K. *Chem. Sci.* **2011**, *2*, 1766–1771. (g) Ramgren, S. D.; Silberstein, A. L.; Yang, Y.; Garg, N. K. *Angew. Chem. Int. Ed.* **2011**, *50*, 2171–2173. (h) Ackermann, L.; Sandmann, R.; Song, W. *Org. Lett.* **2011**, *13*, 1784–1786. (i) Huang, J.-H.; Yang, L.-M. *Org. Lett.* **2011**, *13*, 3750– 3753. (j) Hie, L.; Ramgren, S. D.; Mesganaw, T.; Garg, N. K. *Org. Lett.* **2012**, *14*, 4182–4185.
7. For selected seminal studies on aryl C–C and C–N bond coupling using Ni catalysis, see: (a) Sengupta, S.; Leite, M.; Raslan, D. S.; Quesnelle, C.; Snieckus, V. *J. Org. Chem.* **1992**, *57*, 4066–4068. (b) Wolfe, J. P.; Buchwald, S. L. *J. Am. Chem. Soc.* **1997**, *119*, 6054– 6058. (c) Hamann, B. C.; Hartwig, J. F. *J. Am. Chem. Soc.* **1998**, *120*, 7369–7370. (d) Brenner, E.; Fort, Y. *Tetrahedron Lett.* **1998**, *39*, 5359–5362. (e) Wehn, P. M.; Du Bois, J. *Org. Lett.* **2005**, *7*, 4685– 4688.
8. For solvent selection guides in medicinal chemistry, see: (a) Alfonsi, K.; Colberg, J.; Dunn, P. J.; Fevig, T.; Jennings, S.; Johnson, T. A.; Kleine, H. P.; Knight, C.; Nagy, M. A.; Perry, D. A.; Stefaniak, M. *Green Chem.* **2008**, *10*, 31–36. (b) Henderson, R. K.; Jimeńez-Gonzaĺez, C.; Constable, D. J.

C.; Alston, S. R.; Inglis, G. G. A.; Fisher, G.; Sherwood, J.; Binks, S. P.; Curzons, A. D. *Green Chem.* **2011**, *13*, 854–862.

9. For the ACS Green Chemistry Institute Roundtable Solvent Selection Guide:
http://www.acs.org/content/dam/acsorg/greenchemistry/industriainnovation/roundtable/acs-gci-pr-solvent-selection-guide.pdf
10. *tert*-Amyl alcohol and 2-Me-THF were selected in consultation with the ACS Green Chemistry Institute. *tert*-Amyl alcohol is attractive due to its safety profile, low freezing point (in comparison to *tert*-butanol), and ability to solubilize polar compounds. 2-Me-THF is advantageous because it is obtained from renewable feedstocks and possesses many process chemistry-related over THF. For a discussion of 2-Me-THF, see: Aycock, D. F. *Org. Process Res. Dev.* **2007**, *11*, 156–159.
11. For examples of Pd-catalyzed cross-coupling to access bis-heterocycles, see: (a) Zhao, D.; You, J.; Hu, C. Chem. Eur. J. **2011**, *17*, 5466–5492. (b) Molander, G. A.; Shin, I. *Org. Lett.* **2013**, *15*, 2534– 2537.

Appendix
Chemical Abstracts Nomenclature (Registry Number)

5-Bromopyrimidine; (4595-59-9)
3-Furanylboronic acid; (55552-70-0)
Potassium phosphate tribasic; (7778-53-2)
Bis(tricyclohexylphosphine)nickel(II) chloride; (19999-87-2)
tert-Amyl alcohol; (75-85-4)

Liana Hie was born and raised in Jakarta, Indonesia. She received her B. S. in chemistry from the University of California, Davis (UCD) in 2011. At UCD, she performed research in the laboratory of Prof. Xi Chen. She began her doctoral studies at the University of California, Los Angeles in 2011 under the supervision of Prof. Neil K. Garg. Her research focuses on the development of nickel-catalyzed cross-coupling reactions.

Neil Garg is a Professor of Chemistry at the University of California, Los Angeles. His laboratory develops novel synthetic strategies and methodologies to enable the total synthesis of complex bioactive molecules.

Kristen Flynn was born and raised in New Jersey. She obtained her B.S. in chemistry from Monmouth University in 2016. During her time there, she performed research under the guidance of Prof. Massimiliano Lamberto. She has a research appointment in the Movassaghi group at MIT and will begin her Ph.D. studies at in the fall.

Alyssa Antropow completed her B.S./M.S. in chemistry at Stevens Institute of Technology in 2013. As an undergraduate, she conducted research in the laboratory of Prof. A.K. Ganguly. In 2013, Alyssa began her graduate studies at MIT and joined the laboratory of Prof. Mohammad Movassaghi. Her research is focused on the total synthesis of alkaloid natural products.

Organic Syntheses

Preparation of (S)-Nonafluorobutanesulfinamide

Apiwat Wangweerawong, Scott Kolmar, and Jonathan A. Ellman*[1]

Department of Chemistry, Yale University, New Haven, Connecticut 06520, United States

Checked by Steve Han and Chris Senanayake

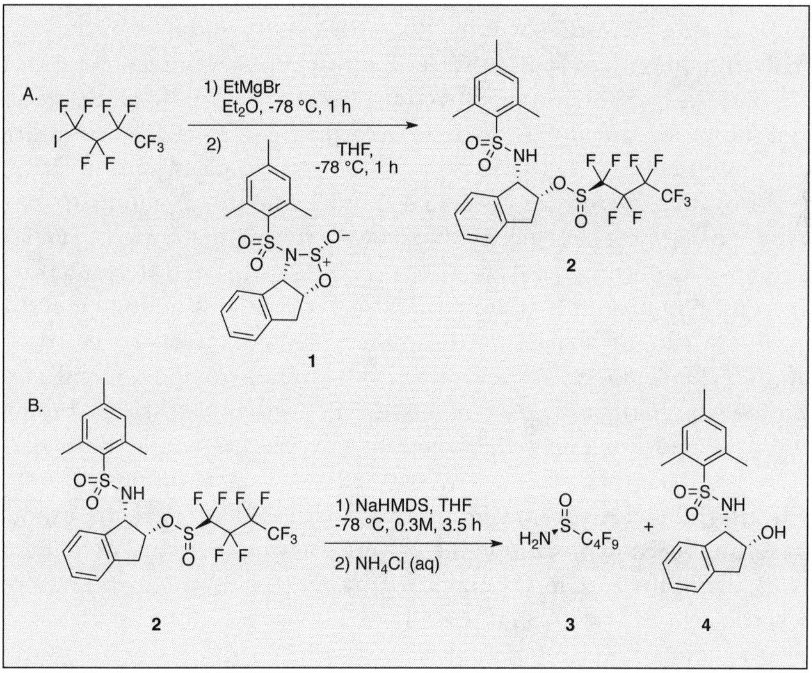

Procedure

A. *(1S,2R)-1-(2,4,6-Trimethylbenzenesulfonamido)-2,3-dihydro-1H-inden-2-yl-(S)-nonafluorobutanesulfinate* (**2**). A flame-dried 250 mL three-necked round-bottomed flask with 24/40 ground glass joints is equipped with a 3 cm Teflon-coated magnetic oval stir bar and fitted with two rubber septa

DOI: 10.15227/orgsyn.093.0319

and a low temperature thermometer as an internal temperature probe (Note 1). The flask is flushed with nitrogen for 10 min via a nitrogen line inlet and then 40 mL of anhydrous diethyl ether (Note 2) is added by syringe. The reaction flask is placed in a dry ice-acetone bath. Nonafluoro-1-iodobutane (3.53 mL, 20.5 mmol, 1.55 equiv) (Note 3) is added by syringe, and the resulting solution is stirred at 700 rpm while an ~3 M ethylmagnesium bromide solution in diethyl ether (6.76 mL, 2.94 M, 19.9 mmol, 1.50 equiv) (Notes 3 and 4) is added dropwise via a syringe over 3 min with the internal temperature maintained below –65 °C. After the addition is complete, the reaction solution turns from a clear light yellow color to a cloudy mixture. The fume hood light is turned off, and the flask is covered with aluminum foil to protect the reaction mixture from light, and the Grignard reagent mixture is stirred for 1 h at –78 °C. While the Grignard reagent mixture is being stirred, to a separate one-necked oven-dried 100 mL round-bottomed flask with a 24/40 ground glass joint is charged with the oxathiazolidine oxide **1** (5.0 g, 13.2 mmol, 1.0 equiv) (Note 5) prepared using the previously reported two-step *Organic Syntheses* procedure.[3] After fitting the flask with a rubber septum and flushing it with nitrogen for 5 min, 40 mL of anhydrous THF (Note 2) is added by syringe before the mixture is shaken until complete dissolution has occurred. The resulting THF solution is transferred via cannula to the 250 mL flask containing the Grignard reagent (1 h after ethylmagnesium bromide addition) in under 6 min while keeping the internal temperature below –60 °C. The 100 mL flask is then washed with an additional 1 mL of anhydrous THF, which is transferred over to the 250 mL flask via cannula. The resulting reaction mixture in the 250 mL flask is stirred at –78 °C for 1 h (Note 6) before the reaction is quenched by a slow dropwise addition of a 4.1M acetic acid in THF (6.5 mL, 26.6 mmol, 2.0 equiv) (Note 3) over 4 min to keep the internal temperature below –60 °C. After warming up to ambient temperature over ca. 30 min, the mixture is transferred to a 250 mL separatory funnel. Additional deionized water (40 mL) and ethyl acetate (20 mL) are used to ensure that all of the product is transferred. The aqueous layer is separated and further extracted with ethyl acetate (3 x 20 mL). The combined organic layer, which has a cloudy appearance, is washed with a saturated aqueous $Na_2S_2O_3$ solution (20 mL) with vigorous shaking until the cloudy appearance changes to clear solution. The organic layer is then washed with saturated aqueous NaCl solution (15 mL) and dried over sodium sulfate (15 g, 10 min). The drying agent is removed by vacuum filtration through a 60 mL medium fritted glass funnel followed by washing the flask and drying agent with ethyl acetate (3 x 10 mL). The filtrate is concentrated using a rotary evaporator (26 °C, <50 mmHg) and then dried with a high vacuum line (1.1 mmHg) to afford the crude product **2** (7.0 g, 88%) as an off-white solid (Note 7). The crude product is of sufficient purity by ^{1}H NMR (Note 8) to be used directly in Step B.

B. *(S)-Nonafluorobutanesulfinamide (3).* A flame dried 250 mL three-necked indented Morton style round bottom flask with 24/40 ground glass joints is equipped with a 3 cm Teflon-coated magnetic oval stir bar and is fitted with two rubber septa and an oven dried 125 mL pressure-equalizing addition funnel with 24/40 ground glass joints and a rubber septum at the top. The flask is flushed with nitrogen via a nitrogen line inlet for 10 min and then charged with 7.0 g of **2** at which point a rubber septum is replaced with a low temperature thermometer as an internal temperature probe (Note 1). To the flask is then added 22 mL of anhydrous THF via syringe (Note 2) followed by stirring at 700 rpm to give a 0.3 M clear yellow solution of **2** in THF. The solution is cooled to an internal temperature of -78 °C using a dry ice-acetone bath. A 2.0 M solution of NaHMDS in THF (17.6 mL, 35.1 mmol, 3 equiv assuming crude **2** is 100% pure) (Notes 3 and 9) is then added dropwise via the 125 mL pressure-equalizing addition funnel over 5 min with the internal temperature maintained below -65 °C. The mixture is stirred for 3.5 h at -78 °C at which time the appearance of the solution changes to a dark brown color. The reaction is monitored by TLC (Note 10). The reaction is quenched by addition of 40 mL of saturated aqueous NH_4Cl solution dropwise via the equipped pressure-equalizing addition funnel over 10 min with the internal temperature maintained below -60 °C. After removing the dry ice-acetone bath, the reaction mixture is allowed to warm to ambient temperature over ~30 min. The biphasic mixture is then stirred at 1000 rpm for 16 h at ambient temperature (Note 11). The mixture is then transferred to a 250 mL separatory funnel with additional deionized water (20 mL) and ethyl acetate (20 mL) used to ensure the complete transfer of crude product. The aqueous layer is separated and further extracted with ethyl acetate (3 x 20 mL). The combined organic layer is washed with saturated aqueous NaCl solution (15 mL) and then dried over sodium sulfate (15 g, 10 min). The drying agent is removed by vacuum filtration through a 60 mL medium fritted glass funnel and washed with three 10 mL portions of ethyl acetate. The filtrate is concentrated using a rotary evaporator (26 °C, <50 mmHg) to afford a brown oil. For purification, the crude material is solubilized in 3 mL of 10% MTBE (by volume) in dichloromethane and loaded onto a 6.5 cm diameter column (35 cm in length) containing 250 mL (about 100 g) of silica gel (Note 12) that has been prepacked with 10% MTBE (by volume) in dichloromethane. The flask is washed with 10% MTBE (by volume) in dichloromethane (3 x 10 mL) with the washings loaded onto the silica gel. Fractions are collected using 18 x 150 mm disposable culture tubes using 10% MTBE (by volume) in dichloromethane as eluent. To recover **4**, fractions containing the *N*-sulfonylaminoindanol **4** and its *O*-silyl-protected derivative are identified by TLC, (Note 13) combined, and subjected to silyl deprotection using TBAF (Note 14). Fractions containing the product sulfinamide **3** are combined and concentrated using a rotary evaporator (26 °C, 100 mmHg). This solid material is subsequently dried with a high vacuum line (1.1 mmHg) to provide 2.50 g (76% over two steps) (Note 15) of enantioenriched (>99% ee) (Note 16) (*S*)-nonafluorobutanesulfinamide **3** as an off-white solid, mp 73–75 °C (Notes 17 and 18).

Notes

1. THERMCO 76mm immersion low temperature thermometer equipped with a thermometer adapter (30.5 cm length, temperature range –100 to 50 °C).
2. THF (HPLC grade) is obtained from Fisher Scientific and diethyl ether (ACS reagent, inhibitor-free) is obtained from Acros Organics. The solvents are purified by passage through activated alumina using a Glass Contour solvent purification system. Chloroform-D ($CDCl_3$, D, 99.8%) is purchased from Cambridge Isotope Laboratories and stored over potassium carbonate.
3. Nonafluoro-1-iodobutane (97%) is obtained from Alfa Aesar and used as received. Ethylmagnesium bromide (3M in diethyl ether) is obtained from Strem Chemicals and titrated before use. Acetic acid (ACS, glacial) is obtained from J.T. Baker and used as received. Sodium bis(trimethylsilyl)amide (2 M NaHMDS in THF) is obtained from Alfa Aesar and used as received. Tetrabutylammonium fluoride (1M in THF) is obtained from Strem Chemicals and used as received.
4. The molarity of the Grignard reagent is determined by titration according to the procedure of Knochel.[2] Specifically, to a flamed-dried 1 dram vial flushed with nitrogen for 5 min and equipped with a magnetic stir bar is weighed out 100 mg of iodine before dissolving in

1.0 mL of a 0.5 M solution of LiCl in THF. The resulting dark brown solution is cooled to 0 °C with an ice bath before ethylmagnesium bromide is added slowly drop by drop to the stirred solution via a 1 mL syringe until the solution turns light yellow and then becomes colorless. The titration is repeated and the average molarity value of the Grignard reagent from the two titrations is obtained.

5. Oxathiazolidine oxide is prepared by a previously reported two-step *Organic Syntheses* procedure.[3]
6. The reaction progress can be monitored by ^{1}H NMR by quenching a 0.1 mL sample of the reaction mixture (obtained via syringe and needle) with acetic acid in THF (3 drops of AcOH in 0.5 mL of ethyl acetate). The resulting mixture is concentrated to dryness using a rotary evaporator before dissolving the material with $CDCl_3$ followed by transfer to a NMR tube. Reaction completion can be monitored by following the disappearance of the triplet of doublet peak at 5.85 ppm, which corresponds to the oxathiazolidine oxide starting material.
7. A duplicate run on the same scale provided 7.5g (94%) of an off white solid. This material was also used in the next step without purification.
8. The crude product **2** is used directly in the next step without further purification and has been characterized as follows: ^{1}H NMR (400 MHz, $CDCl_3$, selected signals) δ: 2.32 (s, 3H), 2.73 (s, 6H), 3.08 – 3.28 (m, 2H), 5.00 (dd, J = 9.6, 4.7 Hz, 1H), 5.21 – 5.26 (m, 2H), 7.00 (s, 2H), 7.12 – 7.18 (m, 1H), 7.20 – 7.24 (m, 2H), 7.24 – 7.32 (m, 1H).
9. At least two equiv of NaHMDS is required for the reaction, one equiv to react with the sulfinate ester group of **1** and the other to deprotonate the sulfonamide group of addition product **2**.
10. The reaction is monitored by TLC (Merck TLC Silica gel 60 F_{254}) using 20% ethyl acetate in hexanes as eluent. The TLC plate is first visualized with a 254 nm UV lamp then with a $KMnO_4$ staining solution. The silyl-protected *N*-sulfonylaminoindanol has an R_f value of 0.70 while the *N*-sulfonylaminoindanol has an R_f value of 0.19.
11. Stirring the biphasic mixture for at least 4 h is necessary to ensure complete cleavage of the trimethylsilyl groups from the sulfinamide product.
12. Silica gel (SiliaFlash® P60, 230-400 mesh) is obtained from Silicycle.
13. Each fraction is analyzed by thin layer chromatography (Merck TLC Silica gel 60 F_{254}) using 10% MTBE in dichloromethane as eluent. The TLC plate is first visualized with a 254 nm UV lamp and then is stained with a $KMnO_4$ staining solution. The reaction product **3** has an R_f value

of 0.31, the *N*-sulfonylaminoindanol **4** has an R_f value of 0.59, and the O-silyl derivative has an R_f value of 0.90.

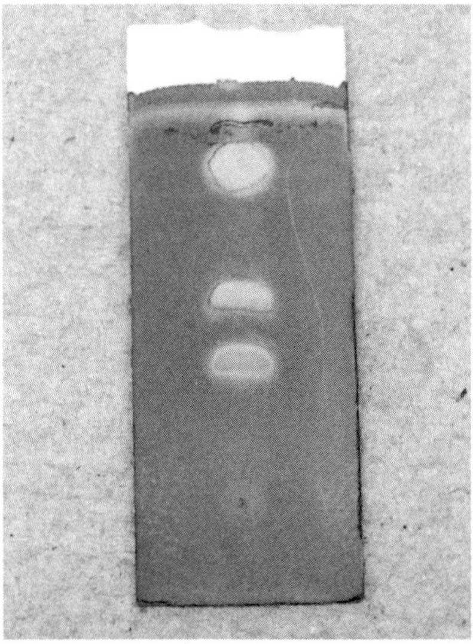

14. The combined fractions 9-43 contain a mixture of *N*-sulfonylaminoindanol **4** and its *O*-silyl derivative and are concentrated using a rotary evaporator (26 °C, 100 mmHg) to afford 12.0 g of dark red-brown oil. The material is transferred to a one-necked 250 mL round bottom flask with a 24/40 ground glass joint using 100 mL of THF (Note 2) to solubilize the material and assist with transfer. To the stirred solution (700 rpm) at ambient temperature is then added 30 mL of 1 M TBAF (29.7 mmol, 1 equiv assuming that the material is 100% the *O*-silyl derivative of **4**) in THF (Note 3). A rubber septum is placed on the flask along with a N_2 inlet. After stirring at ambient temperature for 1 h, the mixture is transferred to a 250 mL separatory funnel. Additional ethyl acetate (50 mL) and deionized water (50 mL) are used to achieve the complete transfer of material. The aqueous layer is separated and further extracted with ethyl acetate (2 x 20 mL). The combined organic layer is washed with deionized water (2 x 30 mL) and saturated aqueous NaCl solution (30 mL) before drying over sodium sulfate (15 g, 10 min). The drying agent is removed by vacuum filtration through a 200 mL medium fritted glass funnel and washed with three 10 mL portions of ethyl acetate. The filtrate is concentrated to dryness using a

rotary evaporator (26 °C, <50 mmHg) before it is triturated with 50 mL of hexanes and vacuum filtered through a Buchner funnel (6.5 cm diameter) equipped with filter paper to afford 10.0 g of a light brown solid material. For purification, a 250 mL Erlenmeyer flask is charged with the crude material and 40 mL of ethyl acetate. The mixture is heated to 50 °C with swirling to dissolve all of the material before 80 mL of hexanes is layered on top. The flask is covered with aluminum foil and then is stored at –20 °C for 16 h. The contents of the flask are vacuum filtered through a Buchner funnel (6.5 cm diameter) equipped with filter paper. The crystalline solid is crushed to a powder using a porcelain mortar and pestle before it is dried with a high vacuum line (1.1 mmHg) to provide 8.0 g (76% recovery over 3 steps) of the *N*-sulfonylaminoindanol **4** as an off-white solid, mp 148–150°C (Note 15). The analytical data for this compound are consistent with previously reported data.[3]

15. A second reaction on identical scale provide 2.45 g (66% over two steps) of the title compound with identical physical properties.
16. The enantiomeric excess of **3** is determined using an Agilent 1100 series HPLC (1.0 mL/min flow rate) equipped with a mutiwavelength detector with monitoring at 230 nm and a Chiralpak IB column. A mixture of 10% ethanol in hexanes is used as the eluent over 30 min: t_R = 7.4 min (*S*) and 11.4 min (*R*).
17. Melting points are obtained using Electrothermal Melting Point Apparatus.
18. The sulfinamide **3** has been characterized as follows: IR (film): 3344, 3241, 3106, 1352, 1230, 1198, 1183, 1135, 1021, 725, 695 cm^{-1}. ^{1}H NMR (400 MHz, $CDCl_3$) δ: 4.73 (s, 2H); ^{13}C{^{1}H} NMR (100 MHz, $CDCl_3$) δ: 111.8 – 105.6 (m); ^{19}F NMR (376 MHz, $CDCl_3$) δ: –125.97 to –126.20 (2F), –122.48 to –123.27 (1F), –121.76 to –121.93 (1F), –121.34 to –121.66 (2F), – 80.77 (tt, *J* = 9.7, 2.3 Hz, 3F); HRMS [M + H] calcd for $C_4H_3F_9NOS$: 283.9792; Found: 283.9769. Anal. Calcd for $C_4H_2F_9NOS$: C, 16.97; H, 0.71; N, 4.95. Found: C, 17.09; H, 0.71; N, 5.34.

Working with Hazardous Chemicals

The procedures in *Organic Syntheses* are intended for use only by persons with proper training in experimental organic chemistry. All hazardous materials should be handled using the standard procedures for work with chemicals described in references such as "Prudent Practices in the Laboratory" (The National Academies Press, Washington, D.C., 2011; the full text can be accessed free of charge at http://www.nap.edu/catalog.php?record_id=12654). All chemical waste should be disposed of in accordance with local regulations. For general guidelines for the management of chemical waste, see Chapter 8 of Prudent Practices.

In some articles in *Organic Syntheses*, chemical-specific hazards are highlighted in red "Caution Notes" within a procedure. It is important to recognize that the absence of a caution note does not imply that no significant hazards are associated with the chemicals involved in that procedure. Prior to performing a reaction, a thorough risk assessment should be carried out that includes a review of the potential hazards associated with each chemical and experimental operation on the scale that is planned for the procedure. Guidelines for carrying out a risk assessment and for analyzing the hazards associated with chemicals can be found in Chapter 4 of Prudent Practices.

The procedures described in *Organic Syntheses* are provided as published and are conducted at one's own risk. *Organic Syntheses, Inc.*, its Editors, and its Board of Directors do not warrant or guarantee the safety of individuals using these procedures and hereby disclaim any liability for any injuries or damages claimed to have resulted from or related in any way to the procedures herein.

Discussion

Chiral α-branched amines are important synthetic targets due to their prevalence in bioactive natural products, pharmaceutical agents and agrochemicals. One of the most frequently employed approaches for the asymmetric synthesis of α-branched amines is the diastereoselective addition of nucleophiles to *N-tert*-butanesulfinyl imines.[4] However, for some types of nucleophiles, additions do not occur because the *N-tert-*

butanesulfinyl imines aren't sufficiently electrophilic. The Liu group has recently investigated *N*-perfluorinated alkanesulfinyl imines for the asymmetric synthesis of amines, including with Strecker reactions,[5,6] vinylogous Mannich reactions,[7] and inverse electron demand aza Diels-Alder cycloadditions.[8] In our efforts to develop asymmetric Rh(III)-catalyzed C–H bond additions to imines we found that *N-tert*-butanesulfinyl aldimines had insufficient electrophilicity for additions to occur. In contrast, Rh(III)-catalyzed C–H bond additions to *N*-perfluorobutanesulfinyl aldimines proceeded in moderate to good yields and with very high diastereoselectivities (95:5 to >99:1 dr).[9,10] We have also found that the *N*-perfluorobutanesulfinyl imino ester derived from ethyl glyoxylate undergoes thermal aza Diels-Alder reactions in good yields and with high stereoselectivities, whereas for the corresponding *N-tert*-butanesulfinyl imino ester, cycloadditions did not occur.[10]

Liu and coworkers were the first to report on the asymmetric synthesis of perfluorinated alkanesulfinamide chiral amine reagents, including perfluorobutanesulfinamide **3**, which is the focus of this procedure.[5] Although this multi-step sequence proceeds in good overall yield and enantiopurity (Scheme 1), it requires two flash column chromatography purification steps, including one for the separation of diastereomers.

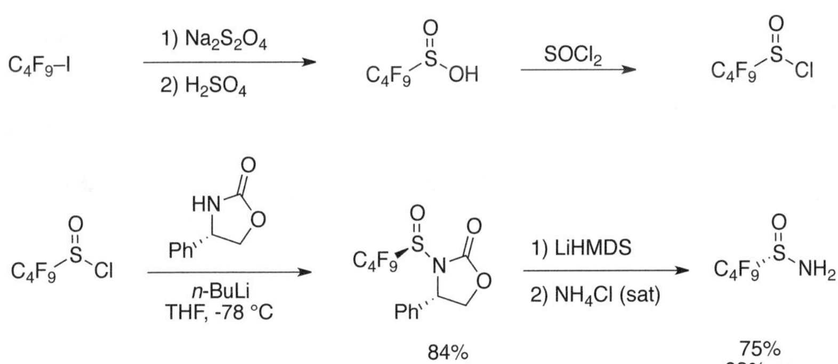

Scheme 1. Liu's Asymmetric Synthesis of Perfluorobutanesulfinamide

To develop a more practical route for the preparation of the perfluorobutanesulfinamide, we focused on Senanayake's chiral sulfinyl transfer auxiliary approach.[11] The two-step synthesis of perfluorobutanesulfinamide from Senanayakes's 2-aminoindanol-derived sulfinyl transfer reagent described in this report proceeds in good overall yield and with excellent enantiomeric purity. The described method uses inexpensive nonafluoro-1-iodobutane as a starting input and requires only one straightforward flash column chromatography. Additionally, the chiral auxiliary **3** can be recovered in good yields.

References

1. Department of Chemistry, Yale University, New Haven, Connecticut 06520, email jonathan.ellman@yale.edu. We thank the National Science Foundation (grant # CHE - 1464129) for support of this work.
2. Krasovskiy, A.; Knochel, P. *Synthesis* **2006**, *5*, 890–891.
3. Ramachandar, T.; Wu, Y.; Zhang, J.; Davis, F. A. *Org. Synth.* **2006**, *83*, 131–139.
4. Robak, M. T.; Herbage, M. A.; Ellman, J. A. *Chem. Rev.* **2010**, *110*, 3600–3740.
5. Liu, L.-J.; Chen, L.-J.; Li, P.; Li, X.-B.; Liu, J.-T. *J. Org. Chem.* **2011**, *76*, 4675–4681.
6. Yang, K.; Liu, L.-J.; Liu, J.-T. *J. Org. Chem.* **2014**, *79*, 3215–3220.
7. Liu, L.-J.; Liu, J.-T. *Tetrahedron* **2014**, *70*, 1236–1245.
8. Li, P.; Liu, L.-J.; Liu, J.-T. *Org. Biomol. Chem.* **2011**, *9*, 74–77.
9. Wangweerawong, A.; Bergman, R. G.; Ellman, J. A. *J. Am. Chem. Soc.* **2014**, *136*, 8520–8523.
10. Wangweerawong, A.; Hummel, J. R.; Bergman, R. G.; Ellman, J. A. *J. Org. Chem.* **2016**, *81*, 1547–1557.
11. Han, Z.; Krishnamurthy, D.; Grover, P.; Fang, Q. K.; Su, X.; Wilkinson, H. S.; Lu, Z.-H.; Magiera, D.; Senanayake, C. H. *Tetrahedron* **2005**, *61*, 6386–6408.

Appendix
Chemical Abstracts Nomenclature (Registry Number)

Perfluorobutanesulfinamide: (*S*)-1,1,2,2,3,3,4,4,4-nonafluorobutane-1-sulfinamide; (1303994-31-1)
Oxathiazolidine oxide: (2R,3aS,8aR)-3-(mesitylsulfonyl)-3,3a,8,8a-tetrahydroindeno[1,2-d][1,2,3]oxathiazole 2-oxide; (473554-02-8)
N-Sulfonylaminoindanol: N-((1S,2R)-2-hydroxy-2,3-dihydro-1H-inden-1-yl)-2,4,6-trimethylbenzenesulfonamide; (473554-01-7)

Apiwat Wangweerawong received his B.S. in Biochemistry in 2011 from Florida State University under the mentorship of Professor Gregory Dudley. Since 2011, he has been a graduate student in the group of Professor Jonathan Ellman working on the asymmetric synthesis of branched amines.

Scott Kolmar obtained his B.S. in Chemistry from Tulane University in 2013. His undergraduate research focused on the pH dependent kinetics of glucosidase enzymes under the supervision of Professor Larry Byers. In 2013, he started his Ph.D. studies with Professor Jonathan Ellman at Yale University working on the synthesis of fluorinated sulfinamides. He is currently studying the mechanisms of small molecule proton-coupled electron transfer reactions in the group of Professor Jim Mayer at Yale University.

Jonathan Ellman earned his B.S. degree from MIT in 1984 and his Ph.D. degree from Harvard as an NSF predoctoral fellow with Professor David Evans in 1989. After an NSF postdoctoral fellowship with Professor Peter Schultz at the University of California at Berkeley, he began his independent career at the same institution in 1992 where he rose to the rank of Professor. Since 2010 he has been the Eugene Higgins Professor of Chemistry and Professor of Pharmacology at Yale University. He is the editor for Volume 88 of *Organic Syntheses*.

Zhengxu (Steve) Han received his B.S degree and Ph.D. in organic chemistry from Lanzhou University with Professor You-Cheng Liu. In 1991, he moved to Tübingen University, Germany as a fellow of the Alexander von Humboldt Foundation with Professor Anton Rieker. After four years of postdoctoral research with Professor Stuart Linn at University California, Berkeley and Professor Lee Magid at the University of Tennessee, Knoxville, he joined the process research group at Sepracor, Inc., in 1998 and then Boehringer Ingelheim (BI) in 2005. He is now a senior research fellow at BI and his research interest center on efficient process and methodology development for asymmetric synthesis.

Preparation of 1-methylimidazole-*N*-oxide (NMI-O)

James I. Murray and Alan C. Spivey*[1]

Department of Chemistry, South Kensington Campus, Imperial College London, United Kingdom, SW7 2AZ

Checked by Moritz Hönig and Erick M. Carreira

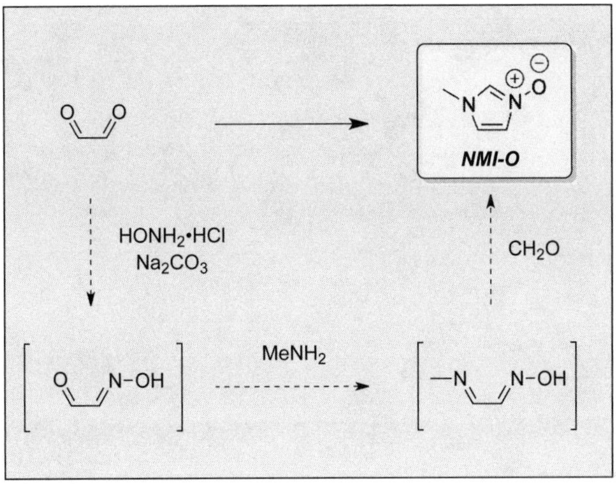

Procedure

1-Methylimidazole-N-oxide (**NMI-O**). A 500 mL, single-necked, round-bottomed flask is charged with an oval 3 cm Teflon coated magnetic stirrer bar. To this flask is added hydroxylamine hydrochloride (11.2 g, 0.161 mol) (Note 1) and sodium carbonate (21.2 g, 0.20 mol, Note 1) followed by distilled water (60 mL) and the resulting white suspension cooled to 0 °C in an ice-water bath (Figure 1A). The headspace in the flask is flushed with nitrogen, the flask fitted with a rubber septum and placed under a positive pressure of nitrogen *via* a syringe needle to a manifold (Figure 1A). After stirring the suspension for 5 min, an aqueous solution of glyoxal (23.0 mL, 40% w/v aq., 8.8 M, 0.20 mol) is added *via* syringe over 5 min resulting in the suspension becoming pale yellow in color (Note 2) (Figure 1B/C). An aqueous solution of methylamine (17.0 mL, 40% w/v aq., 11.6 M, 0.20 mol) is added *via* syringe over 5 min. The addition causes an immediate orange/red coloration of the suspension which darkens progressively during the addition to become a red/brown suspension upon complete addition (Note 3) (Figure 1D/E). An aqueous solution of formaldehyde (15.0 mL, 37% w/v aq., 13.4 M, 0.20 mol) is then added *via* syringe over 5 min during which the brown color is maintained (Note 4) (Figure 1F). The reaction mixture is then stirred for 16 h at room temperature under nitrogen.

Figure 1. A; Initial appearance of reaction mixture, B; During addition of glyoxal, C; 5 minutes after glyoxal addition, D; During addition of $MeNH_2$, E; 5 min after $MeNH_2$ addition, F; During CH_2O addition.

The reaction mixture is cooled to 0 °C with an ice-water bath before adding *conc.* hydrochloric acid (5.3 mL, 37% aq., 12.2 M) (Note 5) over a period of ~ 2 min *via* syringe. The reaction mixture becomes a yellow/orange suspension and this suspension is stirred for a further 2 h at room temperature (Figure 2A).

The reaction mixture is filtered using a Büchner funnel (5 cm diameter) and the filtrate concentrated at reduced pressure on a rotary evaporator (45 °C, 100 mmHg) to afford a brown slurry (Figure 2B). To this residue is added a mixture of dichloromethane/methanol/triethylamine (300 mL, 80:15:5) and after vigorous swirling the resulting mixture is filtered again using a 5 cm diameter Büchner funnel (Note 6). The collected solid residue is washed with a further 100 mL of dichloromethane/methanol/triethylamine (80:15:5 v/v) and the filtrates combined (TLC; 20% methanol in dichloromethane + 0.5% triethylamine, R_f = 0.37, Figure 2C). To the resulting combined brown filtrate solution is added silica gel (15 g) (Note 7) and the mixture concentrated at reduced pressure on a rotary evaporator (45 °C, 100 mmHg) to afford an orange solid. This solid is dry-loaded and purified by flash chromatography (Note 8) to afford the product as brown oil which solidifies upon standing in a fridge or freezer (10.8 g, 68%) (Notes 9 and 10) (Figure 3).

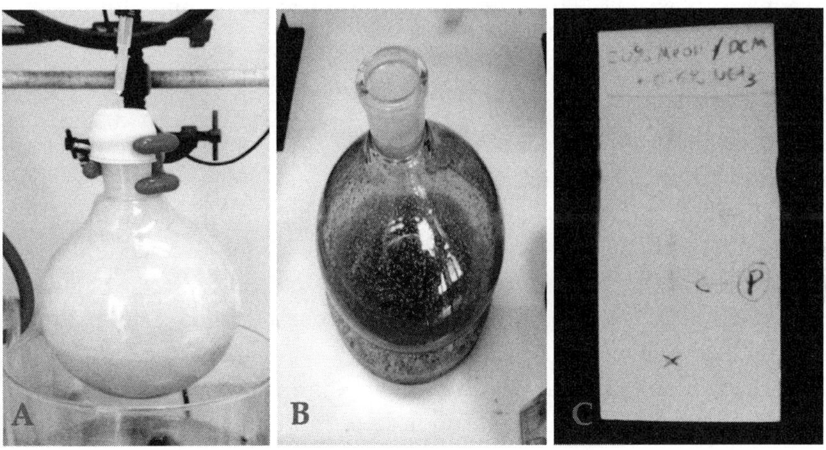

Figure 2. A; After addition of HCl, B; after reaction mixture concentrated, C; TLC of crude reaction mixture P = product.

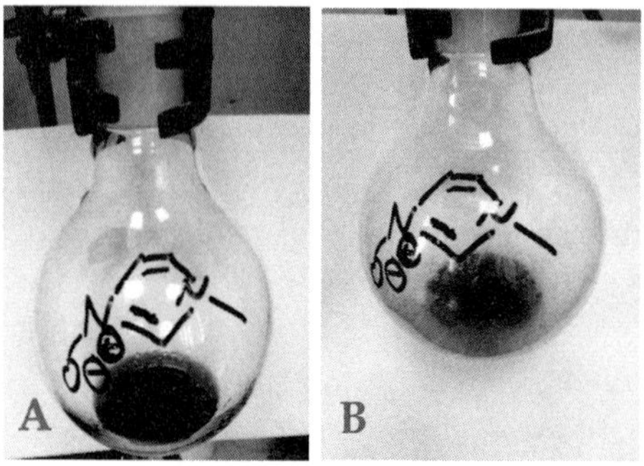

Figure 3. A; Product as an oil directly after concentration of column fractions on rotary evaporator, B; Product as a solid after standing in freezer for 24 h.

Notes

1. Hydroxylamine hydrochloride (HONH2·HCl, 99%) and sodium carbonate (Na2CO3, 99.5%) were purchased from Sigma Aldrich and used as received.
2. Glyoxal solution [(HCO)2, ~40% w/v aq., ~8.8 M] was purchased from Sigma Aldrich and used as received.
3. Methylamine (MeNH2) solution (40% w/v aq., 1.6 M) was purchased from Sigma Aldrich and used as received.
4. Formaldehyde (HCO2H) solution (37% w/v aq., 13.4 M, containing 10-15% MeOH as stabilizer) was purchased from Sigma Aldrich and used as received.
5. Conc. HCl (~37% w/v aq., 12.2 M) was purchased from Sigma Aldrich and used as received.
6. Dichloromethane, methanol and triethylamine were HPLC grade, purchased from Sigma Aldrich and used as received.
7. Silica Gel (high purity grade, pore size 60 Å, 230-400 mesh particle size.
8. Silica gel column (6 cm x 24 cm, 280 g Si). The column is eluted with 10% methanol in dichloromethane + 0.5% triethylamine (1500 mL, first 500 mL discarded) followed by 20% methanol in dichloromethane + 0.5% triethylamine (2000 mL). Fractions were collected in 40 mL test tubes and the desired product is typically obtained in fractions 27–77, which are concentrated by rotary evaporation (45 °C, 100 mmHg) to afford the hygroscopic product as a brown oil. The Rf of the purified product was found to be strongly dependent on the amount of material spotted onto the TLC plate and usually lay in the range of 0.1 to 0.3 (20% methanoi in dichloromethane + 0.5% triethylamine).

9. A reaction run at half-scale provided 6.22 g (78%) of 1-methylimidazole-*N*-oxide.
10. Physical characteristics of 1-methylimidazole-*N*-oxide: low melting point waxy brown solid with 99.0% purity as determined by quantitative ^{1}H NMR using ethylene carbonate as an internal standard. mp 41–44 °C. ^{1}H NMR (400 MHz, $CDCl_3$) δ: 3.66 (s, 3H), 6.68 (s, 1H), 7.10 (s, 1H), 7.84 (s, 1H) (the submitters reported observing ^{4}J coupling of 1.9 Hz in the heterocycle ^{1}H NMR resonances); ^{13}C NMR (100 MHz, $CDCl_3$) δ: 35.7, 117.6, 122.7, 126.9. HRMS (*m/z*) (M, $C_4H_6N_2O$) Calcd 98.0480), Found 98.0475; IR (film) 3142, 2929, 1651, 1551, 1316, 1165, 1090, 1027, 609, 479 cm^{-1}. Elemental analysis could not be secured by either submitters or checkers as a result of the hygroscopic nature of 1-methylimidazole-*N*-oxide. The coordinates from a single crystal X-ray structure determination on the monohydrate of 1-methylimidazole-N-oxide (obtained following atmospheric hydration of a sample of the anhydrous product prepared using this *Organic Syntheses* method) have been deposited at the Cambridge Crystallographic Data Centre: CCDC 1509924.

Working with Hazardous Chemicals

The procedures in *Organic Syntheses* are intended for use only by persons with proper training in experimental organic chemistry. All hazardous materials should be handled using the standard procedures for work with chemicals described in references such as "Prudent Practices in the Laboratory" (The National Academies Press, Washington, D.C., 2011; the full text can be accessed free of charge at http://www.nap.edu/catalog.php?record_id=12654). All chemical waste should be disposed of in accordance with local regulations. For general guidelines for the management of chemical waste, see Chapter 8 of Prudent Practices.

In some articles in *Organic Syntheses*, chemical-specific hazards are highlighted in red "Caution Notes" within a procedure. It is important to recognize that the absence of a caution note does not imply that no significant hazards are associated with the chemicals involved in that procedure. Prior to performing a reaction, a thorough risk assessment should be carried out that includes a review of the potential hazards associated with each chemical and experimental operation on the scale that is planned for the procedure. Guidelines for carrying out a risk assessment

and for analyzing the hazards associated with chemicals can be found in Chapter 4 of Prudent Practices.

The procedures described in *Organic Syntheses* are provided as published and are conducted at one's own risk. *Organic Syntheses, Inc.*, its Editors, and its Board of Directors do not warrant or guarantee the safety of individuals using these procedures and hereby disclaim any liability for any injuries or damages claimed to have resulted from or related in any way to the procedures herein.

Discussion

1-Methylimidazole-*N*-oxide (NMI-O) has been an underused member of the aryl-*N*-oxides, primarily due to its problematic synthesis.[2,3] Tsai *et al.* mistakenly reported the use of NMI-O as an efficient catalyst in the Morita-Baylis-Hillman reaction.[2] However, these authors were subsequently shown by Laus *et al.*[3] to have employed the *N*-methylimidazole (NMI) semiperhydrate (NMI.½H_2O_2) for these studies. The error resulted from the use of hydrogen peroxide (H_2O_2) for the direct oxidation of NMI, which has been repeatedly reported not to yield the desired *N*-oxide.[4] It is likely that NMI-O is concomitantly deoxygenated by H_2O_2 and by peracids during attempted preparation in this fashion.[5] Laus *et al.*[3] demonstrated the first synthesis of authentic NMI-O, requiring an 8 step synthesis from glyoxal.

Our group have reported the use of aryl-*N*-oxides as highly efficient catalysts for the phosphorylation of alcohols,[6] and recently reported the use of 2-aryl-4-dimethylaminopyridine-*N*-oxides for site-selective phosphorylation of polyols and peptides (Scheme 1).[7]

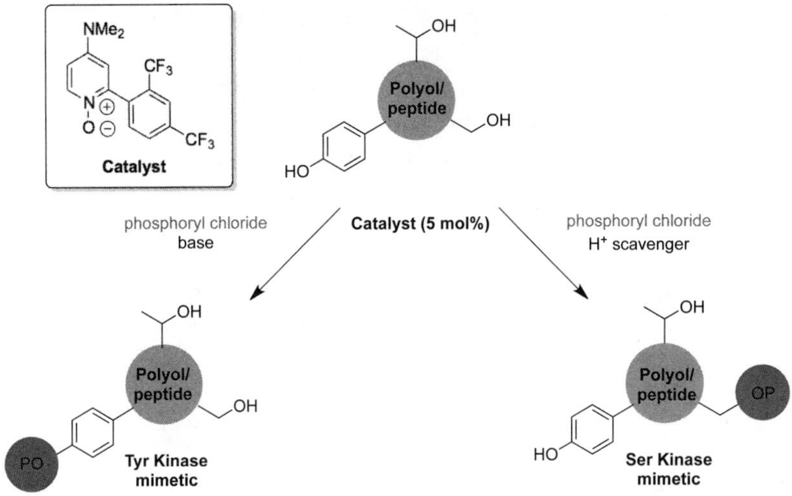

Scheme 1. Aryl-*N*-oxide catalysed site selective phosphorylation of polyols and peptides.

As such, we were interested in the comparative catalytic efficiencies of commonly used *N*-heterocycles and their *N*-oxide counterparts for acylation, sulfonylation and silylation reactions (Scheme 2).[8]

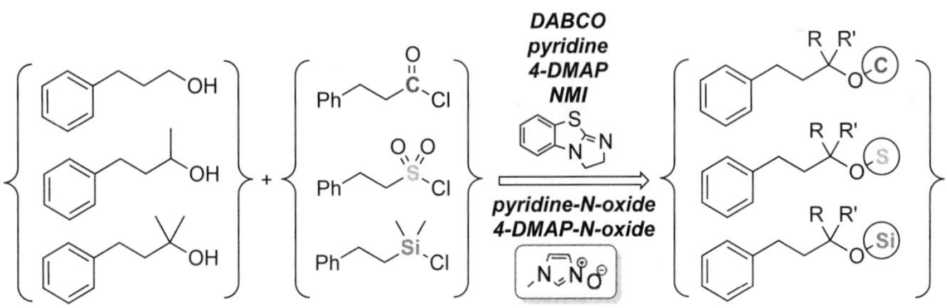

Scheme 2. Comparative study of *N*-heterocycle and *N*-oxide catalysts for the acylation, sulfonylation and silylation of alcohol derivatives.[8]

These studies revealed that whilst *N*-heterocycles were generally more efficient than their *N*-oxide analogues in the acylation process, *N*-oxides were generally more active for the sulfonylation and silylation of alcohol derivatives. In particular, 1-methylimidazole-*N*-oxide (NMI-O) was found to be a highly efficient catalyst both for sulfonylation and silylation. NMI-O was found to be the only amine or *N*-oxide Lewis basic organocatalyst capable of promoting the efficient silylation of tertiary alcohol derivatives with low catalyst loading (2.5 mol%) and under mild reaction conditions (45 °C, CH_2Cl_2) (Scheme 3).[8]

The efficiency of NMI-O as a highly active catalyst for the sulfonylation and silylation of alcohols should mean that it becomes widely used in synthesis. The operationally facile synthesis presented in this *Org. Synth.* preparation allows for multi gram-scale preparation of this highly active Lewis-basic catalyst and so should enable its widespread use and the development of novel derivatives.

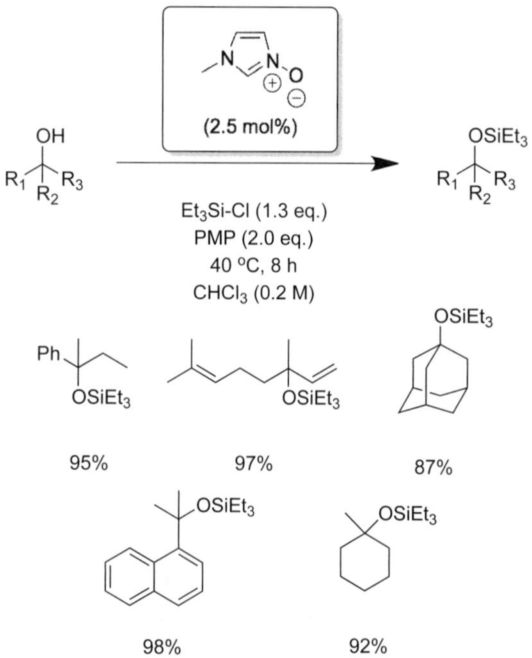

Scheme 3. *N*-Methylimidazole-*N*-oxide (NMIO) catalyzed triethylsilylation of *tert*-alcohols.[8]

References

1. Department of Chemistry, South Kensington Campus, Imperial College London, United Kingdom, SW7 2AZ. Email: a.c.spivey@imperial.ac.uk.
This work was financially supported by the Imperial College Institute of Chemical Biology and the EPSRC. J.I.M gratefully acknowledges the SCI for a Messel Scholarship.
2. Lin, Y.-S.; Liu, C.-W.; Tsai, T. Y. R. *Tetrahedron Lett.* **2005**, *46*, 1859–1861.

3. Laus, G.; Schwarzler, A.; Bentivoglio, G.; Hummel, M.; Kahlenberg, V.; Wurst, K.; Kristeva, E.; Schutz, J.; Kopacka, H.; Kreutz, C.; Bonn, G.; Andriyko, Y.; Nauer, G.; Schottenberger, H. *Z. Naturforsch. Sect. B* **2008**, *63*, 447–464.
4. Nikitina, G. V.; Pevzner, M. S. *Chem. Heterocycl. Compd.* **1993**, *29*, 127–151.
5. Begtrup, M.; Vedsø, P. *J. Chem. Soc., Perkin Trans. 1* **1995**, 243–247.
6. Murray, J. I.; Woscholski, R.; Spivey, A. C. *Synlett* **2015**, 985–990.
7. Murray, J. I.; Woscholski, R.; Spivey, A. C. *Chem. Commun.* **2014**, *50*, 13608–13611.
8. Murray, J. I.; Spivey, A. C. *Adv. Syn. Catal.* **2015**, *357*, 3825–3830.

Appendix
Chemical Abstracts Nomenclature (Registry Number)

$HONH_2 \cdot HCl$: hydroxylamine hydrochloride; (5470-11-1)
Na_2CO_3: sodium carbonate; (497-19-8)
Glyoxal: ethanedial; (107-22-2)
$MeNH_2$: methylamine; (74-89-5)
Formaldehyde (50-00-0)
HCl: hydrochloric acid; (7647-01-0)

James Murray completed his undergraduate studies at Trinity College Dublin under the supervision of Prof. Stephen Connon. During this time, he conducted a one-year placement as a medicinal chemist at GlaxoSmithKline, Stevenage, and as a process chemist at Hoffman-La Roche, Basel. He is currently a 4th year doctoral student in the Spivey Group at Imperial College London where his research is focused on the development of aryl-*N*-oxide catalysts for acylation, sulfonylation and phosphorylation reactions.

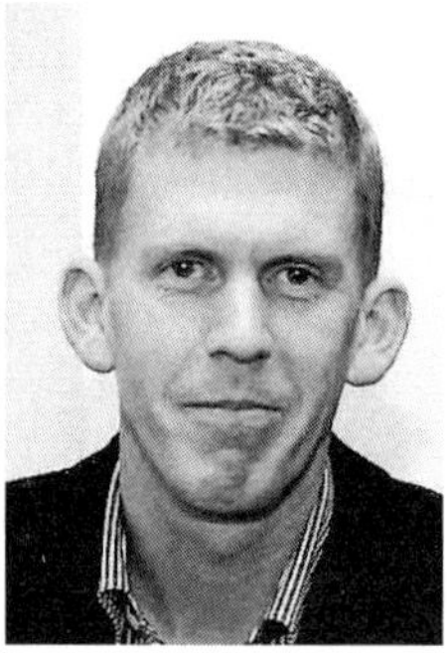

Alan Spivey completed his BSc Chemistry degree at the University of Nottingham in 1988 and DPhil at the University of Oxford in 1991 with Prof. Jack. E. Baldwin. Following post-doctoral studies with Prof. Wolfgang Oppolzer at the Université de Genève, Switzerland and with Prof. Alan R. Battersby at the University of Cambridge, he began his independent career in the Department of Chemistry, University of Sheffield in 1995. In 2003 he moved to Imperial College where he was promoted to Professor in 2008. The focus of the Spivey group's research is the development of useful new synthetic concepts and methods.

Moritz Hönig completed his undergraduate studies at Ludwigs-Maximilians-Universität München in 2014 before he moved to the University of Cambridge for his M.Phil. studies, working in the area of C–H activation in total synthesis under the direction of Professor Matthew J. Gaunt. He is currently a Ph.D. student in the group of Professor Erick M. Carreira at the Eidgenössische Technische Hochschule Zürich. His research focus resides in the total synthesis of natural products.

Organic Syntheses

Palladium-Catalyzed Direct Amination of Allylic Alcohols at Room Temperature

Bao Gao, Lixin Li, Guoying Zhang, and Hanmin Huang*[1]

State Key Laboratory for Oxo Synthesis and Selective Oxidation, Lanzhou Institute of Chemical Physics, Chinese Academy of Sciences, Lanzhou, 730000, China

Checked by Steven A. Loskot and Brian M. Stoltz

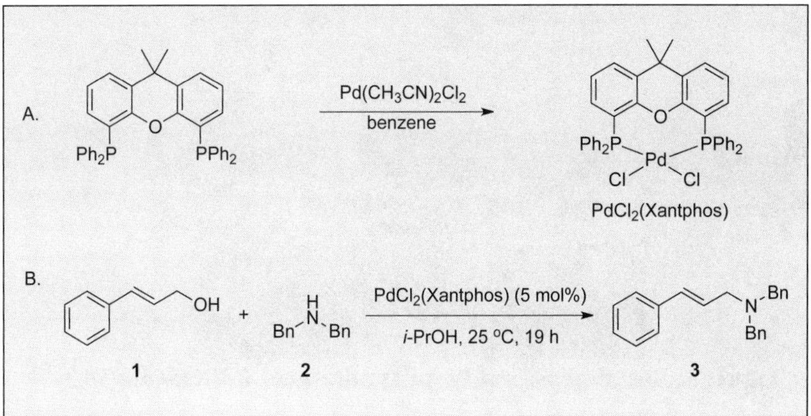

Procedure

A. *Pd(Xantphos)Cl$_2$* An oven-dried, 100-mL Schlenk flask containing a magnetic stir bar (2.5 x 0.8 cm Teflon-coated) (Note 1) is fitted with a reflux conderser and connected to a vacuum line via a one-stopcock adapter in the side arm. The flask is flushed with nitrogen (Note 2) and charged with $Pd(CH_3CN)_2Cl_2$ (1.0 g, 3.86 mmol, 1 equiv) (Note 3), Xantphos (2.46 g, 4.24 mmol, 1.1 equiv) (Note 4) and benzene (80 mL) (Note 5). The reaction mixture is stirred for 48 h at 110 °C (oil bath). After cooling to room temperature, the yellow solid is collected by filtration. The solid is successively washed with benzene (3 x 30 mL) and Et_2O (3 x 30 mL), then

DOI: 10.15227/orgsyn.093.0341

dried under vacuum (1.0 mmHg) for 5 h to give Pd(Xantphos)Cl_2 (2.88 g, 98%) (Note 6) as a yellow solid in 99.2% purity, as determined by quantitative ^{1}H NMR spectroscopy (Note 7).

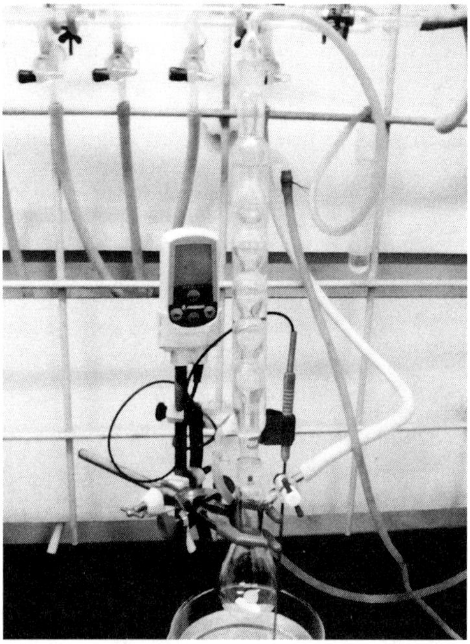

Figure 1. Reaction assembly for synthesis of Pd(Xantphos)Cl_2

B. *(E)-N,N-Dibenzyl-3-phenylprop-2-en-1-amine*. An oven-dried, 100-mL Schlenk flask equipped with a magnetic stir bar (2.5 x 0.8 cm Teflon-coated, ovoid-shaped) is connected to a vacuum line via a one-stopcock adapter in the side arm. The flask is flushed with nitrogen and charged with Pd(Xantphos)Cl_2 (945 mg, 1.25 mmol, 0.05 equiv) and *i*-PrOH (30 mL) (Note 8). The reaction mixture is stirred for one min at room temperature, subsequently (*E*)-3-phenylprop-2-en-1-ol (3.35 g, 25.0 mmol, 1.0 equiv) (Note 9) is added via syringe in one portion (Note 10). Then dibenzylamine (4.8 mL, 25 mmol, 1 equiv) (Note 11) is added via syringe in one portion, and the reaction mixture is stirred for 19 h at room temperature (Notes 12 and 13). The solvent is concentrated in water-aspirator vacuum at (30 mmHg, 40 °C) to obtain the crude product as the viscous brown oil. The resulting residue is purified by column chromatography on silica gel (Note 14) to furnish 7.1 g (91% yield) of ethyl (*E*)-3-(2-acetamido-4-methylphenyl)acrylate as a colorless oil (Notes 15 and 16) with a purity of

98.3%, as determined by quantitative ^{1}H NMR spectroscopy and GC analysis (Notes 17 and 18).

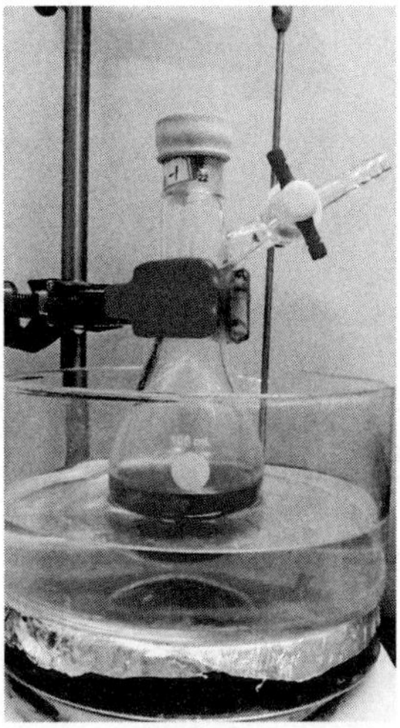

Figure 2. Reaction assembly for synthesis of product 3

Notes

1. All glassware was thoroughly washed and dried in an oven at 100 °C. Teflon-coated magnetic stirring bars were washed with alcohol and dried.
2. This operation is performed by opening the nitrogen inlet from the side arm and flushing the flask for 3 min.
3. $Pd(CH_3CN)_2Cl_2$ was purchased from Sigma-Aldrich, (99% purity, yellow solid) and used as received.
4. Xantphos was purchased from Sigma-Aldrich, (97% purity, white solid) and used as received.
5. Benzene was purchased from Sigma-Aldrich, (99.8% purity, colorless liquid) and used as received.
6. A second reaction on the same scale provided 2.82 g (96%) of Pd(Xantphos)Cl_2.
7. Pd(Xantphos)Cl_2 exhibits the following characteristics: ^{1}H NMR (400 MHz, CD_2Cl_2) δ: 1.87 (s, 6H), 7.06 (td, J = 7.8, 2.5 Hz, 8H), 7.16 – 7.25 (m, 4H), 7.29 – 7.38 (m, 10H), 7.44 (ddd, J = 9.0, 7.7, 1.4 Hz, 2H), 7.73 (dt, J = 7.8, 1.1 Hz, 2H); ^{13}C NMR (101 Hz, CD_2Cl_2) δ: 26.8, 37.4, 37.4, 37.4,

119.9, 119.9, 120.4, 120.4, 125.4, 125.4, 125.5, 125.5, 125.5, 128.4, 128.4, 128.4, 128.5, 128.5, 128.5, 128.7, 129.6, 130.1, 130.4, 130.6, 134.9, 135.0, 136.0, 136.0, 136.0, 154.8, 154.8, 154.9, 154.9; ^{31}P NMR (162 MHz, CD_2Cl_2) δ: 21.8. The purity of product Pd(Xantphos)Cl_2 was determined using ^{1}H QNMR analysis. ^{1}H QNMR was performed using a mixture of Pd(Xantphos)Cl_2 (36.2 mg) and 1,3,5-trimethoxybenzene (15.3 mg) (Alfa Aesar, ≥99% purity, white solid, as an internal standard) in CD_2Cl_2. The purity was calculated according to standard method as 99.7 wt%.

8. *i*-PrOH was purchased from Sigma-Aldrich, (≥99.7% purity, colorless liquid) and used as received.
9. (*E*)-3-Phenylprop-2-en-1-ol (cinnamyl alcohol) was purchased from Sigma-Aldrich, (98% purity, white solid) and used as received.
10. Cinnamyl alcohol was immersed in a water bath at 45 °C for 30 min before use to facilitate its addition via syringe, because of the low melting point of cinnamyl alcohol (30–33 °C). A preheated (45–50 °C) 5-mL glass syringe fitted with a short needle (50 mm) was used in order to avoid solidification of the reagent during the addition. Alternatively, cinnamyl alcohol (3.35 g) could be also dissolved in *i*-PrOH (2 mL) and then added into Schlenk flask as a solution under nitrogen atmosphere.
11. Dibenzylamine was purchased from Sigma-Aldrich, (97% purity, colorless liquid) and used as received.
12. The initially yellow color of the reaction mixture is changed to orange after 10 min stirring, and subsequently becomes orange-yellow after additional 2 h of stirring. The reaction was performed at room temperature in a water bath.
13. The consumption of (*E*)-3-phenylprop-2-en-1-ol is monitored by TLC analysis on silica gel with *n*-hexane:EtOAc (15:1) as eluent. (*E*)-3-Phenylprop-2-en-1-ol (**1**), $R_f = 0.20$; Dibenzylamine (**2**) $R_f = 0.01$; product (**3**) $R_f = 0.85$.
14. The product was purified by flash chromatography on a column (5 x 40 cm) of 100 g of silica gel and eluted with 0.8 L of PE: EtOAc (100:1) followed by 1.0 L of petroleum ether: EtOAc (30:1). The elution was used as received, the boiling point of petroleum ether is 60 °C – 90 °C.
15. The desired product is obtained in fractions 5 through 12, each tube contains 90–100 mL eluent, which are concentrated by rotary evaporation (40 °C, 30 mmHg) and dried under vacuum (1.0 mmHg) for 2 h to give 7.1 g (91%) of **3** as a colorless oil. The product exhibits the following characteristics: ^{1}H NMR (400 MHz, $CDCl_3$) δ: 3.27 (dd, *J* = 6.5, 1.3 Hz, 2H), 3.67 (s, 4H), 6.34 (dt, *J* = 15.9, 6.5 Hz, 1H), 6.58 (dd, *J* = 15.9,

1.5 Hz, 1H), 7.25 – 7.45 (m, 15H); ^{13}C NMR (101 MHz, $CDCl_3$) δ: 55.9, 58.0, 126.3, 126.9, 127.4, 127.8, 128.3, 128.6, 128.9, 132.5, 137.3, 139.7; HRMS (ESI) calcd. for $C_{23}H_{24}N$ [M+H]: 314.1903, found: 314.1911.

16. A second reaction on the same scale provided 7.0 g (90%) of product **3**. When the ratio of **1** and **2** is 1:1.25, **3** was obtained in 83% yield. However, when the load of the catalyst Pd(Xantphos)Cl_2 was reduced to 1.0 mol%, only trace amount of **3** was obtained.
17. The analysis is performed applying gas chromatography (GC). GC conditions: gas chromatography instrument 7890A GC-System from Agilent Technologies equipped with HP-5 column (30 m × 0.32 mm, film 0.25 μm); flow: 1.5 mL/min; injection temperature: 280 °C; temperature profile: initial temperature = 80 °C for 5 min, temperature gradient = 25 °C/min, final temperature = 280 °C for 30 min; Retention time of **3**: 13.028 min.
18. The purity of product **3** was determined using ^{1}H QNMR analysis. ^{1}H QNMR was performed using a mixture of product **3** (31.9 mg) and 1,3,5-trimethoxybenzene (24.0 mg) (Alfa Aesar, ≥99% purity, white solid, as an internal standard) in $CDCl_3$. The purity was calculated according to standard method as 97 wt%.

Working with Hazardous Chemicals

The procedures in *Organic Syntheses* are intended for use only by persons with proper training in experimental organic chemistry. All hazardous materials should be handled using the standard procedures for work with chemicals described in references such as "Prudent Practices in the Laboratory" (The National Academies Press, Washington, D.C., 2011; the full text can be accessed free of charge at http://www.nap.edu/catalog.php?record_id=12654). All chemical waste should be disposed of in accordance with local regulations. For general guidelines for the management of chemical waste, see Chapter 8 of Prudent Practices.

In some articles in *Organic Syntheses*, chemical-specific hazards are highlighted in red "Caution Notes" within a procedure. It is important to recognize that the absence of a caution note does not imply that no significant hazards are associated with the chemicals involved in that procedure. Prior to performing a reaction, a thorough risk assessment should be carried out that includes a review of the potential hazards associated with each chemical and experimental operation on the scale that is planned for the procedure. Guidelines for carrying out a risk assessment and for analyzing the hazards associated with chemicals can be found in Chapter 4 of Prudent Practices.

The procedures described in *Organic Syntheses* are provided as published and are conducted at one's own risk. *Organic Syntheses, Inc.*, its Editors, and its Board of Directors do not warrant or guarantee the safety of individuals using these procedures and hereby disclaim any liability for any injuries or damages claimed to have resulted from or related in any way to the procedures herein.

Discussion

Allylic amines are important compounds that have intrigued chemists due to their unique usage and wide existence in natural products, pharmaceuticals, functional materials, and agrochemicals. In addition, allylamines also serve as attractive precursors in a variety of organic transformations (Figure 3).[2] As such, the development of efficient and sustainable methods for the production of allylic amines is important to the chemical industry and medicinal chemistry, and has attracted a great deal of attention over the past decades.[3]

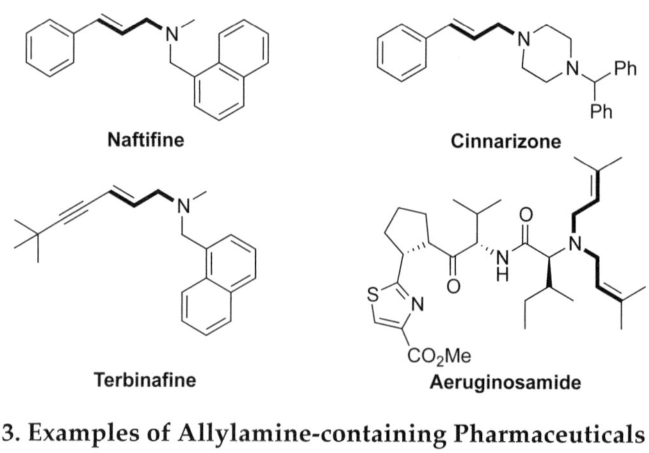

Figure 3. Examples of Allylamine-containing Pharmaceuticals

Among many types of synthetic methods available for constructing the allylamine scaffold,[3] the transition-metal-catalyzed amination of allylic alcohols has proven to be one of the most efficient approaches for the synthesis of allylic amines[4] due to the high atom economy (water is produced as the only by-product) and step economy. In 1999, Yang and Moritani reported the direct coupling of allylic alcohols with amines through Pd-catalyzed amination, wherein the Lewis acid ($Ti(O\textit{i}Pr)_4$) as promoter was added to enhance the leaving ability of the hydroxy group in the presence of molecular sieves at the high temperature (Scheme 1a).[5]

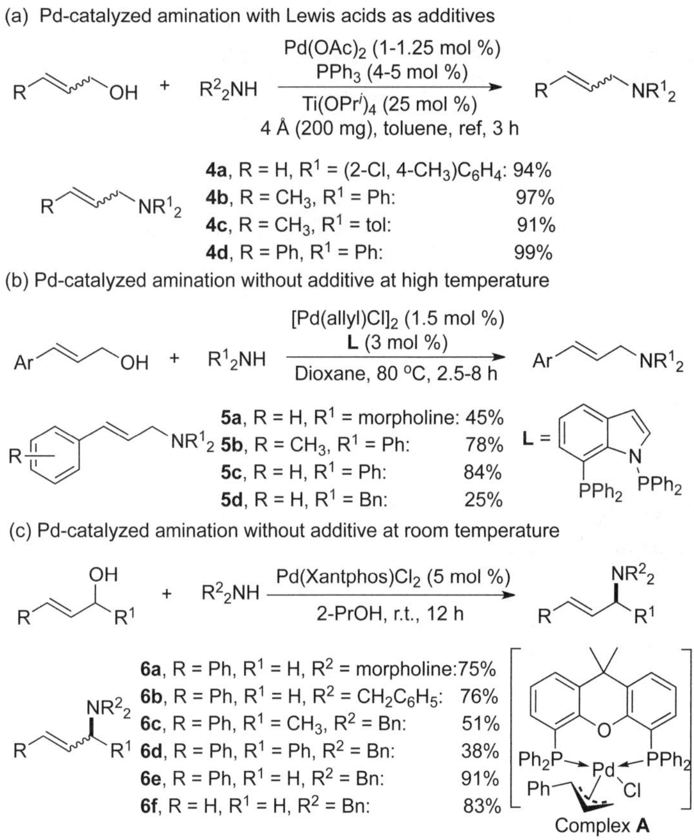

Scheme 1. Palladium(II)-catalyzed Amination of Allyl Alcohols

Notable progress was made by using $[Pd(allyl)Cl]_2$ complexes with 1,7-bis(diphenylphosphino)-1H-indole. The reactions proceeded smoothly in

1,4-dioxane at 80 °C without any additives; however, use of this bisphosphine ligand, which is not easily prepared, is still quite rare (Scheme 1b).[6] Our research group found that the use of the less expensive and readily available Pd(Xantphos)Cl_2 catalyst enabled the direct amination of allylic alcohols with amines at room temperature in the absence of additives (Scheme 1c).[7] This method is compatible with a variety of functional groups and can be used to prepare a wide range of linear allylic amines in good to excellent yields with high stereoselectivity. Moreover, this method was utilized to synthesize the antihistamine pharmaceutical cinnarizine.[8]

In summary, a simple palladium(II) complex has been identified to be an efficient catalyst for the direct amination of allylic alcohols with amines via C-O bond cleavage. This simple reaction can be performed at room temperature and can be used for synthesis of a broad range of linear allylamines, which are important for natural product synthesis and drug discovery. Considering the practical importance of this atom- and step-economical amination reaction, significant further applications are expected.

References

1. State Key Laboratory for Oxo Synthesis and Selective Oxidation, Lanzhou Institute of Chemical Physics, Chinese Academy of Sciences, Lanzhou, 730000, China. E-mail: hmhuang@licp.cas.cn. This work was supported by the National Natural Science Foundation of China (21222203 and 21172226).
2. (a) Towse, G. *J. Laryngol. Otol.* **1980**, *94*, 1009; (b) Shupak, A.; Doweck, I.; Gordon, C. R.; Spitzer, O. *Clin Pharmacol Ther.* **1994**, *55*, 670; (c) Johannsen, M.; Jørgensen, K. A. *Chem. Rev.* **1998**, *98,* 1689; (d) Chen, Z.; Ye, T. *New J. Chem.* **2006**, *30*, 518.
3. (a) Cheik, R. B.; Chaabouni, R.; Laurent, A.; Mison, P.; Nafti, A. *Synthesis* **1983**, 685; (b) Overman, L. E.; Carpenter, N. E. *Org. React.* **2005**, *66*, 1; (c) Miyabe, H.; Takemoto, Y. *Synlett.* **2005**, 1641; (d) Takeuchi, R.; Kezuka, S. *Synthesis* **2006**, 3349; (e) Hartwig, H. F.; Stanley, L. M. *Acc. Chem. Res.* **2010**, *43*, 1461; (f) Collet, F.; Lescot, C.; Dauban, P. *Chem. Soc. Rev.* **2011**, *40*, 1926; (g) Ramirez, T. A.; Zhao, B.; Shi, Y. *Chem. Soc. Rev.* **2012**, *41*, 931.
4. (a) Kimura, M.; Futamata, M.; Shibata, K.; Tamaru, Y. *Chem. Commun.* **2003**, 234; (b) Qin, H.; Yamagiwa, N.; Matsunaga, S.; Shibasaki, M.

Angew. Chem. Int . Ed. **2007**, *46*, 409; (c) Defieber, C.; Ariger, M. A.; Moriel, P.; Carreira, E. M. *Angew. Chem. Int. Ed.* **2007**, *46*, 3139; (d) Mora, G.; Deschamps, B.; van Zutphen, S.; Goff, X. F. L.; Ricard, L.; Floch, P. L. *Organometallics* **2007**, *26*, 1846; (e) Utsunomiya, M.; Miyamoto, Y.; Ipposhi, J.; Ohshima, T.; Mashima, K. *Org. Lett.* **2007**, *9*, 3371; (f) Ohshima, T.; Miyamoto, Y.; Ipposhi, J.; Nakahara, Y.; Utsunomiya, M.; Mashima, K. *J. Am. Chem. Soc.* **2009**, *131*, 14317; (g) Nishikata, T.; Lipshutz, B. H. *Org. Lett.* **2009**, *11*, 2377; (h) Tao, Y.; Wang, B.; Wang, B.; Qu, L.; Qu, J. *Org. Lett.* **2010**, *12*, 2726; (i) Roggen, M.; Carreira, E. M. *J. Am. Chem. Soc.* **2011**, *132*, 11917; (j) Hikawa, H.; Yokoyama, Y. *J. Org. Chem.* **2011**, *76*, 8433; (k) Das, K.; Shibuya, R.; Nakahara, Y.; Germain, N.; Ohshima, T.; Mashima, K. *Angew. Chem. Int. Ed.* **2012**, *51*, 150; (l) Wagh, Y. S.; Sawant, D. N.; Dhake, K. P.; Bhanage, B. M. *Catal. Sci. Technol.*, **2012**, *2*, 835.

5. Yang, S. -C.; Hung, C.-W. *J. Org. Chem.* **1999**, *64*, 5000.
6. Ghosh, R.; Sarkar, A. *J. Org. Chem.* **2011**, *76*, 8508.
7. Wang, M.; Xie, Y.; Li, J.; Huang, H. *Synlett* **2014**, *25*, 2781.
8. (a) Towse, G. *J. Laryngol. Otol.* 1**980**, *94*, 1009; (b) Singh, B. N. *Br. J. Clin. Pharmacol.* **1986**, *21*, 109S; (c) Shupak, A.; Doweck, I.; Gordon, C. R.; Spitzer, O. *Clin. Pharmaclo. Ther.* **1994**, *55*, 670.

Appendix
Chemical Abstracts Nomenclature (Registry Number)

Dibenzylamine; (103-49-1)
1,3,5-Trimethoxybenzene; (621-23-8)
i-PrOH: Isopropyl alcohol; (67-63-0)
(*E*)-3-Phenylprop-2-en-1-ol; (104-54-1)
$PdCl_2(CH_3CN)_2$: Bis(acetonitrile)palladium(II) chloride; (14592-56-4)
Xantphos: 9,9-Dimethyl-4,5-bis(diphenylphosphino)xanthene; (161265-03-8)

Hanmin Huang was born in Hubei, China, and completed his M.S. degree at the Huazhong University of Science & Technology. He obtained his Ph.D. degree in 2003 at the Dalian Institute of Chemical Physics, Chinese Academy of Sciences (CAS), under the supervision of Professor Huilin Chen and Professor Zhuo Zheng. He then moved to Nagoya University and worked as a JSPS postdoctoral research fellow with Professor Masato Kitamura. In 2008, he initiated his independent research in the Lanzhou Institute of Chemical Physics, CAS. In March 2016, he moved to the University of Science and Technology of China as a full professor. His current research interests are focused on organometallic chemistry and the development of new and efficient synthetic methodologies for green organic synthesis.

Bao Gao was born in 1987 in Shanxi, China. He received his B.S. in chemistry at Yanbian University in 2010 and completed his M.S in chemistry on organic functional materials at Lanzhou University in 2013. He is now pursuing his Ph.D. in the group of Prof. Hanmin Huang at University of Chinese Academy of Sciences (2013-present). His current research interests include organometallic chemistry and carbonylative reaction.

Lixin Li was born in 1987 in Henan, China. He received his B.S. degree and M.S. degree from Henan University in 2010 and 2013 respectively, Now he is pursuing his Ph.D. at the University of Chinese Academy of Sciences (2014-present) under the supervision of Professor Hanmin Huang. His current research interests are focused on palladium-catalyzed the cleavage of C-N bonds.

Guoying Zhang was born in 1986 in Shandong, China. He received his B.S. from University of Jinan in 2009 and received M.S. from Wenzhou University in 2012.He is now pursuing his Ph.D. in thc group of Prof. Hanmin Huang at University of Chinese Academy of Sciences (2012-present). His current research interests include organometallic chemistry and catalysis.

Steven A. Loskot was born in San Jose, CA in 1991. He received hi B.S. degree in Biochemistry from Seattle University in 2014 where he conducted research for Professor Joseph M. Langenhan. He is now pursuing his graduate studies at the California Institute of Technology under the guidance of Professor Brian M. Stoltz. His graduate research focuses on the total synthesis of novel natural products.

Organic Syntheses

Preparation of 1,1-Difluoroallenes by Difluorovinylidenation of Carbonyl Compounds

Kohei Fuchibe, Masashi Abe, Ken Oh, and Junji Ichikawa*[1]

Division of Chemistry, Faculty of Pure and Applied Sciences, University of Tsukuba, Tsukuba, Ibaraki 305-8571, Japan

Checked by Shogo Sato, Keiichiro Sakata, and Keisuke Suzuki

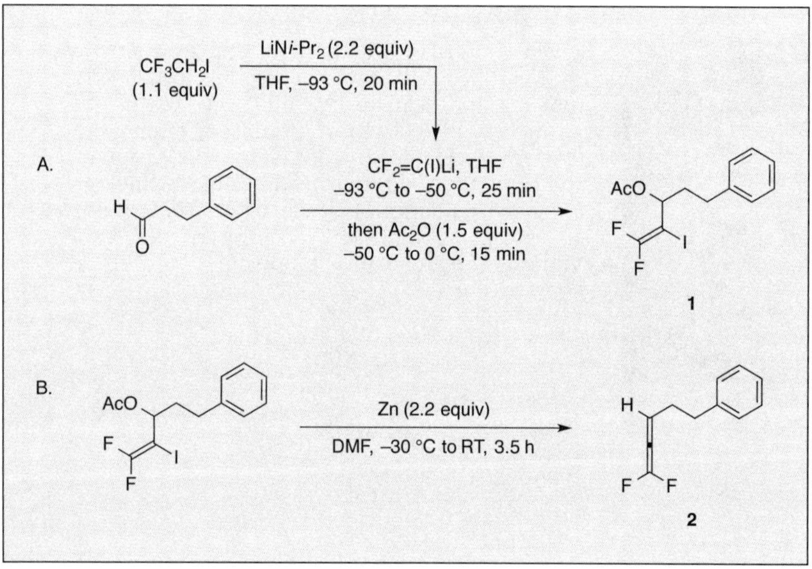

Procedure

A. *1,1-Difluoro-2-iodo-5-phenylpent-1-en-3-yl acetate* (**1**). A flame-dried, 500-mL, three-necked, round-bottomed flask (Figure 1, a) is equipped with a Teflon-coated magnetic stirring bar (b), an internal thermometer (c), a dropping funnel (100 mL, d), and a three-way stopcock (e) fitted with an argon inlet. The probe of the thermometer is inserted to the round-bottomed flask through a septum (f1). A septum (f2) is placed on the dropping funnel.

The flask is charged with diisopropylamine (16.4 mL, 11.8 g, 117 mmol, 2.24 equiv) (Note 1) and THF (80 mL) (Note 2) via the dropping funnel, and the reaction mixture is chilled to 0 °C (bath temperature). Butyllithium (71.3 mL, 1.64 M in hexane, 117 mmol, 2.24 equiv) (Note 3) is added to the THF solution of diisopropylamine over 40 min via the dropping funnel at 0 °C. The inside of the dropping funnel is rinsed with THF (5 mL). The reaction mixture is stirred at 0 °C for an additional 10 min, and then chilled (internal temperature: –93 °C; bath temperature: –95 °C) using an ethanol/liq. N_2 bath (Note 4). 1,1,1-Trifluoro-2-iodoethane (5.66 mL, 12.1 g, 57.4 mmol, 1.1 equiv) (Note 5) is added to the THF solution of LDA over 15 min by a syringe through the septum (f1), during which the internal temperature is kept below –90 °C. The reaction mixture is stirred for 30 min. A THF (20 mL) solution of 3-phenylpropanal (6.87 mL, 7.00 g, 52.2 mmol) (Note 6) is added over 15 min via the dropping funnel. The inside of the dropping funnel is rinsed with THF (5 mL). After the addition of the aldehyde, the reaction temperature is raised (internal temperature: –60 to –50 °C; bath temperature: –50 °C) over 20 min, followed by an additional 15 min stirring. When complete consumption of the aldehyde is confirmed by TLC analysis (Note 7), acetic anhydride (7.54 mL, 8.14 g, 79.8 mmol, 1.53 equiv) (Note 8) is added by a syringe through the septum (f1) over 20 min, keeping the internal temperature below –50 °C. After the addition is finished, the ethanol/liq. N_2 bath is changed to an ice bath and the reaction mixture is stirred for an additional 15 min. A spot-to-spot formation of acetate (**1**) is confirmed by TLC analysis (Note 9). A saturated aq. NH_4Cl (350 mL) is added over 1 min at 0 °C (bath temperature). The products are extracted with ethyl acetate (3 × 20 mL) using a 500-mL separatory funnel. The combined organic layers are washed with brine (50 mL) and dried over anhydrous Na_2SO_4 (ca. 90 g). After removal of Na_2SO_4 by paper filtration, the solvent is removed by rotary evaporation (30 °C, 110 mmHg). The residue is kept under reduced pressure (room temp, 3 mmHg) to remove the solvent completely. A crude acetate (**1**) is obtained as a dark brown oil (approx. 20.9 g) in 91% purity (Note 10) and used for the next step without purification. Column chromatography (Note 11) is used when material of higher purity is required (Note 12).

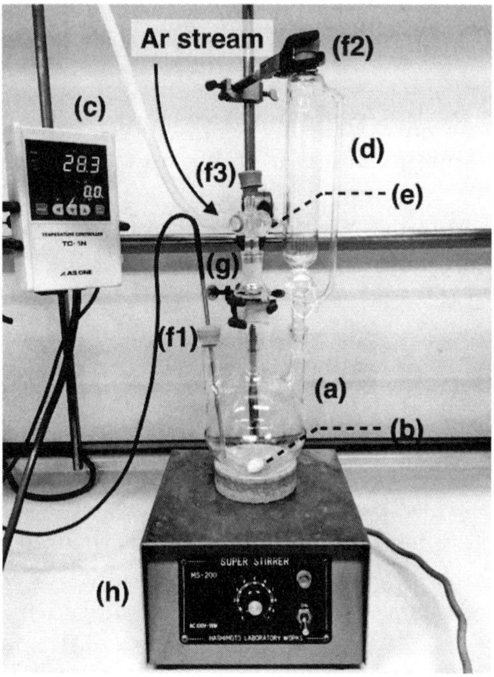

Figure 1. Set-up for Steps A and B (a: round-bottomed flask, 500 mL; b: magnetic stirring bar; c: thermometer; d: dropping funnel; e: three-way stopcock; f1–3: rubber septum; g: adaptor; h: magnetic stirrer)

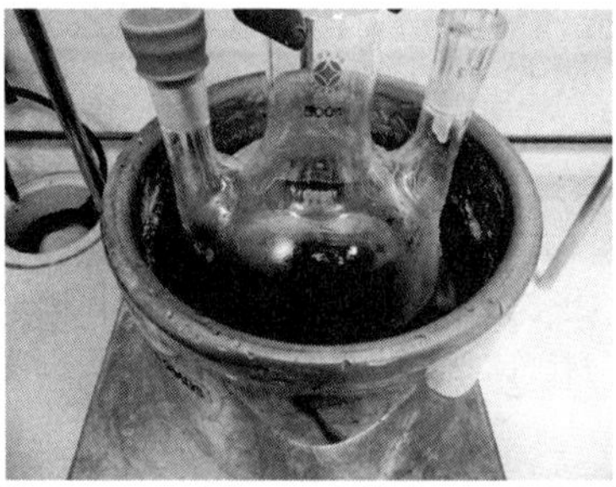

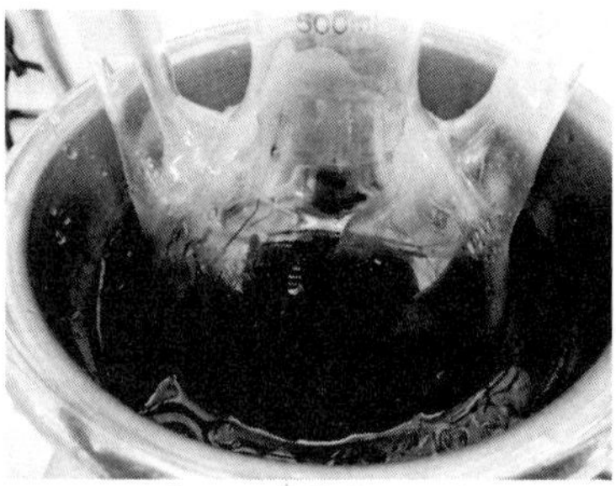

Figure 2. After addition of CF_3CH_2I **Figure 3.** After addition of Ac_2O

Figure 4. Crude acetate (1)

B. *1,1-Difluoro-5-phenylpenta-1,2-diene* (**2**). Zinc dust is activated by the following procedure. A 200-mL Erlenmeyer flask is charged with zinc dust (50 g) under air (Note 13). The zinc dust is washed with 2 M aq. HCl at room temperature (3 × 80 mL) (Note 14). The zinc dust is then washed with deionized water (8 × 60 mL) (Note 15) until the pH value of the water becomes 7 (Note 16). The zinc dust is further washed with methanol (8 × 60 mL) (Note 17) and ether (8 × 60 mL) (Note 18). The obtained active zinc dust is dried under reduced pressure (room temp, 6 mmHg) and stored under argon.

A flame-dried, 500-mL, three-necked, round-bottomed flask (Figure 1, a) is equipped with a Teflon-coated magnetic stirring bar (b), an internal thermometer (c), a dropping funnel (100 mL, d), and a three-way stopcock (e) fitted with an argon inlet. The probe of the thermometer is inserted to the round-bottomed flask through a septum (f1). A septum (f2) is placed on the dropping funnel.

The flask is charged with zinc dust (6.84 g, 105 mmol, 2.2 equiv) (Note 19). Dimethylformamide (DMF) (153 mL) (Note 20) is added via the dropping funnel and the suspension is chilled to –20 °C (bath temperature). A DMF (26 mL) solution of the crude acetate (**1**, 20.9 g) is added over 10 min via the dropping funnel. The inside of the dropping funnel is rinsed with DMF (5 mL). The reaction mixture is warmed to room temperature and stirred for 70 min (Note 21). A spot-to-spot formation of 1,1-difluoroallene (**2**) (Note 22) is confirmed by TLC analysis. Unreacted zinc dust is removed by passing through a small pad of celite (7 cm × 1 cm) (Note 23) and the celite is washed with ether (50 mL). Saturated aq. NH_4Cl (50 mL) and water (20 mL) are added and the products are extracted with ether (3 × 20 mL) using a 500-mL separatory funnel. The combined organic layers are washed with brine (50 mL) and dried over anhydrous Na_2SO_4 (ca. 30 g). After removal of Na_2SO_4 by paper filtration, the solvent is removed by rotary evaporation (30 °C, 260 mmHg).

The residue (14.0 g) is charged on a column (5 cm × 30 cm) of silica gel (100 g), and eluted with pentane (60 mL fractions collected in test tubes) (Note 24). Product (**2**) is obtained in fractions 4–9. These fractions are combined and the solvent is removed by rotary evaporation (30 °C, 290 mmHg) (Note 25). The obtained oil is kept under reduced pressure for a short period of time (room temp, 3 mmHg, 3 min) to remove the solvent completely. 1,1-Difluoroallene (**2**) is obtained as a colorless oil (6.18 g, 72% yield based on crude acetate (**1**) and 66% yield based on 3-phenypropanal) (Notes 26, 27 and 28).

Figure 5. After addition of crude acetate

Figure 6. Product (2)

Notes

1. Diisopropylamine (99.0+%) was purchased from Tokyo Chemical Industry Co., Ltd. and was used after distillation from NaOH.
2. THF (anhydrous; Kanto Chemical Co., Inc.) was purified under argon using a solvent purification unit (Wako Pure Chemical Co., Inc.)
3. Butyllithium (hexane solution, 1.6 M) was purchased from Nakalai Tesque, Inc. and titrated with diphenylacetic acid in THF at room temperature.
4. During the preparation of CF_2=C(I)Li and the addition of aldehyde, the internal temperature must be kept below –90 °C to prevent the decomposition of the vinyllithium through elimination of LiF.
5. 1,1,1-Trifluoro-2-iodoethane (95+%) was purchased from Kanto Chemical Co., Inc. and used after distillation from CaH_2.

6. 3-Phenylpropanal (90+%) was purchased from Kanto Chemical Co., Inc. and used after distillation from $CaSO_4$.
7. TLC analysis is performed on silica gel with 20:1 hexane-AcOEt eluent (1 cm × 5 cm, glass-backed, Merck KGaA, TLC silica gel 60 F_{254}, 1.05715.0001). 3-Phenylpropanal has Rf = 0.31 and 1,1-difluoro-2-iodo-5-phenylpent-1-en-3-ol has Rf = 0.19.
8. Acetic anhydride (97+%) was purchased from Kanto Chemical Co., Inc. and used after distillation from *N,N*-diethylaniline.
9. TLC analysis is performed on silica gel with 20:1 hexane-AcOEt eluent. 1,1-Difluoro-2-iodo-5-phenylpent-1-en-3-yl acetate (**1**) has $R_f = 0.41$.
10. ^{19}F NMR analysis of an aliquot (55.0 mg) of the crude acetate (**1**, 20.9 g) using α,α,α-trifluorotoluene (19.6 mg) as internal reference indicates that acetate (**1**) is obtained in 91% yield. α,α,α-Trifluorotoluene (98+%) was purchased from Tokyo Chemical Industry Co., Ltd.
11. If necessary, acetate (**1**) can be purified by column chromatography. For example, the crude acetate (**1**, 21.0 g) obtained in another run from 3-phenylpropanal (6.87 mL, 7.00 g, 52.2 mmol) is dissolved in 20:1 hexane–AcOEt (ca. 30 mL) and charged on a column (5 cm × 45 cm) of silica gel (120 g) and eluted with 20:1 hexane-AcOEt (60 mL fractions collected in test tubes). Acetate (**1**) is obtained in fractions 3–15. These fractions are combined and the solvent is removed by evaporation (40 °C, 40 mmHg; RT, 4 mmHg), which affords pure acetate (**1**) as a colorless oil (15.74 g, 82% yield) in 97.1% purity (Note 10).
12. 1,1-Difluoro-2-iodo-5-phenylpent-1-en-3-yl acetate (**1**) is bench-stable but should be stocked in a refrigerator (0 °C). The purified compound has the following spectroscopic properties: ^1H NMR (600 MHz, $CDCl_3$, $SiMe_4$) δ: 1.86–1.92 (m, 1H), 2.05–2.11 (m, 1H), 2.07 (s, 3H), 2.56–2.61 (m, 2H), 4.96–4.99 (m, 1H), 7.17–7.22 (m, 3H), 7.29 (dd, *J* = 7.4, 7.6 Hz, 2H); ^{13}C NMR (151 MHz, $CDCl_3$, $SiMe_4$) δ: 20.9, 31.0, 36.1, 53.9 (dd, *J*= 25, 26 Hz), 68.9 (d, *J* = 3 Hz), 126.3, 128.3, 128.6, 140.3, 154.1 (dd, *J* = 286, 286 Hz), 169.6; ^{19}F NMR (565 MHz, $CDCl_3$, C_6F_6) δ: 89.2 (d, *J* = 22 Hz, 1F), 90.2 (d, *J* = 22 Hz, 1F); IR (ATR): 3028, 2934, 1747, 1717, 1269, 1225, 1027, 699 cm^{-1}; HRMS (ESI-TOF): $[M+Na]^+$ calcd. for $C_{13}H_{13}F_2IO_2Na$: 388.9826, found: 388.9821.
13. Zinc dust (96+%) was purchased from Tokyo Chemical Industry Co., Ltd.
14. In the first wash, zinc dust is stirred for 15 min in 2 M aq. HCl and the aq. HCl is then removed by decantation. In the second and third

washes, zinc dust is stirred for 30 sec. Conc. HCl was purchased from Kishida Chemical Co., Ltd.

15. Zinc dust is stirred for 30 sec in deionized water and the water is then removed by decantation.
16. The test paper was purchased from Advantec Toyo Kaisha, Ltd. (UNIV, pH 1–11).
17. Zinc dust is stirred for 30 sec in methanol. Methanol (99%) was purchased from Nakalai Tesque, Inc. and used as received.
18. Zinc dust is stirred for 30 sec in diethyl ether. Diethyl ether (anhydrous; Kanto Chemical Co., Inc.) was purified under argon using a solvent purification unit (Wako Pure Chemical Co., Inc.)
19. Active zinc dust is weighed under air and quickly put into the reaction vessel.
20. DMF (99%) was purchased from Nakalai Tesque, Inc. and used after distillation from CaH_2.
21. After removing the cold bath, the internal temperature slowly rises to about 26 °C and then declines to room temperature.
22. TLC analysis is performed on silica gel with 20:1 hexane–AcOEt eluent. 1,1-Difluoro-5-phenylpenta-1,2-diene (**2**) has $R_f = 0.97$.
23. Celite was purchased from Wako Pure Chemical Industries Ltd. (No. 500).
24. Pentane was purchased from Nacalai Tesque, Inc. and used as received.
25. Removal of the solvent by rotary evaporation should be stopped when condensation of the solvent is not observed. Otherwise, yield of the product (**2**) decreases by evaporation.
26. The reaction was checked two additional times on identical scale and provided 5.95 g (63%) and 6.20 g (66%) of the product with the yield based on 3-phenypropanal.
27. Isolated difluoroallene (**2**) was analyzed by ^{1}H NMR spectroscopy as 98.9 wt% using ethylene carbonate as an internal standard. The submitters report that difluoroallene was observed as a single peak by GC analysis (GC-14B instrument, Shimadzu Corporation), but this analysis was not attempted by the Checkers. The submitter's GC instrument was equipped with Rtx-5 (Restek Corporation) column (30 m × 0.25 mm, film thickness 0.25 μm). Retention time (**2**): t = 7.40 min with N_2 flow rate of 1.9 mL/min. Temperature profile: T = 60 °C/2 min, 60–280 °C/11 min, then 280 °C/30 min.
28. 1,1-Difluoro-5-phenylpenta-1,2-diene (**2**) undergoes slow decomposition at room temperature, which therefore should be stored

in a refrigerator (–19 °C). This compound has the following spectroscopic properties: ^{1}H NMR (600 MHz, $CDCl_3$, $SiMe_4$) δ: 2.54–2.60 (m, 2H), 2.81 (t, J = 7.6 Hz, 2H), 6.47 (tt, J = 6.1, 2.4 Hz, 1H), 7.18–7.22 (m, 3H), 7.30 (dd, J = 7.3, 7.3 Hz, 2H); ^{13}C NMR (151 MHz, $CDCl_3$, $SiMe_4$) δ: 33.85, 33.86, 121.5 (t, J = 5.5 Hz), 126.3, 128.4, 128.5, 140.7, 152.9 (t, J = 259.2 Hz), 170.2 (t, J = 36.0 Hz); ^{19}F NMR (565 MHz, $CDCl_3$, C_6F_6) δ: 60.027 (s), 60.035 (s); IR (ATR): 3029, 2930, 2861, 2014, 1463, 1197, 747, 699 cm^{-1}.

Working with Hazardous Chemicals

The procedures in *Organic Syntheses* are intended for use only by persons with proper training in experimental organic chemistry. All hazardous materials should be handled using the standard procedures for work with chemicals described in references such as "Prudent Practices in the Laboratory" (The National Academies Press, Washington, D.C., 2011; the full text can be accessed free of charge at http://www.nap.edu/catalog.php?record_id=12654). All chemical waste should be disposed of in accordance with local regulations. For general guidelines for the management of chemical waste, see Chapter 8 of Prudent Practices.

In some articles in *Organic Syntheses*, chemical-specific hazards are highlighted in red "Caution Notes" within a procedure. It is important to recognize that the absence of a caution note does not imply that no significant hazards are associated with the chemicals involved in that procedure. Prior to performing a reaction, a thorough risk assessment should be carried out that includes a review of the potential hazards associated with each chemical and experimental operation on the scale that is planned for the procedure. Guidelines for carrying out a risk assessment and for analyzing the hazards associated with chemicals can be found in Chapter 4 of Prudent Practices.

The procedures described in *Organic Syntheses* are provided as published and are conducted at one's own risk. *Organic Syntheses, Inc.*, its Editors, and its Board of Directors do not warrant or guarantee the safety of individuals using these procedures and hereby disclaim any liability for any injuries or damages claimed to have resulted from or related in any way to the procedures herein.

Discussion

1,1-Difluoroallenes have been synthesized via either the formation of the second C–C double bond in fluorinated alkenes or the rearrangement of the C–C triple bonds in fluorinated alkynes.[2] The former involves elimination reactions of (i) trifluoromethylalkenes and (ii) difluoroalkenes, while the latter involves substitution reactions of difluoropropargyl bromides. The substitution reactions are further classified into two types, wherein the propargyl moieties serve as (iii) nucleophiles or (iv) electrophiles.

Table 1. Synthesis of 1,1-Difluoroallenes

Entry	Carbonyl compound	1,1-Difluoroallene 2	1, 2 / %
1	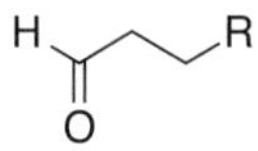	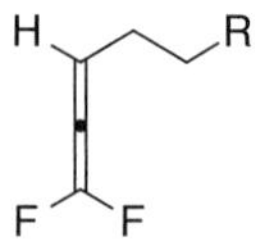	82, 86 (R = Ph)
2			83, 82 (R = 1-Naph)
3			84, 87 (R = n-C_7H_{15})
4	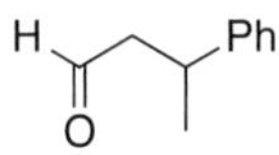	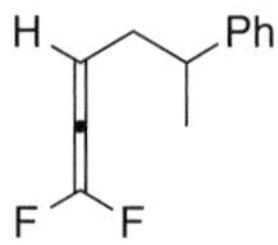	81, 95
5	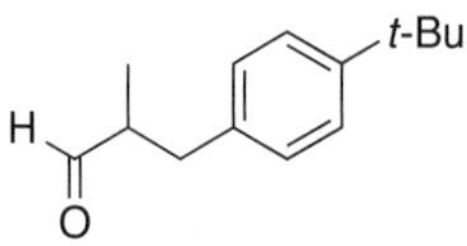	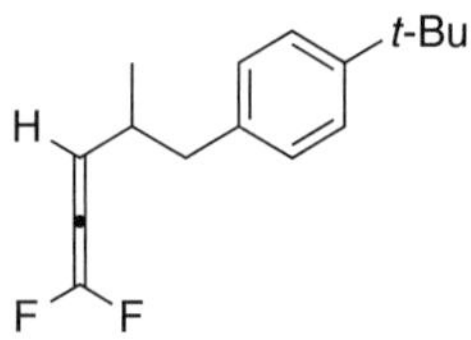	87, 92
6	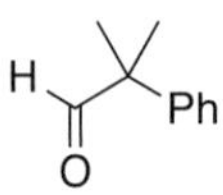	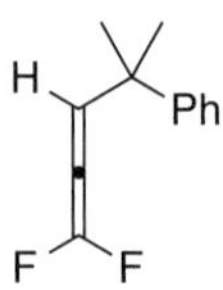	83, 93
7	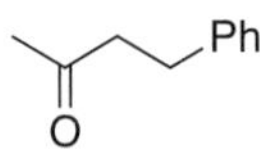	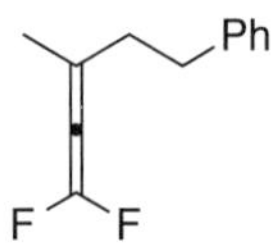	80,[a] 86

[a] Acetylation was performed with isopropenyl acetate (14 equiv)/2 mol%

The present 1,1-difluoroallene synthesis is classified as the type (ii) synthesis and provides a highly general route to the variously substituted 1,1-difluoroallenes because of the wide availability of starting aldehydes and ketones (Table 1).[3] Not only linear aldehydes (Entries 1–3) but also branched aldehydes (Entries 4 and 5), including tertiary alkanals (Entry 6), afford the corresponding 1,1-difluoroallenes in high yields. Ketones are also efficiently converted to the corresponding 1,1-difluoroallenes (Entry 7). It is noteworthy that both ester and pyridine moieties are tolerant to the reaction conditions and the corresponding difluoroallenes are obtained in high yields (Scheme 1).

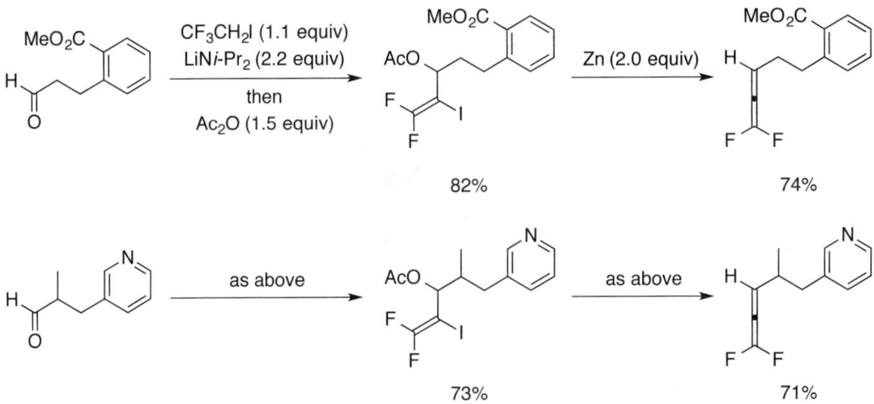

Scheme 1. Synthesis of 1,1-Difluoroallenes Bearing Ester and Pyridine Moieties

This difluoroallene synthesis is a modification of our former 1,1-dibromo-2,2-difluoroethene-based protocol (Scheme 2).[4] Since there is a short supply of dibromodifluoroethene because of its ozone depletion property, trifluoroiodoethane is adopted instead as a starting material.

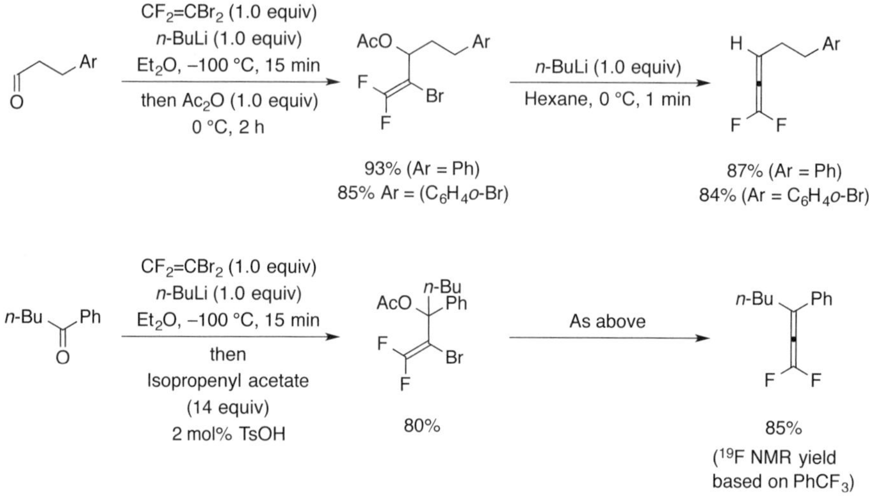

Scheme 2. Dibromodifluoroethene-Based Synthesis of 1,1-Difluoroallenes

1,1-Difluoroallenes are versatile intermediates that undergo nucleophilic additions, substitutions, and cycloadditions.[1] In addition, difluoroallenes serve as an appropriate platform for the synthesis of polycyclic aromatic hydrocarbons (PAHs).[5] 1,1-Difluoroallenes, bearing an aryl group and an alkyl substituent adjacent to the difluoroallene moiety, are readily prepared from arylacetonitriles by half-reduction followed by difluorovinylidenation (Scheme 3). Treatment of the 1,1-difluoroallenes with 2 mol% of $InBr_3$ facilitates the domino Friedel–Crafts-type cyclization/1,2-alkyl migration (or deprotonation) to afford regioselectively fluorinated naphthalenes in high yields. Difluoroallenes bearing an aryl group and a cyclopentene moiety similarly undergo the domino Friedel–Crafts-type cyclization/alkyl migration sequence, resulting in ring formation and ring expansion. Subsequent one-pot dehydrogenation with 2,3-dichloro-5,6-dicyano-*p*-benzoquinone (DDQ) efficiently provides pinpoint-fluorinated higher-order PAHs (Scheme 4), which are promising organic materials for printable electronic devices.[6]

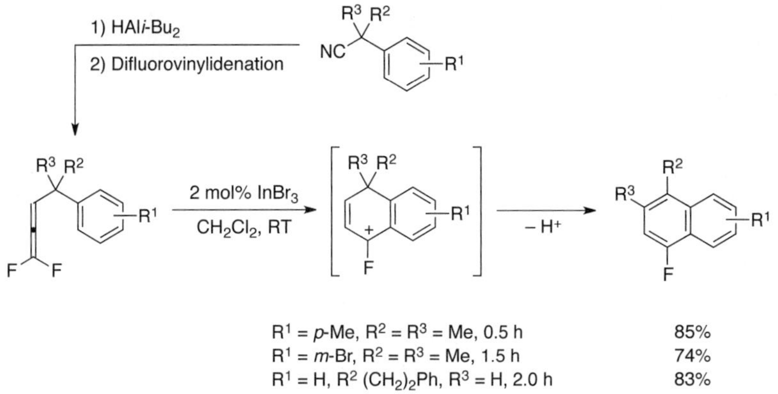

Scheme 3. Domino Synthesis of Pinpoint-Fluorinated Naphthalenes

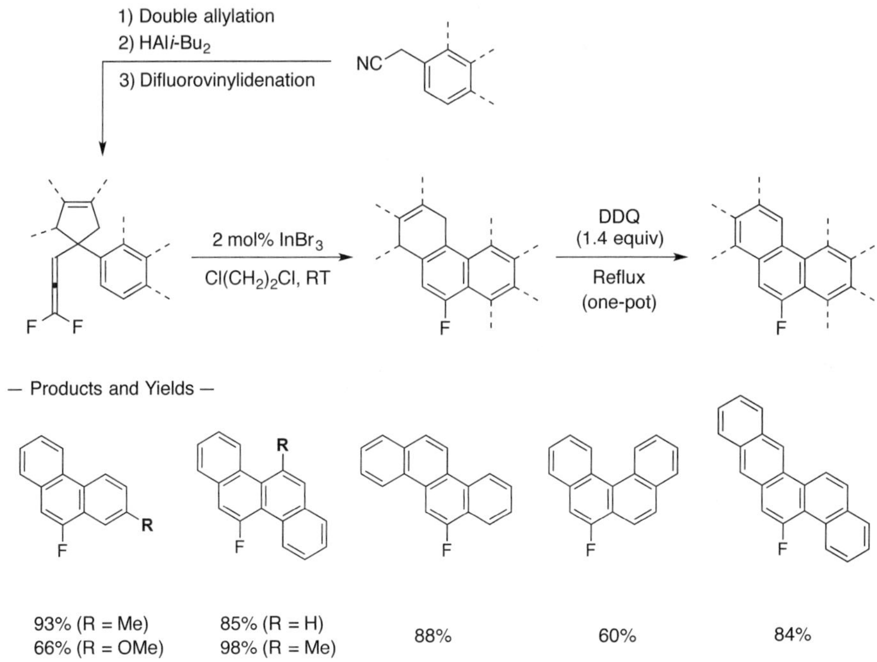

Scheme 4. Domino Synthesis of Pinpoint-Fluorinated PAHs

References

1. Division of Chemistry, Faculty of Pure and Applied Sciences, University of Tsukuba, Tsukuba, Ibaraki 305–8571, Japan. E-mail: junji@chem.tsukuba.ac.jp. This work was partially supported by Tosoh F-Tech, Inc.
2. Fuchibe, K.; Ichikawa, J. *1,1-Dihaloallenes* in Web of Science Knowledge Updates 2014/2, Thieme, Stuttgart, **2014**, pp. 217–231.
3. (a) Oh, K.; Fuchibe, K.; Ichikawa, J. *Synthesis* **2011**, 881–886. (b) Oh, K.; Fuchibe, K.; Yokota, M.; Ichikawa, J. *Synthesis* **2012**, *44*, 857–861.
4. Yokota, M.; Fuchibe, K.; Ueda, M.; Mayumi, Y.; Ichikawa, J. *Org. Lett.* **2009**, *11*, 3994–3997.
5. (a) Fuchibe, K.; Mayumi, Y.; Zhao, N.; Watanabe, S.; Yokota, M.; Ichikawa, J. *Angew. Chem. Int. Ed.* **2013**, *52*, 7825–7828. (b) Fuchibe, K.; Mayumi, Y.; Yokota, M.; Aihara, H.; Ichikawa, J. *Bull. Chem. Soc. Jpn.* **2014**, *87*, 942–949.
6. Fuchibe, K.; Morikawa, T.; Shigeno, K.; Fujita, T.; Ichikawa, J. *Org. Lett.* **2015**, *17*, 1126–1129.

Appendix
Chemical Abstracts Nomenclature (Registry Number)

1,1-Difluoro-2-iodo-5-phenylpent-1-en-3-yl acetate: Benzenepropanol, α-(2,2-difluoro-1-iodoethenyl)-, 1-acetate; (**1**) (1309570-95-3)
Diisopropylamine: 2-Propanamine, *N*-(1-methylethyl)-; (108-18-9)
Butyllithium: Lithium, butyl-; (109-72-8)
1,1,1-Trifluoro-2-iodoethane: Ethane, 1,1,1-trifluoro-2-iodo-; (353-83-3)
3-Phenylpropanal: Benzenepropanal; (104-53-0)
Acetic anhydride: Acetic acid, 1,1'-anhydride; (108-24-7)
1,1-Difluoro-5-phenylpenta-1,2-diene: Benzene, (5,5-difluoro-3,4-pentadien-1-yl)-; (**2**) (1186222-29-6)
Zinc; (7440-66-6)

Kohei Fuchibe was born in Fukui, Japan in 1974. He received his B.Sc. in 1999 and Ph.D. in 2002 from the University of Tokyo (Prof. K. Narasaka). He joined Gakushuin University as a Research Associate in 2002 and was shifted to an Assistant Professor in 2007. He moved to University of Tsukuba as a Lecturer in 2007 and was promoted to an Associate Professor in 2011. His research interests involve organic synthetic reactions catalyzed by transition metals or organic small molecules.

Masashi Abe was born in the state of New Hampshire, USA in 1991. He received his B.Sc. in 2014 from University of Tsukuba and is pursuing a master's degree at the same university (Prof. J. Ichikawa). He is majoring in synthetic organic chemistry and now working on the development of synthetic methods for polycyclic aromatic hydrocarbons.

Ken Oh (Jian Wang) was born in Shandong, China in 1970, graduated from Wuhan University in 1991, moved to Japan as a research student in 1998, received his M.Sc. in 2001 from the University of Tokyo (Prof. K. Kitazawa) and Ph.D. in 2012 from University of Tsukuba (Prof. J. Ichikawa). He joined Shandong Non-metallic Materials Institute in 1991. He worked as a Senior Chemist for DuPont-Mitsui Fluorochemicals Co., Ltd. during 2005-2012. From 2012, he worked as a Chief Senior Researcher in Zeon Corporation. His research interests involve organic synthesis of functional molecules (from small to macro), especially organofluorine compounds.

Junji Ichikawa was born in Tokyo, Japan in 1958. He received his B.Sc. in 1981 and Ph.D. in 1986 from the University of Tokyo (Prof. T. Mukaiyama). He joined Kyushu University as an Assistant Professor in 1985. In 1989, he was a postdoctoral research associate in Harvard University (Prof. E. J. Corey) and then worked in Kyushu Institute of Technology as a Lecturer and an Associate Professor. In 1999, he moved to the University of Tokyo, as an Associate Professor. He was appointed Professor in University of Tsukuba in 2007. His research interests lie in the area of synthetic methodology based on the properties of metals and fluorine.

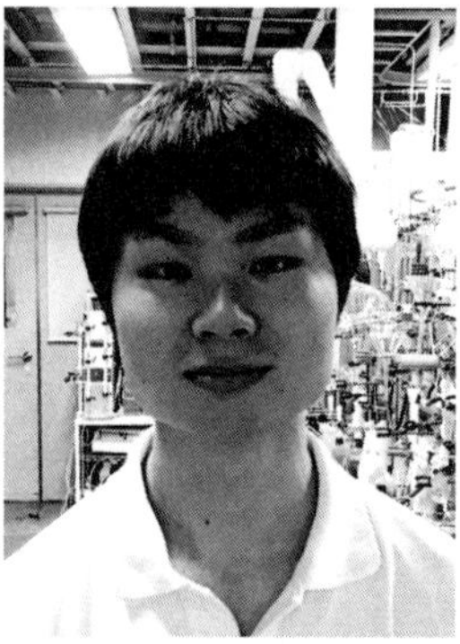

Shogo Sato was born in Kanagawa, Japan in 1990. He received his B.Sc. degree from Aoyama Gakuin University (Prof. H. Sugimura) in 2013. In the same year, he joined the research group of Prof. Keisuke Suzuki at Tokyo Institute of Technology. His research focuses on the synthesis of natural products.

Keiichiro Sakata was born in Tokyo, Japan in 1992. He received his B.S. degree in 2015 from Waseda University (Prof. S. Hosokawa) and pursing a master's degree at Tokyo Institute of Technology (Prof. K. Suzuki). He is majoring in synthetic organic chemistry and now focuses on synthetic methods of natural products.

Rhodium-Catalyzed Addition of Carboxylic Acids to Terminal Alkynes towards *Z*-Enol Esters

Submitted by Stephanie Ganss,[a] Julia Pedroni,[a] Alexandre Lumbroso,[a] Günther Leonhardt-Lutterbeck,[a] Antje Meißner,[b] Siping Wei,[b] Hans-Joachim Drexler,[b] Detlef Heller,[b] and Bernhard Breit[a*]

[a] Institut für Organische Chemie
Albert-Ludwigs-Universität Freiburg
Albertstraße 21, 79104 Freiburg i. Brsg. (Germany)

[b] Leibniz-Institut für Katalyse e.V.
Albert-Einstein-Straße 29a, 18059 Rostock (Germany)

Checked by Sonia Rodriguez and Chris H. Senanayake

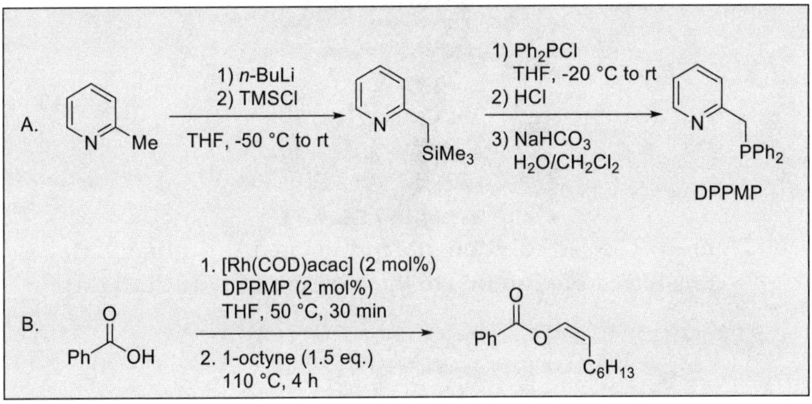

Procedure

> *Caution! A protection shield should be placed in front of the reaction set-up used for the preparation of (Z)-1-benzoyloxy)-1-octene.*

A. 2-(Diphenylphosphino-methyl) pyridine (DPPMP). A dry argon-flushed (Notes 1 and 2) 250 mL round-bottomed Schlenk-flask (Note 3)

DOI: 10.15227/orgsyn.093.0367
Published on the Web 11/3/2016

containing a magnetic stirring bar (oval, 30 x 16 mm) is charged with THF (75 mL) (Notes 4 and 5). 2-Picoline (7.40 mL, 6.98 g, 75.0 mmol) (Note 6) is added by syringe (Note 5) at rt. The solution is cooled to –50 °C (Note 7). Subsequently, a solution of ^{n}BuLi (30.0 mL, 2.5 M in hexane, 75.0 mmol, 1 equiv) (Note 8) is added dropwise by syringe over 1 h (Note 9). The reaction mixture is then stirred at –20 °C (Note 7) for 1.5 to 2 h. Meanwhile, a solution of TMSCl (9.50 mL, 8.13 g, 75.0 mmol, 1.0 equiv) (Note 10) in THF (10 mL) (Notes 4 and 5) is prepared in a dry argon-flushed (Notes 1 and 2) 250 mL round-bottomed Schlenk-flask (Note 3) containing a magnetic stirring bar (oval, 30 x 16 mm) and cooled to –20 °C (Note 7). The solution of deprotonated 2-picoline is then transferred *via* cannula (Note 11) to the TMSCl solution over a period of 2 h (Figure 1). After complete addition, the

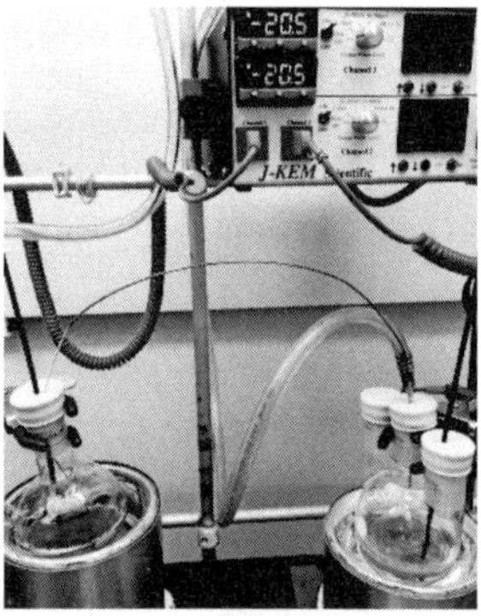

Figure 1. The red solution of deprotonated 2-picoline (left) is transferred *via* cannula to the solution of TMSCl (right)

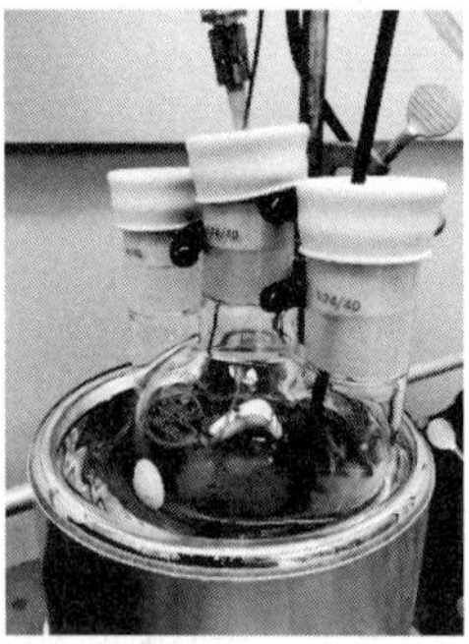

Figure 2. Reaction mixture after complete addition of deprotonated 2-picoline to TMSCl

flask of the 2-picoline solution is rinsed with THF (10 mL) (Notes 4 and 5) and this solution is also transferred by cannula. The reddish reaction mixture (Note 12) is allowed to warm to rt and stirred overnight (Figure 2). The reaction flask is equipped with a distillation apparatus (Note 13), and solvent and unreacted TMSCl are removed by distillation at ambient pressure. The residue is transferred by syringe to a 50 mL round-bottomed flask and fractionally distilled at 90 °C under reduced pressure of 20 mmHg (Note 14) (Figure 3) to afford 2-[(trimethylsilyl)methyl]pyridine as colorless oil (9.57 g, 57.9 mmol, 77%) (Note 15).

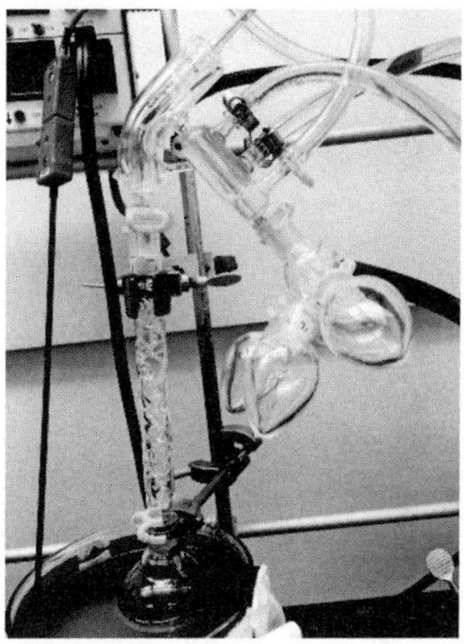

Figure 3. Distillation apparatus for the purification of 2-[(trimethylsilyl)methyl]pyridine

A dry argon-flushed (Notes 1 and 2) 250 mL round-bottomed Schlenk-flask (Note 3) containing a magnetic stirring bar (oval, 30 x 16 mm) is charged by syringe with 2-[(trimethylsilyl)methyl]pyridine (8.83 g, 53.4 mmol) and THF (45 mL) (Notes 4 and 5). The solution is cooled to –20 °C (Note 7) and PPh_2Cl (10.6 mL, 13.0 g, 58.8 mmol, 1.1 equiv) (Note 16) is added by syringe (Note 5) over 10 min. The reaction mixture is stirred at –20 °C for 1 h, allowed to warm to rt and stirred overnight. Then HCl (2.0 M in Et_2O, 30 mL, 60 mmol, 1.1 equiv) (Note 5) (Note 17) is added over 5 min at 0 °C (Note 18). The mixture is stirred for 1 h before the solvent is carefully removed while being connected to oil pump vacuum (0.15 mmHg). The residue is recrystallized (Note 19) (Figures 4, 5, and 6) from EtOH (Notes 20 and 21) and Et_2O (Note 21) to afford 2-((diphenylphosphino)-

methyl)pyridine hydrochloride (DPPMP•HCl) as acicular crystals (12.0 g, 38.2 mmol, 72%) (Note 22).

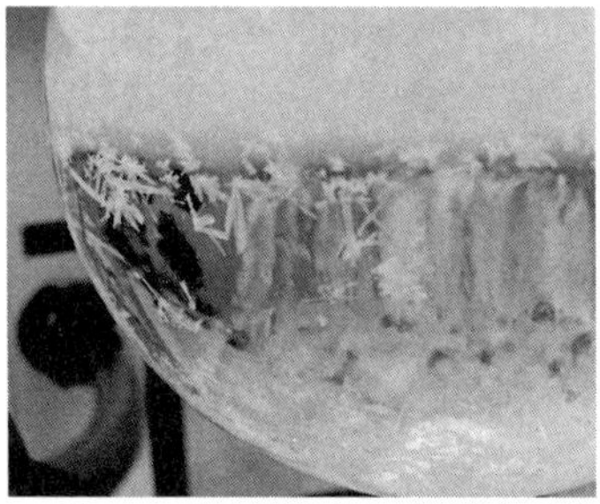

Figure 4. Crystal formation of DPPMP•HCl in EtOH overlaid by Et_2O

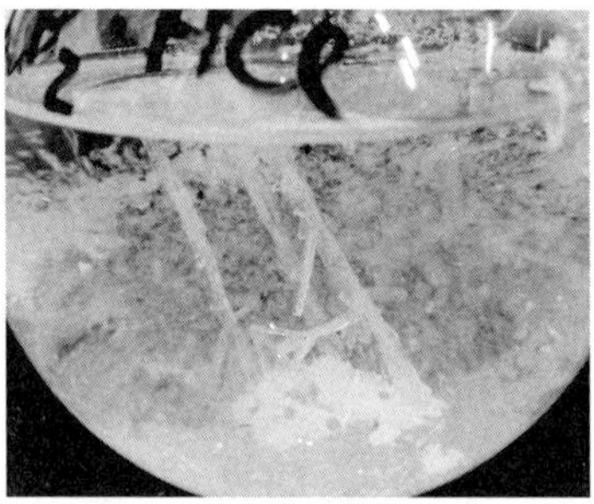

Figure 5. Crystals of DPPMP•HCl in EtOH/Et_2O

Figure 6. Crystals of DPPMP•HCl after removal of the solvents

A dry, argon-flushed (Notes 1 and 2) 50 mL round-bottomed Schlenk-flask (Note 3) containing a magnetic stirring bar (cylindric, 15 x 4.5 mm) is charged with $NaHCO_3$ (368 mg, 4.38 mmol, 2.2 equiv) (Note 23). Distilled water (25 mL) (Note 20) is added by syringe and the mixture stirred at rt until the salt is completely dissolved. Meanwhile, an argon-flushed 100 mL round-bottomed Schlenk-flask containing a magnetic stirring bar (cylindric,

20 x 6 mm) is charged with DPPMP•HCl (627 mg, 2.00 mmol) (Note 24). Dichloromethane (33 mL) (Note 25) is added by syringe to give a clear colorless solution. The aqueous solution of $NaHCO_3$ is added by syringe and the biphasic mixture is vigorously stirred for 30 min. The organic layer is transferred by syringe to an argon-flushed 50 mL round bottomed Schlenk-flask (Note 3). The organic layer is washed with distilled water (10 mL) (Note 21), carefully concentrated and dried under oil pump vacuum (0.15 mmHg). The free ligand 2-((diphenylphosphino)-methyl)pyridine is obtained as a colorless solid (546 mg, 1.97 mmol, 99%) (Notes 26 and 27).

B. *(Z)-1-(Benzoyloxy)-1-octene*. A dry argon-flushed (Notes 1 and 2) 120 mL Teflon screw cap pressure vessel (Note 28) containing a magnetic stirring bar (cylindric, 20 x 6.0 mm) is charged with [Rh(COD)acac] (248 mg, 0.800 mmol, 2.0 mol%) (Note 29), 2-(diphenylphosphino-methyl) pyridine (DPPMP, 222 mg, 0.8 mmol, 2.0 mol%) and benzoic acid (4.88 g, 40.0 mmol) (Note 30) in a glove box. Degassed anhydrous THF (62 mL) (Notes 4 and 5) is added, the flask sealed (Note 31), and immediately immersed in a pre-heated oil bath (bath temperature 50 °C) outside of the glove box. The yellow mixture (Figure 7) is stirred for 30 min and then allowed to cool down to rt (Note 32). 1-Octyne (8.85 mL, 6.61 g, 60.0 mmol,

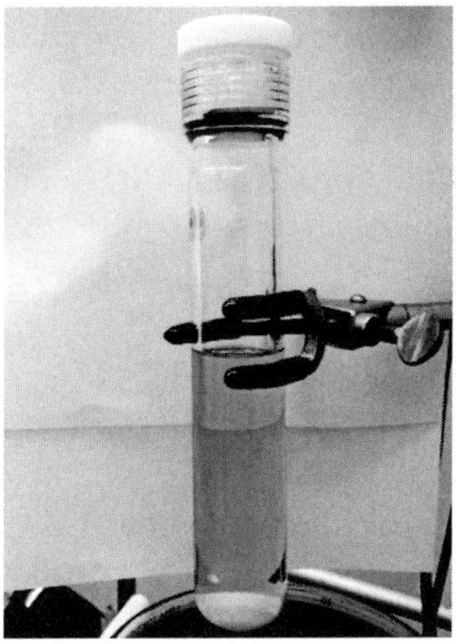

Figure 7. Reaction mixture before addition of 1-octyne

1.5 equiv) (Note 33) is added in the glove box, the flask sealed (Note 31), and immediately immersed in a pre-heated oil bath (bath temperature 110 °C) outside of the glove box. The mixture turns orange (Figure 8) while

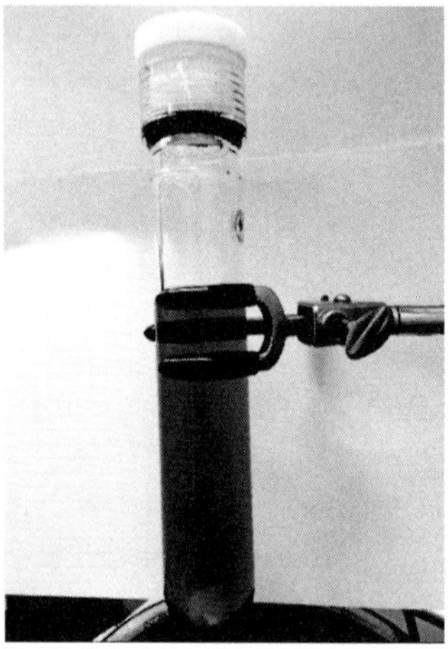

Figure 8. Reaction mixture after addition of 1-octyne and heating

being stirred for 4 h. After cooling to rt for 25 min, the mixture is filtered through a pad of silica gel (30 x 70 mm) (Note 34) and eluted with ethyl acetate (3 x 20 mL) (Note 35) to separate the catalyst. The solvent is removed under reduced pressure first on a rotary evaporator (225 mmHg down to 15 mmHg, bath temperature 40 °C), then under oil pump vacuum (0.2 mmHg, room temperature) and the crude product (Note 36) purified by column chromatography on silica gel (100 x 120 mm, eluent pentane : dichloromethane = 5:1, R_f = 0.3) (Note 37). The solvents are removed under reduced pressure first on a rotary evaporator (600 mmHg down to 15 mmHg, bath temperature 40 °C), then under oil pump vacuum (0.2 mmHg, room temperature) to furnish (Z)-1-(benzoyloxy)-1-octene (8.67 g, 37.3 mmol, 93%) (Note 38) as a light orange oil (Note 39).

Notes

1. Vessels are dried by heating with the heat gun for 2 minutes while connected to the oil pump vacuum (0.15 mmHg).
2. Argon 5.0 from Sauerstoffwerke Friedrichshafen.
3. Equipped with a rubber septum and connected to argon.
4. THF (anhydrous, >99.9%, inhibitor-free) was purchased from Sigma Aldrich.
5. A syringe purchased from Henke Sass Wolf is flushed with argon three times and used for the addition.

6. 2-Picoline (98% purity) was purchased from Aldrich and distilled at 75 °C under 150 mmHg. The freshly distilled colorless 2-picoline was removed under an Argon counter flow. A yellowish oil remained in the flask.
7. Cooling was performed using a dry ice/ethanol bath.
8. ^{n}BuLi (2.5 M in hexane) was purchased from Sigma Aldrich. The concentration was checked by titration using *N*-(*o*-tolyl)pivalamide (Suffert's reagent) prior to use.[2]
9. The previously colorless solution turns orange and finally ruby red.
10. Chlorotrimethylsilane (≥98.0%) was purchased from Sigma Aldrich and distilled at 57 °C under argon prior to use.
11. The picoline flask is attached to a continuous argon flow. A stainless steel cannula is used for the transfer.
12. The reaction mixture turns yellowish then reddish after complete addition of 2-picoline.
13. The distillation apparatus consists of a Claisen stillhead, a fused Liebig condenser (10 cm) and a fused vacuum connection. Solvent and TMSCl are distilled off at about 65 °C (oil bath temperature 85 °C) at ambient pressure. The distillates are collected in a 100 mL flask.
14. A Vigreux column (10 cm) is attached and 2-[(trimethylsilyl)methyl]-pyridine is distilled at 90 °C (oil bath temperature 150 °C) and 22 mmHg. The distillates are separated *via* a multi limb delivery adapter and collected in suitable flasks (Figure 3).
15. A second run on the same scale provided 9.74 g (79%) of the same material. 2-[(trimethylsilyl)methyl]pyridine exhibits the following spectroscopic properties: ^{1}H NMR (400 MHz, $CDCl_3$) δ: 0.00 (s, 9H), 2.33 (s, 2H), 6.90–6.98 (m, 2H), 7.47 (dt, *J* = 7.6, 2.0 Hz, 1H), 8.41 (bd, *J* = 5.2 Hz, 1H); ^{13}C NMR (101 MHz, $CDCl_3$) δ: –1.70, 30.31, 119.10, 122.12, 135.78, 149.01, 161.35; ESI-HRMS (*m*/*z*) calcd. for $C_9H_{16}NSi$ $[M+H]^+$ 166.10465, found 166.1039.
16. Chlorodiphenylphosphine (96% purity) was purchased from Sigma Aldrich and distilled at 90 °C (oil bath temperature 150 °C) at 0.004 mmHg.
17. HCl (2.0 M in Et_2O) was purchased from Sigma Aldrich and used as received.
18. Cooling is performed using an ice-water bath.
19. The residue is dissolved in a minimum amount of EtOH (70 to 100 mL) and carefully overlaid with Et_2O (EtOH:Et_2O = 1:1) (Figure 4). After

crystallization overnight at rt (Figure 5), the mother liquor is transferred *via* cannula to another 250 mL round-bottomed Schlenk-flask and 2-((diphenylphosphino)methyl)-pyridine hydrochloride is afforded in acicular crystals (8.90 g, 28.4 mmol) (Figure 6). The mother liquor was concentrated under oil pump vacuum (0.15 mmHg) and the residue recrystallized overnight at rt from a minimum amount of EtOH (50 mL) and carefully overlaid with Et_2O (EtOH:Et_2O = 1:2) to give more DPPMP•HCl (2.30 g, 7.3 mmol).

20. Absolute anhydrous EtOH was purchased from Sigma Aldrich.
21. The solvent is degassed by bubbling argon through the solvent using a cannula for 30 min while being stirred.
22. 2-((Diphenylphosphino)methyl)pyridine hydrochloride exhibits the following spectroscopic and physical properties: NMR samples should be prepared in degassed solvents. The solvent (~3 mL) is degassed by bubbling argon through the solvent for 5-10 min. ^{1}H NMR (400 MHz, CD_2Cl_2) δ: 4.19 (s, 2H), 7.34–7.39 (m, 1H), 7.40–7.45 (m, 6H), 7.55–7.61 (m, 4H), 7.61–7.67 (m, 1H), 8.11 (dt, J = 8.0, 1.6 Hz, 1H), 8.51 (dd, J = 5.8, 1.0 Hz, 1H); ^{13}C NMR (101 MHz, CD_2Cl_2) δ: 33.3, 123.8, 127.2, 128.8, 129.7, 133.1, 135.3, 140.4, 144.4, 155.7; ^{31}P NMR (121 MHz, CD_2Cl_2) δ: –4.66; ESI-HRMS (*m/z*) calcd. for $C_{18}H_{17}NOP$ $[M\text{-}Cl+O+H]^+$ 294.10422, found 294.1036.
23. $NaHCO_3$ was purchased from Fisher Chemical and used as received.
24. After being charged with the DPPMP•HCl the flask is evacuated and filled with argon three times, equipped with a septum and an argon line.
25. Anhydrous dichloromethane from Sigma Aldrich and used as received after degassing by bubbling argon for 15 minutes.
26. 2-((Diphenylphosphino)methyl)pyridine exhibits the following spectroscopic and physical properties: mp 57–58 °C; ^{1}H NMR (400 MHz, CD_2Cl_2) δ: 3.54 (s, 2H), 6.89-6.99 (m, 2H), 7.21-7.26 (m, 6H), 7.31-7.42 (m, 5H), 8.33-8.38 (m, 1H); ^{13}C NMR (101 MHz, CD_2Cl_2) δ: 38.5, 120.9, 123.6, 128.4, 128.7, 132.8, 136.0, 138.3, 149.2, 158.2; ^{31}P NMR (121 MHz, CD_2Cl_2) δ: –11.12.; ESI-HRMS (*m/z*) calcd. for $C_{18}H_{17}NP$ $[M+H]^+$ 278.10931, found 278.1095.
27. The ligand DPPMP is sensitive to oxidation when being stored over a longer period of time. Therefore, the authors recommend releasing the required amount of free ligand from the HCl salt shortly before the catalysis. The pure ligand should be stored under argon. The oxide

may be separated by filtration over silica gel under inert atmosphere using degassed dichloromethane.

28. Pressure vessel was purchased from Chem Glass.
29. [Rh(COD)acac] (elemental Rh (ICP): 32.99%, $[Rh(COD)Cl]_2 < 0.1\%$) was purchased from Alfa Aesar and used as received.
30. Benzoic acid (>99.5% purity) was purchased from Sigma Aldrich and used as received.
31. The cap was screwed back on tightly. A protection shield should be placed in front of the reaction apparatus.
32. The cooling required 15 minutes.
33. 1-Octyne (98% purity) was purchased from Alfa Aesar and filtered over a pad of basic alumina (30x 20 mm) prior to use. The basic alumina was purchased from Sigma Aldrich.
34. Silica gel 60, 230-400 mesh (Fisher Chemical) is used as stationary phase.
35. Ethyl acetate of technical grade is used after evaporation and recollection of the solvent in the receiving flask of a rotary evaporator (200 mmHg, bath temperature 50-60 °C).
36. The ratio *anti*-Markovnikov-*Z* (AM-*Z*):*anti*-Markovnikov-*E* (AM-*E*):Markovnikov (M) = 92:4:4 is determined by ^{1}H NMR analysis.[3] An enyne-byproduct from homocoupling of 1-octyne is observed.
37. The product is purified by flash chromatography through silica gel using a Teledyne Isco CombiFlash Rf (120 g column, from Silicycle) with hexane:dichloromethane = 5:1 as eluent. The product is typically found in fractions 11-39 *via* TLC analysis on silica gel (hexane:dichloromethane = 5:1, $R_f = 0.3$, visualization with $KMnO_4$ stain).
38. A second run on the same scale provided 8.49 g (91%) of the same products. The enyne-byproduct was separated and not observed in the ^{1}H NMR product spectrum after column chromatography. A ratio AM-*Z*:AM-*E*:M = 94:1:4 was determined by ^{1}H NMR analysis. Quantitative NMR (500 MHz, $CDCl_3$, 1,3,5-trimethoxybenzene (≥ 99% purity, Sigma Aldrich)) delivers a purity of >96 wt% with a ratio of AM-*Z*:AM-*E*:M = 95:1:4. ^{1}H NMR (500 MHz, $CDCl_3$) δ: 0.84–0.91 (m, 3H), 1.26–1.40 (m, 6H), 1.40–1.49 (m, 2H), 2.29 (dq, *J* = 7.5, 1.5 Hz, 2H), 7.24–7.28 (m, 1H), 7.45–7.50 (m, 2H), 7.62–7.57 (m, 1H), 8.09–8.13 (m, 2H); ^{13}C NMR (125 MHz, $CDCl_3$) δ: 14.1, 22.6, 24.6, 28.9, 29.2, 31.7, 115.0,

128.5, 129.5, 129.9, 133.4, 134.2, 163.6; ESI-HRMS (*m/z*) calcd. for $C_{15}H_{24}NO_2$ $[M+NH_4]^+$ 250.18015, found 250.1798.

39. If product with higher purity is desired, the product (7.47 g) is distilled at 75 °C at 0.4 mmHg to provide a colorless oil (7.02 g, 94%). Quantitative NMR (500 MHz, $CDCl_3$, 1,3,5-trimethoxybenzene (≥ 99% purity, Sigma Aldrich)) delivers a purity of >96 wt% with a ratio of AM-Z:AM-*E*:M = 95:1:4. The distillation apparatus consists of a 25 mL round-bottomed flask connected to a Claisen stillhead, a fused Liebig condenser (5.5 cm) with fused vacuum adapter and a multi-limb delivery adapter with three 10 mL flasks (Figure 9).

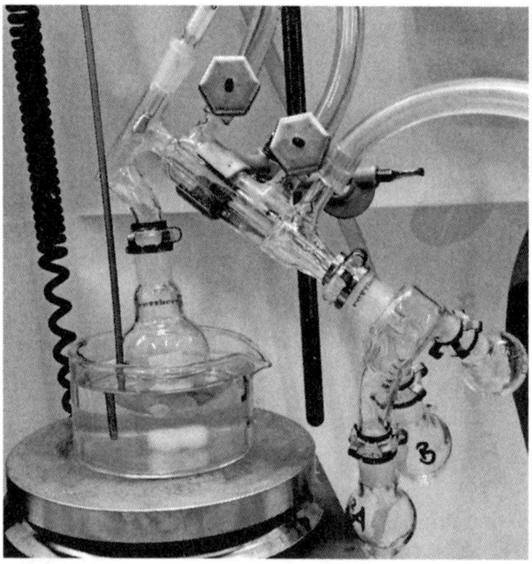

Figure 9: Distillation apparatus for further purification of the product

Working with Hazardous Chemicals

The procedures in *Organic Syntheses* are intended for use only by persons with proper training in experimental organic chemistry. All hazardous materials should be handled using the standard procedures for work with chemicals described in references such as "Prudent Practices in the Laboratory" (The National Academies Press, Washington, D.C., 2011; the full text can be accessed free of charge at http://www.nap.edu/catalog.php?record_id=12654). All chemical waste should be disposed of in accordance with local regulations. For general guidelines for the management of chemical waste, see Chapter 8 of Prudent Practices.

In some articles in *Organic Syntheses*, chemical-specific hazards are highlighted in red "Caution Notes" within a procedure. It is important to recognize that the absence of a caution note does not imply that no significant hazards are associated with the chemicals involved in that procedure. Prior to performing a reaction, a thorough risk assessment should be carried out that includes a review of the potential hazards associated with each chemical and experimental operation on the scale that is planned for the procedure. Guidelines for carrying out a risk assessment and for analyzing the hazards associated with chemicals can be found in Chapter 4 of Prudent Practices.

The procedures described in *Organic Syntheses* are provided as published and are conducted at one's own risk. *Organic Syntheses, Inc.*, its Editors, and its Board of Directors do not warrant or guarantee the safety of individuals using these procedures and hereby disclaim any liability for any injuries or damages claimed to have resulted from or related in any way to the procedures herein.

Discussion

Enol esters are versatile building blocks in organic synthesis, finding application as substrates in a wide range of reactions such as acylations,[4] asymmetric epoxidations,[5] cyclopropanations[6] and hydroformylations[7] besides being important monomers for polymerisation.[8] Moreover, the enol ester motif is found in several biologically active natural products such as Phenochalasin B[9] and Grenadadiene.[10] Since the first report on a ruthenium-catalyzed addition of carboxylic acids to an alkyne by Shvo *et al.* in 1983,[11] several other intermolecular methodologies were developed employing iridium,[12] rhenium,[13] palladium,[14] gold[15] as well as an intramolecular rhodium-catalyzed synthesis of alkylidene lactones.[16]

While these mostly lead to Markovnikov (M) or *anti*-Markovnikov (AM-*E*) addition, our recently reported atom- and redox-economic rhodium-catalyzed hydro-oxycarbonylation[17] leads selectively to the *Z*-enol ester (AM-Z) in high yields in a reaction time of 16-24 hours. Recently, the methodology was successfully applied in natural product synthesis by Burke *et al.* in the total synthesis of Patulolide C.[18]

Following our first report employing the $[Rh(COD)Cl]_2$/DPPMP catalyst system, further investigations have led to a more active

Rh(COD)acac/DPPMP catalyst system, which furnishes the desired Z-enol esters in equally high yields and selectivities however with a reduced reaction time of only 4 hours.[19] Furthermore, this novel catalyst system employed in the herein reported protocol was found to be compatible with a wide range of functionalities (Table 1).

Table 1. Further Z-enol esters prepared using this protocol[20]

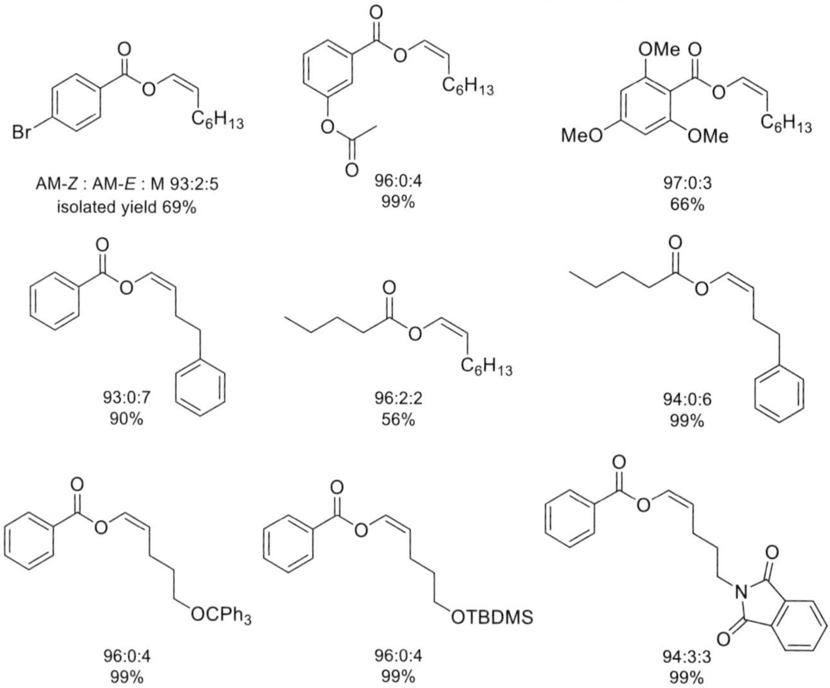

As a result of the milder reaction conditions, this improved methodology is highly suitable for the synthesis of Z-enol esters containing sensitive moieties such as the enol ester derived from *p*-bromo benzoic acid. In addition, the compatibility with commonly employed protecting groups such as trityl, TBDMS and *N*-phthalimide reinforces the applicability of our methodology in total synthesis.

In conclusion, the catalyst systems developed by us show complementary functional group compatibility, allowing for a broad application of the rhodium-catalyzed hydro-oxycarbonylation in organic synthesis.

References

1. Institut für Organische Chemie, Albert-Ludwigs-Universität Freiburg, Albertstr. 21, 79104 Freiburg im Breisgau. (Germany). E-mail: bernhard.breit@chemie.uni-freiburg.de. We would like to thank the DFG, the International Research Training Group "Catalysis and Catalytic Reactions for Organic Synthesis" (IRTG 1038), the Krupp Foundation for financial support as well as Umicore, BASF and Wacker for generous gifts of chemicals. E-mail: bernhard.breit@chemie.uni-freiburg.de, detlef.heller@catalysis.de.
2. Suffert, J. *J. Org. Chem.* **1989**, *54*, 509–510.
3. (a) AM-*E*: Huang, F.; Quach, T. D.; Batey, R. A. *Org. Lett.* **2013**, *15*, 3150-3153; (b) M: Lee, P. O.; Kang, D.; Choi, S.; Kim, S. *Org. Lett.* **2011**, *13*, 3470–3473.
4. For selected examples, see: (a) Bruneau, C.; Neveux, M.; Kabouche, Z.; Ruppin, C.; Dixneuf, P. H. *Synlett* **1991**, 755–763. (b) Tu, W.; Floreancig, P. E. *Angew. Chem., Int. Ed.* **2009**, *48*, 4567–4571. (c) Quirin, C.; Kazmaier, U. *Eur. J. Org. Chem.* **2009**, *48*, 371–377. (d) Lin, M.-H.; RajanBabu, T. V. *Org. Lett.* **2000**, *2*, 997–1000. (e) Grasa, G. A.; Güveli, T.; Singh, R.; Nolan, S. P. *J. Org. Chem.* **2003**, *68*, 2812–2819; (f) Verzijl, G. K. M.; de Vries, J. G.; Broxterman, Q. B. *Tetrahedron: Asymmetry* **2005**, *16*, 1603–1610. (g) Tu, W.; Floreancig, P. E. *Angew. Chem.* **2009**, *121*, 4637–4641; *Angew. Chem., Int. Ed.* **2009**, *48*, 4567–4571.
5. (a) Sharpless, K. B.; Michaelson, R. C. *J. Am. Chem. Soc.* **1973**, *96*, 6136-6137. (b) Lurain, A. E.; Maestri, A.; Kelly, A. R.; Carroll, P. J.; Walsh, P. J. *J. Am. Chem. Soc.* **2004**, *126*, 13608–13609. (c) Lurain, A. E., Carroll, P. J.; Walsh, P. J. *J. Org. Chem.* **2005**, *70*, 1262–1268. (d) Oppolzer, W.; Radinov, R. N.; De Brabander, J. *Tetrahedron Lett.* **1995**, *36*, 2607–2610. (e) Salvi, L.; Jeon, S.-J.; Fisher, E. L.; Carroll, P. J.; Walsh, P. J. *J. Am. Chem. Soc.* **2007**, *129*, 16119–16125.
6. For a review, see Pellissier, H. *Tetrahedron* **2008**, *64*, 7041-7095. For relevant examples, see (a) Kim, H. Y.; Lurain, A. E.; García-García, P.; Carroll, P. J.; Walsh, P. J. *J. Am. Chem. Soc.* **2005**, *127*, 13138–13139. (b) Kim, H. Y.; Salvi, L.; Carroll, P. J.; Walsh, P. J. *J. Am. Chem. Soc.* **2009**, *131*, 954–962. (c) Oppolzer, W.; Radinov, R. N. *J. Am. Chem. Soc.* **1993**,

115, 1593–1594. (d) Fournier, J. F.; Mathieu, S.; Charette, A. B. *J. Am. Chem. Soc.* **2005**, *127*, 13140–13141.

7. (a) Thomas, P. J.; Axtell, A. T.; Klosin, J.; Peng, W.; Rand, C. L.; Clark, T. P.; Landis, C. R.; Abboud, K. A. *Org. Lett.* **2007**, *9*, 2665–2668. (b) Risi, R. M.; Burke, S. D. *Org. Lett.* **2012**, *14*, 1180–1182.
8. Montheard, J. P.; Camps, M.; Seytre, G.; Guillet, J.; Dubois, J. C. *Angew. Makromol. Chem.* **1978**, *72*, 45–55.
9. (a) Tomoda, H.; Namatame, I.; Tabata, N.; Kawaguchi, K.; Si, S.; Omura, S. J. Antibiot. **1999**, *52*, 857–861. (b) Quang, D. N.; Hashimoto, T.; Tanaka, M.; Baumgartner, M.; Stadler, M.; Asakawa, Y. *J. Nat. Prod.* **2002**, *65*, 1869–1874.
10. (a) Sitachitta, N.; Gerwick, W. H. *J. Nat. Prod.* **1998**, *61*, 681–684. (b) Bisogno, T.; Melck, D.; De Petrocellis, L.; Bobrov, M. Y.; Gretskaya, N. M.; Bezuglov, V. V.; Sitachitta, N.; Gerwick, W. H.; Di Marzo, V. *Biochem. Biophys. Res. Commun.* **1998**, *248*, 515–522.
11. Rotem, M.; Shvo, Y. *Organometallics* **1983**, *2*, 1689–1691.
12. Nakagawa, H.; Okimoto, Y.; Sakaguchi, S.; Ishii, Y. *Tetrahedron Lett.* **2003**, *44*, 103–106.
13. Hua, R.; Tian, X. *J. Org. Chem.* **2004**, *69*, 5782–5784.
14. Smith, D. L.; Goundry, W. R. F.; Lam, H. W. *Chem. Commun.* **2012**, *48*, 1505–1507.
15. Chary, B. C.; Kim, S. *J. Org. Chem.* **2010**, *75*, 7928–7931.
16. (a) Chan, D. M. T.; Marder, T. B.; Milstein, D.; Taylor, N. J. *J. Am. Chem. Soc.* **1987**, *109*, 6385–6388. (b) Elgafi, S.; Field, L. D.; Messerle, B. A. *J. Organomet. Chem.* **2000**, *607*, 97–104. (c) Mas-Marzá, E.; Sanaú, M.; Peris, E. *Inorg. Chem.* **2005**, *44*, 9961–9967.
17. Lumbroso, A.; Vautravers, N. R.; Breit, B. *Org. Lett.* **2010**, *12*, 5498–5501.
18. Risi, R. M.; Burke, S. D. *Org. Lett.*, **2012**, *14*, 1180–1182.
19. Wei, S.; Pedroni, J.; Meißler, A.; Lumbroso, A.; Drexler, H.-J.; Heller, D.; Breit, B. *Chem. Eur. J.* **2013**, *19*, 12067–12076.
20. These results were obtained for reactions performed at a 0.45 mmol scale. For the full characterization data see: Wei, S.; Pedroni, J.; Meißler, A.; Lumbroso, A.; Drexler, H.-J.; Heller, D.; Breit, B. *Chem. Eur. J.* **2013**, *19*, 12067–12076.

Appendix
Chemical Abstracts Nomenclature (Registry Number)

Benzoic acid: (65-85-0)
1-Octyne: (629-05-0)
[Rh(COD)acac]: (Acetylacetonato)(1,5-cyclooctadiene) rhodium (I); (12245-39-5)
n-BuLi: *n*-Butyllithium; (109-72-8)
2-Picoline: (109-06-8)
TMSCl: Chlorotrimethylsilane; (75-77-4)
PPh_2Cl: Chlorodiphenylphosphine; (1079-66-9)
1,3,5-Trimethoxybenzene: (621-23-8)

Stephanie Ganss studied chemistry in Darmstadt (Germany) and Lausanne (Switzerland) and received her diploma degree from the Technical University Darmstadt in 2010. She completed her diploma thesis in the group of Professor Waser at the EPF Lausanne (Switzerland) working on palladium catalyzed electrophilic acetylination of indoles using hypervalent iodine compounds. In 2016, she completed her Ph.D. under the supervision of Professor Breit at the Albert-Ludwigs University Freiburg (Germany). Her current research focuses on the application of rhodium catalyzed CH functionalization and the development of macrolactonization reactions.

Julia Pedroni was born in 1984 in Campinas, Brazil. During her undergraduate studies at the Albert-Ludwigs-Universität Freiburg she worked in the group of Professor Breit on Rh-catalysis. Since July 2013, she is working in the group of Professor Cramer at the EPF Lausanne as a Ph.D. student.

Alexandre Lumbroso studied chemistry at the University of Nantes (France). He completed his Ph.D. under the supervision of Professor Jean-Paul Quintard in 2009, working on the utilization of organostannanes for the stereoselective synthesis of imino sugar derivatives. His postdoctoral studies in the group of Prof. Dr. Bernhard Breit at the Albert-Ludwigs-University Freiburg (Germany) focused on the development of new atom-economic regio-and stereoselective reactions using homogeneous catalysis. He is now working as laboratory leader at Syngenta (Switzerland).

Günter Leonhardt-Lutterbeck successfully completed his apprenticeship for lab-technician at the Universtiy of Freiburg in 1986. He then joined the group of Professor Horst Prinzbach working on the synthesis of pagodan. Since 2001, he is part of the group of Professor Breit mainly focusing on ligand syntheses for hydroformylation.

Antje Meißner studied Biology and Chemistry at the University of Rostock (Germany) and received her degree in Chemistry in 2010. She is currently doing her doctorate under the supervision of Prof. Dr. Detlef Heller. Her present research focuses on the systematic investigations into the in situ synthesis of dinuclear rhodium diphosphine complexes and the implications of the results in homogeneous catalysis.

Siping Wei received a Ph.D. from the Sichuan University in 2011, and then worked as a post-doctoral fellow/research scientist at Leibniz-Institut für Katalyse e. V. at the University of Rostock (LIKAT Rostock, Germany). He has been an associate professor at Luzhou Medical College since 2014. His current research interest focuses on developing new NHC-Metal Complexes for homogeneous catalysis.

Hans-Joachim Drexler graduated in Chemistry at the University of Rostock (Germany) under the supervision of Prof. Hans-Jürgen Holdt in 1997. Since 1999 he is a scientific co-worker at the Leibniz-Institut für Katalyse e.V., Rostock. His research interests are in the fields of organometallic chemistry, asymmetric homogeneous catalysis and especially X-ray analysis. The successful cooperation with industrial partners is documented by several joint projects and patents.

Detlef Heller studied Chemistry at the University of Greifswald (Germany), where he also obtained his Ph.D. under the supervision of Prof. Kurt Madeja. Since 1987 he is a researcher at the Leibniz-Institut für Katalyse e.V., Rostock, since 1998 as group leader. In 1999 he completed his Habilitation in the field of physical chemistry. His research interests within organometallic chemistry deal with the kinetics, mechanisms and deactivation phenomena of homogeneous catalysis, with a special emphasis on rhodium complexes towards the design of efficient general in situ techniques for catalyst generation. Beside he is very fond of the history of Chemistry from a philatelic point of view.

Bernhard Breit studied chemistry at the University of Kaiserslautern, Germany, where he obtained his doctorate in 1993 with Professor Regitz. After postdoctoral training with Professor Trost at Stanford University, he worked in Marburg, Germany, with Professor R. W. Hoffmann to obtain his habilitation in 1998. In 1999 he was appointed as an Associate Professor at the University of Heidelberg, Germany. Since 2001 he has been a Full Professor of Organic Chemistry at the Albert-Ludwigs-Universität Freiburg i. Brsg. His research interests focus on the exploration and development of catalytic methods for organic synthesis. He is author of 200 publications.

Sonia Rodriguez studied Chemical Sciences at the Universitat Autònoma de Barcelona, Spain, where she received her Ph.D. in Organic Chemistry in 2000 under the supervision of Profs. Pedro de March and Marta Figueredo. After industrial experience at Finnovequim and Deripol, she broadened her synthetic background during a postdoctoral appointment with Prof. Peter Wipf at the University of Pittsburgh from 2001 to 2003. Subsequently, she joined the department of Chemical Development at Boehringer Ingelheim Pharmaceuticals, Inc., where she has been a member of the Catalysis group since 2007. Her research interests include the development and application of catalytic asymmetric reactions in the development of new economical processes.

Organic Syntheses

Copper-Catalyzed Semi-Reduction of Alkynes

Nick Cox,[1] Hester Dang,[2] Aaron M. Whittaker,[3] and Gojko Lalic[4*]

Department of Chemistry, University of Washington, Seattle, Washington 98195, United States

Checked by Miyu Furuta and John L. Wood

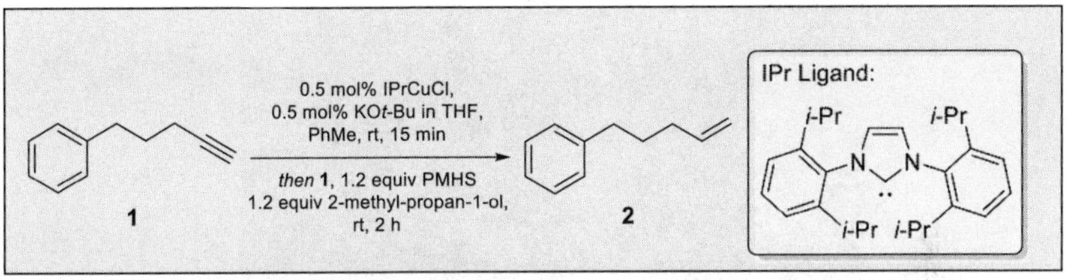

Procedure

4-Penten-1-ylbenzene (**2**). A 500 mL round-bottomed Airfree reaction flask with a 24/40 jointed neck and side-arm equipped with valve (Flask A) is charged with 20.6 g of 3 Å molecular sieves and sealed with a silicone-greased 24/40 glass stopper (Figure 1), then flame-dried under vacuum for approximately 10 min (Note 1). Flask A is then refilled with dry nitrogen, the glass stopper is replaced with a rubber septum, and Flask A is allowed to cool to room temperature. To Flask A is added a Teflon-coated magnetic stir bar (5 cm × 8 mm). Anhydrous toluene (PhMe) (128 mL) (Note 2), 4-pentyn-1-ylbenzene (**1**) (6.83 mL, 45.0 mmol, 1.00 equiv) (Note 3) and polymethylhydrosiloxane (PMHS) (3.25 mL, 54.0 mmol, 1.20 equiv of hydride) (Note 4) are added sequentially by syringe, followed by isobutanol (5.00 mL, 54.0 mmol, 1.20 equiv), also by syringe (Note 5). The contents of Flask A are stirred at 500 rpm at ambient temperature for approximately 40 min. While the solution in Flask A is stirring, a second 500 mL round-bottomed Airfree reaction flask with a 24/40 jointed neck and side-arm equipped with valve (Flask B) is charged with a Teflon-coated magnetic stir

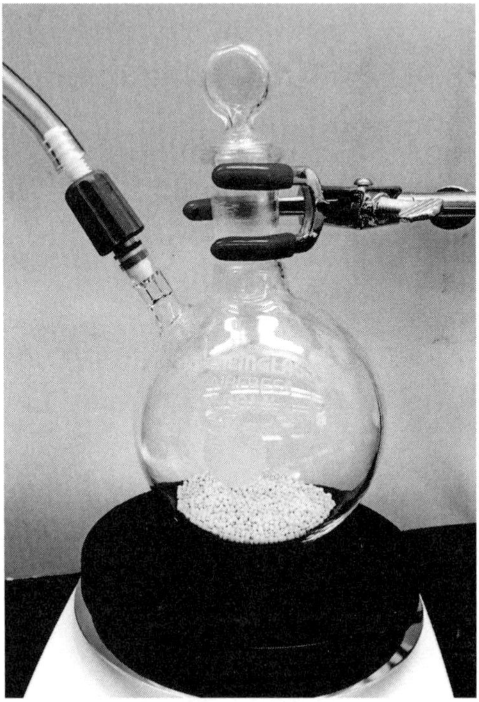

Figure 1. Apparatus Assembly for Flask A

bar (5 cm × 8 mm) and sealed with a silicone-greased 24/40 glass stopper, then flame-dried under vacuum. Flask B is then refilled with dry nitrogen, the glass stopper is replaced with a rubber septum, and Flask B is allowed to cool to room temperature. To Flask B is added chloro[1,3-bis(2,6-diisopropylphenyl)imidazol-2-ylidene]-copper(I) (IPrCuCl) (109 mg, 0.22 mmol, 0.0050 equiv). Anhydrous PhMe (16 mL) and potassium *tert*-butoxide (KO*t*-Bu) (as 1 M solution in THF, 230 μL, 0.23 mmol, 0.0050 equiv) are added sequentially by syringe (Note 6). The contents of Flask B are stirred at 300 rpm at ambient temperature for 5 min, at which point the solution transitions from a slightly cloudy suspension to a clear, colorless solution (Figure 2). Additional anhydrous PhMe (16 mL) is added by syringe, and the contents of Flask B are allowed to stir for an additional 15 min. At this point, the stir rate of flask B is increased to 500 rpm, and the solution from Flask A is transferred over 10 min via cannula to Flask B. The resultant mixture undergoes a small amount of foaming and changes from a yellow to an ash color (Figure 3). The sieves that remain in Flask A are then rinsed several times with anhydrous PhMe (4 x 16 mL), transferring the solvent to Flask B by cannula after each rinse. The stir rate is increased to 700 rpm, and the contents of Flask B are allowed to stir at ambient temperature for 70 min, at which point the alkyne (**1**) is completely consumed based on GC analysis (Note 7).

Figure 2. Color Transition Observed in Flask B

Figure 3. Addition of the solution of Flask A to Flask B

Filtration through silica gel is used to remove copper salts from the crude reaction products. A 60 mL Büchner filter funnel with a coarse frit is placed atop a 500 mL round-bottomed flask and packed with celite (1 cm) and then 30 g of silica gel (Note 8). The contents of Flask B are vacuum-filtered through the silica gel plug, which is then rinsed with 200 mL of hexanes. The flask is charged with a Teflon-coated magnetic stir bar (5 cm × 8 mm), and connected to a distillation head equipped with a 6 cm Vigreux column and a 1 L round-bottomed collection flask (Figure 4). The filtrate is concentrated to a volume of approximately 15 mL (Note 9 and 10).

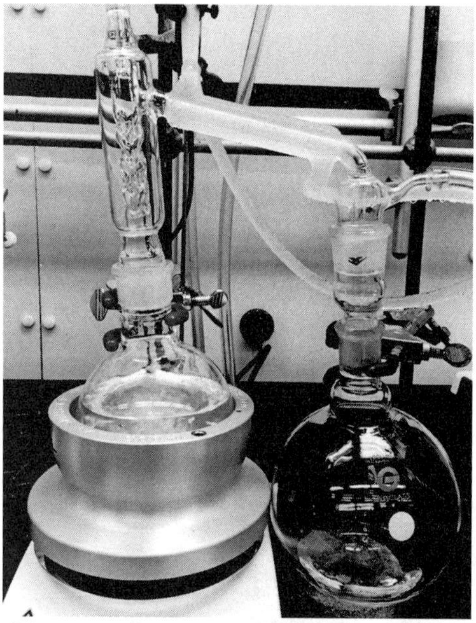

Figure 4. Concentration of the filtrate

The remaining residue is transferred to a 25 mL round-bottomed flask via pipette and the flask was rinsed with hexanes (2 x 3 mL). The 25 mL flask containing the product is equipped with a Teflon-coated magnetic stir bar (2.5 cm × 1 cm), a short path distillation apparatus and a 10 mL pear-shaped collection flask (Figure 5). Distillation (Note 11) is carried out under reduced pressure (20 mmHg) to afford compound (**2**) (5.60 g, 85%) (Note 12) as a clear, colorless liquid (Note 13 and 14).

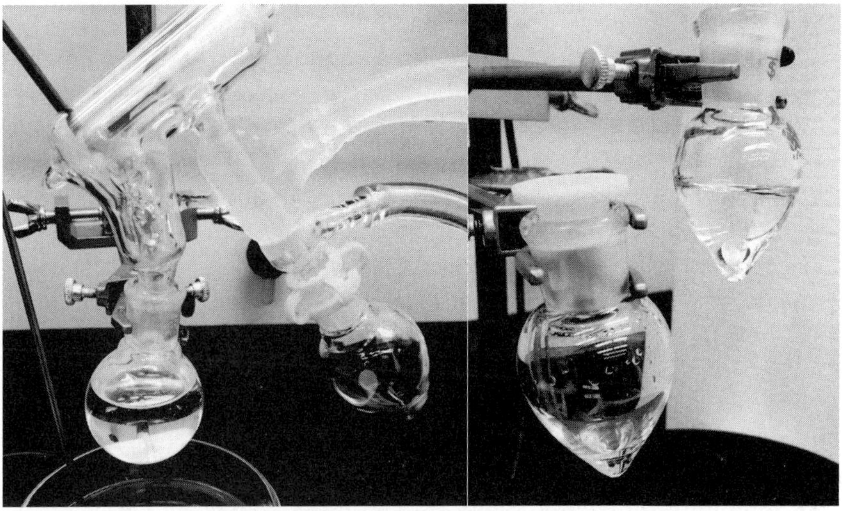

Figure 5. Distillation apparatus and pure final products

Notes

1. The submitters note that gently agitating the sieves by manually swirling the flask over the flame facilitates the removal of residual moisture. Heating was continued until the pressure in the vacuum line dropped below 0.05 mmHg.
2. The Submitters purchased anhydrous toluene (99.8%) in a Sure/Seal™ bottle from Sigma-Aldrich and used it as received. The Checkers used toluene (99.9%), purchased from EMD Millipore and passed it through a purification column (Glass Contour Solvent Systems by SG Water, USA) immediately prior to use.
3. 4-Pentyn-1-ylbenzene (98%) was purchased from GFS Chemicals and used as received.
4. PMHS was purchased from Acros Organics and used as received. The yellow color observed after addition of PMHS results from the formation of copper hydride. The Submitters found foaming is likely due to the formation of hydrogen gas, which can result from a reaction between copper hydride and residual moisture present in the reaction mixture.
5. Isobutanol, anhydrous (99.5%), was purchased from Sigma-Aldrich and used as received.

6. IPrCuCl (98%) was purchased from Strem Chemicals and used as received. The Submitters found this complex can be stored on the bench at ambient temperature and can be handled in air. KO*t*-Bu (1.0 M solution in THF) was purchased from Sigma-Aldrich and used as received.
7. GC analysis: The Submitters performed this analysis on a Shimadzu GC-2010 with a flame ionization detector and a SHRXI-5MS column (15 m x 0.25 mm x 0.25 μm). Temperature program: 60 °C hold 2 min, ramp to 160 °C at 13 °C/min, ramp to 250 °C at 30 °C/min, hold 5.50 min; t_R(**1**) = 5.71 min, t_R(**2**) = 5.40 min. The Checkers used an Agilent Technologies 7820A with flame ionization detector and an Agilent HP-5MS column (30 m x 0.25 mm x 0.25 μm). Temperature program: 60 °C hold 2 min, ramp to 160 °C at 13 °C/min, ramp to 250 °C at 30 °C/min, hold 5.50 min; t_R(**1**) = 8.00 min, t_R(**2**) = 7.69 min.
8. The Submitters purchased Celite 545 from J.T. Baker and used it as received. Silica gel (60 Å pore size, 230-400 mesh) was purchased from Agela Technologies and used as received. The Checkers purchased Celite 545 from Sigma-Aldrich and used it as received. Silica gel (60 Å pore size, 230-400 mesh) was purchased from SiliCycle and used as received.
9. The distillation is carried out under reduced pressure (61 °C/200 mmHg) with a bath temperature of 100 °C.
10. The Submitters note, at this stage, the major impurity consists of PMHS reaction byproducts, which are removed by a second silica gel filtration. The Submitters concentrated the filtrate using rotary evaporation (28 °C, 300 mmHg ramped down to 35 mmHg) to remove as much PhMe as possible. The Submitters found that for the isolation of non-polar compounds that are more mobile on silica gel than the PMHS-derived byproducts, it is essential to remove as much PhMe as possible prior to the next step of the purification sequence. The Submitters report that complete removal of PhMe can be acheived by re-suspending the crude reaction concentrate in 500 mL of hexanes and repeating the concentration using rotary evaporation (repeat if necessary). In attempting this procedure the checkers noted the loss of some product and thus employed distillation.
11. The distillation was performed as follows. Stirring was started and the pressure was gradually reduced to 20 mmHg. After vigorous bubbling due to loss of hexanes subsided, the distillation pot was heated to 105–125 °C. Two fractions were collected. The first fraction was of material

distilling from 62– 82 °C (5.16 g, 97.0% pure by QNMR). After removal of the vacuum and replacing the receiving flask, a 2nd fraction of material distilling from 82–83 °C at 20 mmHg was collected (0.44 g, 99.4% pure by QNMR) for a combined yield of 85% yield.

12. In a subsequent run it was found that collecting a single fraction distilling from 65–86 °C at 20 mmHg (bath temp. 100– 125 °C) furnished 5.56 g (84%) of product that was 97.3% pure by QNMR.
13. The Submitters performed the purification as follows: The desired alkene product (**2**) can be isolated by filtration through silica gel. A 350 mL Buchner filter funnel with a coarse frit is packed with celite (1 cm) followed by 180 g of silica gel, and then a layer of sand (1 cm). The crude reaction products are then dissolved in 33 mL of hexanes and loaded evenly over the top of the sand layer. An additional 33 mL of hexanes is used to recover remaining material from the flask, which is then poured evenly over the top of the sand layer. The silica gel plug is then washed with 1,200 mL of hexanes. The filtrate is collected in two 1 L round-bottomed flasks and concentrated using rotary evaporation (28 °C, 300 mmHg ramped down to 35 mmHg) to afford a clear, colorless liquid. Compound (**2**) is obtained in 87–92% yield (5.70–6.04 g) as a clear, colorless liquid. The purity is >99%(GC).
14. 4-Penten-1-ylbenzene (**2**) has the following properties: ^{1}H NMR (400 MHz, CD_2Cl_2) δ: 1.70 (tt, *J* = 7.7, 6.8 Hz, 2H). 2.09 (dtt, *J* = 7.7, 6.8, 1.5 Hz, 2H), 2.62 (t, *J* = 7.7 Hz, 2H), 4.97 (ddt, *J* = 10.4, 2.1, 1.5 Hz, 1H), 5.03 (ddt, *J* = 17.2, 2.1, 1.5 Hz, 1H), 5.85 (ddt, *J* = 17.2, 10.4, 6.8 Hz, 1H), 7.25–7.29 (m, 2H), 7.14-7.20 (m, 3H). ^{13}C NMR (100 MHz, $CDCl_3$) δ: 30.8, 33.4, 35.5, 114.8, 125.8, 128.4, 128.6, 138.8, 142.6. IR (neat, cm^{-1}): 3064, 3027, 2930, 2857, 1640, 1603, 1496, 1453, 1030, 991, 908, 741, 634. Elemental Analysis calculated for $C_{11}H_{14}$: C, 90.35; H, 9.65, Found: C, 90.16; H, 9.67. ESI-MS calculated for [M+H]+ 147.1168, found 147.1167. The purity was determined to be >97.0% by quantitative NMR spectroscopy with ethylene carbonate as the internal standard.

Working with Hazardous Chemicals

The procedures in *Organic Syntheses* are intended for use only by persons with proper training in experimental organic chemistry. All hazardous materials should be handled using the standard procedures for work with chemicals described in references such as "Prudent Practices in the Laboratory" (The National Academies Press, Washington, D.C., 2011; the full text can be accessed free of charge at http://www.nap.edu/catalog.php?record_id=12654). All chemical waste should be disposed of in accordance with local regulations. For general guidelines for the management of chemical waste, see Chapter 8 of Prudent Practices.

In some articles in *Organic Syntheses*, chemical-specific hazards are highlighted in red "Caution Notes" within a procedure. It is important to recognize that the absence of a caution note does not imply that no significant hazards are associated with the chemicals involved in that procedure. Prior to performing a reaction, a thorough risk assessment should be carried out that includes a review of the potential hazards associated with each chemical and experimental operation on the scale that is planned for the procedure. Guidelines for carrying out a risk assessment and for analyzing the hazards associated with chemicals can be found in Chapter 4 of Prudent Practices.

The procedures described in *Organic Syntheses* are provided as published and are conducted at one's own risk. *Organic Syntheses, Inc.*, its Editors, and its Board of Directors do not warrant or guarantee the safety of individuals using these procedures and hereby disclaim any liability for any injuries or damages claimed to have resulted from or related in any way to the procedures herein.

Discussion

The semi-reduction of alkynes to form alkenes is one of the most important reactions in organic chemistry. Traditional methods for achieving this transformation, such as Lindlar[5] or P-2 Nickel catalysts,[6] are complicated by a variety of factors including the *E*/*Z* isomerization of the alkene products[7] and over-reduction of the desired alkene to form alkanes.[8] These difficulties limit the practical use of these catalysts, as reactions must be carefully monitored to maximize conversion while minimizing over-reduction and isomerization. Furthermore, because the unreacted starting materials, alkene isomers, and over-reduced alkanes have very similar physical properties, purification of the desired alkene can be challenging.

Because of these complications, much effort has been devoted to the development of more selective techniques for alkyne semi-reduction. Several attempts have been made to modulate the reactivity of solid-state catalysts by using various poisons or by modifying the catalyst surface.[9] Other approaches using homogeneous catalysts have demonstrated improved selectivity using a variety of palladium,[8, 10] and copper[11] complexes in the reduction of arylacetylenes. Although significant progress has been made, these methods still suffer from one or more problems associated with isomerization, over-reduction, poor reactivity, limited functional group tolerance, the need for activating functional groups, high temperatures, or high pressures of molecular hydrogen. As a consequence, the Lindlar catalyst is still the preferred method for alkyne semi-reduction despite its drawbacks. The limitations of these methods provide motivation for the development of a convenient technique for the selective semi-reduction of alkynes that displays broad functional group compatibility with no over-reduction or isomerization.

The copper-catalyzed technique described here is an extension of previous work developed in our laboratory[12] and that of Tsuji.[11] This work is based on an earlier report by Sadighi,[13] who described the transfer of hydride from silicon to copper, and subsequent hydrocupration of an alkyne by copper hydride in a series of stoichiometric experiments. We showed that the resulting alkenyl copper species underwent protonation in the presence of a proton source to form an alkene and copper alkoxide complex. These observations led to our development of a copper-catalyzed semi-reduction of alkynes that is highly functional-group-tolerant, suffers from no over-reduction or isomerization of the alkene, and in the case of internal alkynes, is completely selective for the formation of the *Z* stereoisomer.[12a] As shown in Table 1, this approach can be used to selectively reduce both internal and terminal alkynes to the corresponding alkene in the presence of diverse functional groups, such as nitro, cyano, imide, carbamate, sulfonate, ester, silyl ether, hydroxyl, aryl iodide, aryl bromide, and alkyl and aryl chlorides. Importantly, there is no need for the alkyne to be adjacent to an activating group, no over-reduction to the alkane is detected, and the reduction of internal alkynes affords only the *Z* alkene.

Table 1. Representative alkenes obtained from the copper-catalyzed semi-reduction of internal and terminal alkynes (previous work).

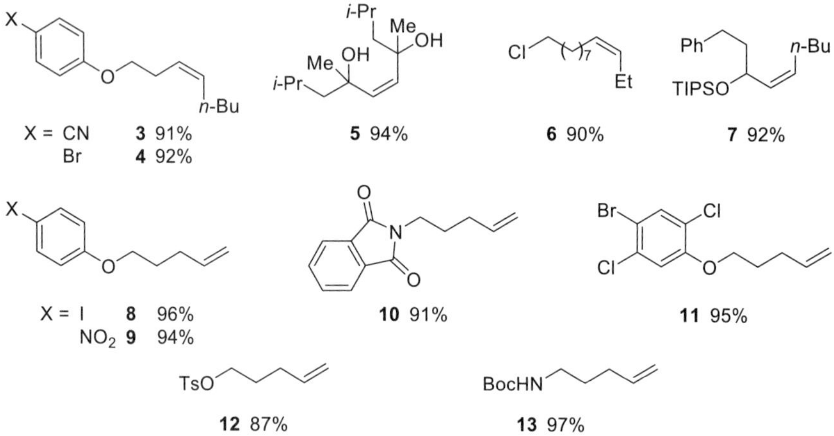

Previous conditions[12a] for internal alkyne reduction: 2.0 mol% IPrCuO*t*-Bu, 2.0 equiv PMHS, 2.5 equiv *t*-BuOH, toluene, 45 °C, 8 h. Previous conditions[12a] for terminal alkyne reduction: 0.5 mol% IPrCuO*t*-Bu, 1.2 equiv PMHS, 1.2 equiv *i*-BuOH, toluene, rt, 1 h.

Following our initial report,[12a] several additional developments have been made using NHC-Cu catalysts in the semi-reduction of alkynes. A variation using a $Cu(OAc)_2 \bullet H_2O$ pre-catalyst and IPr•HCl salt was provided by Zhong,[14] which avoids the need to prepare the air-unstable IPrCuO*t*-Bu catalyst. An alkoxide-tethered NHC-Cu catalyst was developed by Teichert,[15] which enabled the efficient semihydrogenation of arylacetylenes using H_2 as the terminal reducing agent. Using a similar approach, Sawamura[16] recently described a procedure for alkyne semihydrogenation using H_2, the SIPrCuO*t*-Bu catalyst, and a NaO*t*-Bu additive. Other recent examples of alkyne semi-reduction have been reported using palladium,[17] nickel,[18] and gold[19] catalysts. A transition-metal free approach to the *E*-selective semi-reduction of diarylalkynes using $Na_2S \bullet H_2O$ has also been reported.[20]

Based on our initial report,[12a] we developed a more practical procedure[12b] that allows the reaction to be carried out with a commercially available pre-catalyst, without the need for an inert-atmosphere glovebox, and with an improved purification scheme that in most cases allows the desired alkenes to be isolated with two simple filtration steps (Scheme 1). IPrCuO*t*-Bu is air-sensitive and is not commercially available; this challenge has prompted study of suitable pre-catalysts in our laboratory and by Zhong.[14] We found that the active catalyst can be efficiently generated *in situ* using the commercially available IPrCuCl complex and potassium *tert*-butoxide. Although the reaction is sensitive to the presence of moisture, briefly exposing the starting materials to molecular sieves and using standard air-free techniques offers a convenient alternative to rigorously drying each reagent or using a moisture and air-free glovebox.

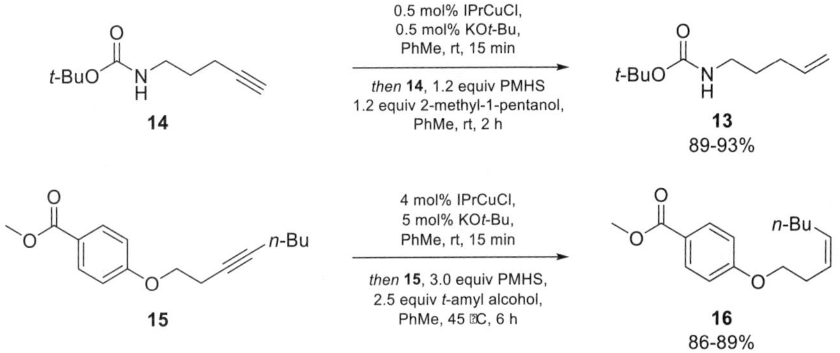

Scheme 1. Modified procedures for the glovebox-free semi-reduction of terminal and internal alkynes

The optimal reaction and purification conditions depend upon two factors: (1) the degree of substitution of the alkyne, and (2) the polarity of the alkene products. For the reduction of terminal alkynes, primary alcohols, such as isobutanol or 2-methyl-1-pentanol, should be used as the proton source (Scheme 1, top). For the reduction of internal alkynes, best results are achieved by increasing the catalyst loading to 4%, raising the temperature to 45 °C, and when using tertiary alcohols, such as *tert*-butanol or *tert*-amyl alcohol as the proton source (Scheme 1, bottom). With both terminal and internal alkynes, the choice of the alcohol depends on the polarity of the substrate and the purification procedure that will be used (see below).

The optimized purification scheme allows isolation the desired alkene products by two simple filtration steps in most cases. First, filtering the reaction mixture through a short plug of silica gel facilitates the removal of copper salts. After concentrating to remove toluene, filtration through a second plug of either silica gel or alumina allows for clean separation of

PMHS-derived polymer byproducts. For the semi-reduction of hydrophobic alkynes (Rf ≥ 0.3 in hexanes), more polar alcohols should be used as the proton source in the reaction: isobutanol for the reduction of terminal alkynes, and *tert*-butanol for the reduction of internal alkynes. Additionally, the second filtration step should be carried out using silica gel, as described in the Procedure section above. It should be noted that discretion should be applied in the amount of solvent used for the second filtration, as the polymer byproducts will elute following the desired alkenes.

Table 2. Summary of differences in procedure for the semi-reduction of alkynes based on substitution and polarity.

	Rf ≥ 0.3 in hexanes	Rf < 0.3 in hexanes
terminal alkyne	• 0.5 mol% catalyst • room temperature • isobutanol • 2nd filtration on silica gel	• 0.5 mol% catalyst • rt • 2-methyl-1-pentanol • 2nd filtration on alumina
internal alkyne	• 4 mol% catalyst • 45 °C • *tert*-butanol • 2nd filtration on silica gel	• 4 mol% catalyst • 45 °C • *tert*-amyl alcohol • 2nd filtration on alumina

For the semi-reduction of alkynes bearing polar functional groups (Rf < 0.3 in hexanes), less polar alcohols should be used as the proton source in the reaction: 2-methyl-1-pentanol for terminal alkynes, and *tert*-amyl alcohol for the reduction of internal alkynes. Additionally, the second filtration should be performed using alumina. We speculate that under these conditions, the resulting PMHS-derived byproducts are significantly more mobile on alumina than the desired alkene products, allowing for improved separation. Although slightly different protocols are used depending upon substitution and polarity of the alkyne, these adjustments do not significantly affect the procedure for reaction setup or purification. For clarity, these differences are summarized in Table 2.

In summary, these procedures offer a convenient and inexpensive method for the semi-reduction of alkynes to alkenes that is compatible with a wide variety of functional groups. These reactions reach full conversion without any over-reduction of the alkene to the alkane, avoiding the need for careful monitoring of reaction progress. In the case of internal alkynes, the semi-reduction is completely *Z*-selective and no isomerization is detected. Finally, this transformation can be achieved using commercially available reagents without the need for a glovebox, and the products can be easily isolated by filtration through silica or alumina.

References

1. Stanford ChEM-H, Stanford University, Stanford, California 94305, United States.
2. Department of Chemistry, Yale University, New Haven, Connecticut, 06520, United States.
3. Department of Process Chemistry, Merck & Co. Inc, Rahway, New Jersey 07065, United States.
4. Department of Chemistry, University of Washington, Seattle, Washington, 98195, United States. Email: lalic@chem.washington.edu, Phone (206) 616-8493. Funding by NSF is gratefully acknowledged (CAREER Award #1254636).
5. (a) Lindlar, H. *Helv. Chim. Acta* **1952,** *35*, 446–450; (b) Lindlar, H.; Dubuis, R. *Org. Synth. Coll.* **1973,** *5*, 880.
6. (a) Brown, C. A. *Chem. Commun.* **1970**, 139–140; (b) Brown, C. A.; Ahuja, V. K. *J. Org. Chem.* **1973,** *38* (12), 2226–2230.
7. (a) Ulan, J. G.; Maier, W. F.; Smith, D. A., *J. Org. Chem.* **1987,** *52* (14), 3132–3142; (b) Molnár, Á.; Sárkány, A.; Varga, M. *J. Mol. Catal. A: Chem.* **2001,** *173* (1–2), 185–221; (c) Oger, C.; Balas, L.; Durand, T.; Galano, J.-M. *Chem. Rev.* **2013,** *113* (3), 1313–1350.
8. (a) van Laren, M. W.; Elsevier, C. J. *Angew. Chem. Int. Ed.* **1999,** *38* (24), 3715–3717; (b) Kluwer, A. M.; Koblenz, T. S.; Jonischkeit, T.; Woelk, K.; Elsevier, C. J. *J. Am. Chem. Soc.* **2005,** *127* (44), 15470–15480.
9. (a) Nishio, R.; Sugiura, M.; Kobayashi, S. *Org. Biomol. Chem.* **2006,** *4* (6), 992–995; (b) Panpranot, J.; Phandinthong, K.; Sirikajorn, T.; Arai, M.; Praserthdam, P. *J. Mol. Catal. A: Chem.* **2007,** *261* (1), 29–35; (c) Dominguez-Dominguez, S.; Berenguer-Murcia, A.; Pradhan, B. K.; Linares-Solano, A.; Cazorla-Amoros, D. *J. Phys. Chem. C* **2008,** *112* (10), 3827–3834; (d) Hori, J.; Murata, K.; Sugai, T.; Shinohara, H.; Noyori, R.; Arai, N.; Kurono, N.; Ohkuma, T. *Adv. Synth. Catal.* **2009,** *351* (18), 3143–3149; (e) Weerachawanasak, P.; Mekasuwandumrong, O.; Arai, M.; Fujita, S.-I.; Praserthdam, P.; Panpranot, J. *J. Catal.* **2009,** *262* (2), 199–205;

(f) Armbrüster, M.; Kovnir, K.; Behrens, M.; Teschner, D.; Grin, Y.; Schlögl, R. *J. Amer. Chem. Soc.* **2010,** *132* (42), 14745–14747; (g) Yabe, Y.; Yamada, T.; Nagata, S.; Sawama, Y.; Monguchi, Y.; Sajiki, H. *Adv. Synth. Catal.* **2012,** *354* (7), 1264–1268.

10. (a) Hauwert, P.; Maestri, G.; Sprengers, J. W.; Catellani, M.; Elsevier, C. J. *Angew. Chem. Int. Ed.* **2008,** *47* (17), 3223–3226; (b) Drost, R. M.; Bouwens, T.; van Leest, N. P.; de Bruin, B.; Elsevier, C. J. *ACS Catalysis* **2014,** *4* (5), 1349–1357.
11. Semba, K.; Fujihara, T.; Xu, T.; Terao, J.; Tsuji, Y. *Adv. Synth. Catal.* **2012,** *354* (8), 1542–1550.
12. (a) Whittaker, A. M.; Lalic, G. *Org. Lett.* **2013,** *15* (5), 1112–1115; (b) Cox, N.; Dang, H.; Whittaker, A. M.; Lalic, G. *Tetrahedron* **2014,** *70* (27–28), 4219–4231.
13. Mankad, N. P.; Laitar, D. S.; Sadighi, J. P. *Organometallics* **2004,** *23* (14), 3369–3371.
14. Wang, G.-H.; Bin, H.-Y.; Sun, M.; Chen, S.-W.; Liu, J.-H.; Zhong, C.-M. *Tetrahedron* **2014,** *70* (12), 2175–2179.
15. Pape, F.; Thiel, N. O.; Teichert, J. F. *Chem. Eur. J.* **2015,** *21* (45), 15934–15938.
16. Wakamatsu, T.; Nagao, K.; Ohmiya, H.; Sawamura, M. *Organometallics* **2016,** *35* (10), 1354–1357.
17. (a) Planellas, M.; Guo, W.; Alonso, F.; Yus, M.; Shafir, A.; Pleixats, R.; Parella, T. *Adv. Synth. Catal.* **2014,** *356* (1), 179–188; (b) Slack, E. D.; Gabriel, C. M.; Lipshutz, B. H. *Angew. Chem. Int. Ed.* **2014,** *53* (51), 14051–14054; (c) Verho, O.; Zheng, H.; Gustafson, K. P. J.; Nagendiran, A.; Zou, X.; Bäckvall, J. E. *Chem. Cat. Chem.* **2016,** *8* (4), 773–778; (d) Zhong, J.-J.; Liu, Q.; Wu, C.-J.; Meng, Q.-Y.; Gao, X.-W.; Li, Z.-J.; Chen, B.; Tung, C.-H.; Wu, L.-Z. *Chem. Commun.* **2016,** *52* (9), 1800–1803.
18. Richmond, E.; Moran, J. *J. Org. Chem.* **2015,** *80* (13), 6922–6929.
19. (a) Vasilikogiannaki, E.; Titilas, I.; Vassilikogiannakis, G.; Stratakis, M. *Chem. Commun.* **2015,** *51* (12), 2384–2387; (b) Wagh, Y. S.; Asao, N. *J. Org. Chem.* **2015,** *80* (2), 847–851; (c) Liang, S.; Hammond, G. B.; Xu, B. *Chem. Commun.* **2016,** *52* (35), 6013–6016.
20. Chen, Z.; Luo, M.; Wen, Y.; Luo, G.; Liu, L. *Org. Lett.* **2014,** *16* (11), 3020–3023.

Appendix
Chemical Abstracts Nomenclature (Registry Number)

Chloro[1,3-bis(2,6-di-i-propylphenyl)imidazol-2-ylidene]copper(I); (578743-87-0)
Isobutanol (2-methyl-1-propanol); (78-83-1)
4-Pentyn-1-ylbenzene (1823-14-9)
Potassium *tert*-butoxide; (865-47-4)
Polymethylhydrosiloxane; (9004-73-3)

Gojko Lalic grew up in Belgrade, Serbia where he completed his B.Sc. in chemistry. After he obtained a Ph.D. from Harvard University under the mentorship of Prof. Matthew Shair, he completed a postdoctoral fellowship with Prof. Robert Bergman at UC Berkeley. After another postdoctoral fellowship with Prof. E. J. Corey, he joined the faculty at the University of Washington as an assistant professor in fall 2008.

Nick Cox received his B.S. in Chemical Engineering from the University of Washington in 2007. After spending a year in industry, he returned to the University of Washington, where he completed his Ph.D. in Chemistry in 2013 under the mentorship of Professor Gojko Lalic. Nick is currently working as a postdoctoral scholar in medicinal chemistry at Stanford ChEM-H under the supervision of Professor Chaitan Khosla.

Hester Dang received her B.S. in Chemistry from Syracuse University in 2009. In 2010, she joined the University of Washington to pursue her Ph.D. in chemistry. As a member of Professor Gojko Lalic's group, she developed and investigated metal-catalyzed reactions. In the summer of 2016, Hester received her Ph.D. and has joined Seth Herzon's group as postdoctoral fellow at Yale University.

Aaron Whittaker received his B.S. from Northern Arizona University in 2008 where he studied the toxicity of depleted uranium under the supervision of Prof. Diane Stearns. He then moved to the University of Washington to develop copper-catalyzed transformations with Prof. Gojko Lalic where he received his Ph.D. in 2013. Aaron then did postdoctoral research on nickel catalyzed transfer hydrogenations with Prof. Vy Dong. He is currently a Senior Scientist in the Process Chemistry department at Merck Research Laboratories.

Miyu Furuta received her B.S. degree (2012) and her Ph.D. degree (2016) from Keio University under the direction of Professor Takeshi Sugai. She is a JSPS research fellow from 2015 and currently a post-doctoral fellow in the laboratory of Professor John L. Wood at Baylor University.

Organic Syntheses

Synthesis of Cyclobutanes by Lewis Acid-Promoted Ketene-Alkene [2+2] Cycloadditions

Christopher M. Rasik, Eleni M. Salyers, and M. Kevin Brown[1*]

Department of Chemistry, Indiana University, 800 E. Kirkwood Avenue, Bloomington, Indiana 47405, United States

Checked by Charel Prost, Estíbaliz Merino, and Cristina Nevado

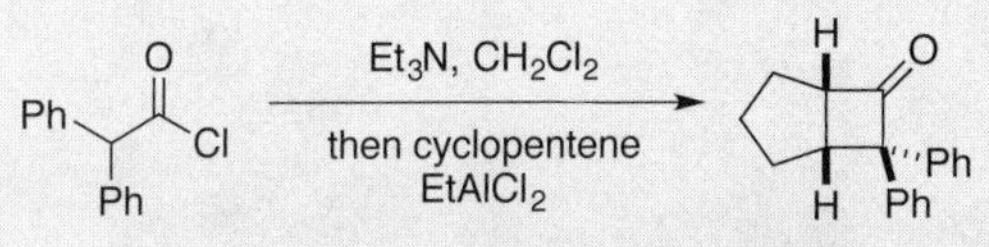

1 2

Procedure

A. *7,7-Diphenylbicyclo[3.2.0]heptan-6-one* (**2**). A two-necked, 500 mL round-bottomed flask equipped with a magnetic stir bar (1.9 x 5.0 cm, Teflon-coated, egg-shaped) is flame-dried under vacuum and then allowed to cool. The flask is backfilled with nitrogen and diphenylacetyl chloride (**1**, 9.68 g, 37.8 mmol, 1.00 equiv, 90% technical grade), is weighed out on the benchtop, and added (Note 1). The flask is evacuated and backfilled with nitrogen three times, and then fitted with a rubber septum and Schlenk nitrogen line. Dry dichloromethane (38.0 mL) and triethylamine (5.4 mL, 38.7 mmol, 1.02 equiv) are added by syringe through the septum and the mixture became yellow and heterogeneous (Figure 1) within minutes (Notes 2, 3, and 4). The reaction is stirred at room temperature for 30 min. The flask

DOI: 10.15227/orgsyn.093.0401
Published on the Web 11/9/2016

Figure 1. Reaction Mixture after Addition of Triethylamine

is placed in a dry ice/acetone bath (-78 °C) and cyclopentene (7.0 mL, 79.2 mmol, 2.1 equiv) is added down the walls of the flask by syringe through the septum (Note 5). The rubber septum is then quickly replaced with a 250 mL addition funnel equipped with a rubber septum that had been oven-dried and flushed with nitrogen. Ethylaluminum dichloride (95.0 mL, 95.0 mmol, 2.51 equiv, 1 M in hexanes) is added by syringe through the septum to the addition funnel (Notes 6 and 7). The solution of ethylaluminum dichloride is added dropwise to the flask over 50 min with continuous stirring and cooling (Notes 8, 9, and 10). After the addition, the reaction is stirred for 1 h at –78 °C (Figure 2).

Figure 2. Reaction Mixture after Addition of Ethylaluminum Dichloride

The addition funnel is removed and triethylamine (5 mL) is added down the walls of the flask over 5 min to quench the reaction. Deionized water (20 mL) is added directly to the reaction dropwise over 15 min. The flask is removed from the dry ice/acetone bath and the reaction is stirred until it reached room temperature (Note 11). As the solution warmed the color changed from brown to a light yellow. When the reaction reaches room temperature (~30 min) diethyl ether (100 mL) is added to the flask followed by 1 mL portion-wise addition of 1M HCl (100 mL) over 10 minutes with pipette (Notes 12 and 13) (Figure 3). The mixture stirred until it became mostly homogenous (~15 min). The biphasic mixture is then transferred to a 1000 mL separatory funnel. Diethyl ether (10 mL) and 1M HCl (10 mL) are used to rinse the round-bottomed flask and are then added to the separatory funnel. The aqueous layer is separated from the organic layer. The aqueous layer is washed with diethyl ether (100 mL). The layers are separated again; the organic phases are combined, and then are washed with 1M sodium hydroxide (100 mL) (Note 14). The separated organic layer is dried with sodium sulfate (50 g) over 10 min, gravity filtered through filter paper into a 500 mL round-bottomed flask and concentrated via rotary evaporator (ca. 20 mmHg, water bath temperature 30 °C) (Note 15).

Figure 3. Reaction Mixture after Addition of HCl

The crude residue is purified by a combination of silica gel flash column chromatography and recrystallization. Column chromatography using EtOAc and hexanes as eluent (Notes 16, 17 and 18) yielded the title compound as an air and moisture stable, light yellow amorphous solid, containing a small amount of impurities.

The chromatographed mixture of product and impurities is purified by recrystallization using hexanes and diethyl ether. To the crude solid in a 250 mL round-bottomed flask is added hexanes (50 mL) and diethyl ether (10 mL), and the resulting solution is swirled vigorously and gently warmed with a heat gun until all solids dissolved. The flask is loosely capped and left to sit at room temperature for 24 h while white needles form (Note 19). The mother liquor is decanted away into a 100 mL round-bottomed flask and the crystals are collected via vacuum filtration on a fritted Büchner funnel. The resulting crystals are transferred to a pre-weighed 250 mL round-bottomed flask and dried at 0.1 mmHg for 3 h to yield **2** (5.29 g, 20.16 mmol, 53%) as air and moisture stable white needles (Note 20) (Figure 4).

Figure 4. White Crystalline Product

An additional crop of crystals could be obtained by a second recrystallization. The mother liquor is concentrated in a 100 mL round-bottomed flask via rotary evaporator (ca. 20 mmHg, water bath temperature 30 °C) to yield a white solid. To the crude solid is added hexanes (15 mL) and diethyl ether (3 mL) and the resulting solution is swirled vigorously and gently heated with a heat gun until all solids dissolved. The flask is loosely capped and left to sit at room temperature for 24 h while white needles form. The mother liquor is decanted away and the crystals are collected via vacuum filtration on a fritted Büchner funnel. The resulting crystals are transferred to a pre-weighed 100 mL round-bottomed flask and dried at 0.1 mmHg for 3 h to yield **2** (1.09 g, 4.15 mmol, 11%, total of 64% yield) (Notes 20 and 21).

Notes

1. Diphenylacetyl chloride (90% technical grade) is purchased from Sigma Aldrich and used as received.
2. Dichloromethane was collected in an oven-dried (140 °C for 24 h) 250 mL Schlenk flask, which had been purged with argon atmosphere by three evacuation-backfill cycles from a dry solvent system (Innovative Technology).

3. Triethylamine (>99.5%) was purchased from Sigma Aldrich and used as received.
4. Upon the addition of triethylamine the solution becomes a canary yellow color and a white solid forms within minutes.
5. Cyclopentene (97%) was purchased from Alfa Aesar and used as received.
6. Ethylaluminum dichloride (1 M in hexane) was purchased from Sigma Aldrich and used as received.
7. Excess ethylaluminum dichloride (>1 equiv) is necessary due to product inhibition and reaction with Et_3NHCl. For details see Ref 3a.
8. Transfer and addition of ethylaluminum dichloride should be performed carefully, in small portions, since it reacts violently with water. The transfer of ethylaluminum dichloride to the addition funnel typically requires ~15 min.
9. The rate of the addition was held at approximately one drop every two seconds.
10. During the addition, the reaction smokes and the mixture becomes a dark red/black color.
11. After the bath is removed, vigorous stirring is maintained to prevent excess bubbling/bumping.
12. The quenching process must be completed very slowly to avoid excessive bubbling and bumping.
13. Hydrogen chloride (12 M aqueous) was purchased from Merck and was used as received.
14. Sodium hydroxide was purchased from Honeywell and was used as received.
15. Sodium sulfate was purchased from Merck and was used as received.
16. Thin layer chromatograph was performed on silica gel 60 F_{254} TLC plate (Merck TLC Silica Gel 60 Glass Plates, with 20% EtOAc in hexanes as eluent. The following R_f value was observed (visualized under 254 nm UV light and $KMnO_4$): R_f(product): 0.68 Picture of TLC plate is shown below (A: mixture after column chromatography; B: product after recrystallization (Figure 5).

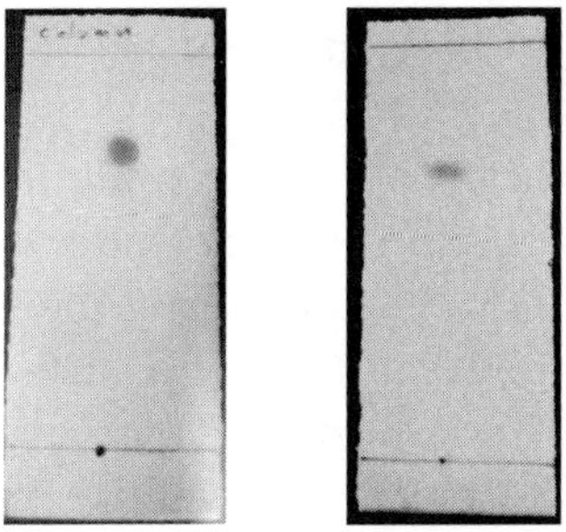

A B

Figure 5. TLC Analysis of A) Product Mixture after Column and B) Product after Recrystallization

17. Silica gel (pore size 60 Å, 230-400 mesh particle size, 40-63 μm particle size) was purchased from Sigma Aldrich (catalog number: 60737) and used as received.
18. Dichloromethane (10 mL) was added to the crude mixture. The solution of crude product was loaded onto a column packed with silica gel slurry in hexanes. Column is 9 cm in diameter, height of silica gel is 20 cm. Gradient elution using hexanes (50 mL), 5% EtOAc in hexanes (100 mL), 10% EtOAc in hexanes (200 mL), 15% EtOAc in hexanes (240 mL), 30% EtOAc in hexanes (580 mL), 40% EtOAc in hexanes (320 mL), and 50% EtOAc in hexanes (800 mL) were collected in 65 mL fractions. All fractions containing product (including mixed fractions) were collected.
19. Crystals typically appeared within 1 h.
20. The product was isolated as a white needles and had the following characteristics: mp = 90–92 °C; ^{1}H NMR (400 MHz, $CDCl_3$) δ: 1.21–1.36 (m, 1H), 1.51–1.70 (m, 3H), 1.77–1.87 (m, 1H), 2.15 (dd, J = 11.2, 6.3 Hz, 1H), 3.70–3.75 (m, 2H), 7.18–7.21 (m, 2H), 7.26–7.30 (m, 4H), 7.32–7.35 (m, 2H), 7.44–7.47 (m, 2H); ^{13}C NMR (101 MHz, $CDCl_3$) δ: 26.4, 29.4, 29.5, 41.6, 61.88, 75.9, 126.5, 126.8, 127.0, 128.4, 128.7, 140.1, 141.7, 214.4. IR (neat): 2950, 2932, 2864, 1760, 1490, 1445, 1148, 1079, 1031, 834, 743, 695, 662, 604, 567, 529, 512, 473 cm$_{-1}$. HRMS (MM: EI): m/z calcd for $C_{19}H_{19}O_{1}$ $[M]^{+}$: 263.1430. Found: 263.1432. Anal. Calcd for $C_{19}H_{18}O_{1}$: C, 86.99; H, 6.92; Found: C, 87.08; H, 6.87.
21. A reaction performed on half-scale provided 3.30 g (67%) of the desired product.

Working with Hazardous Chemicals

The procedures in *Organic Syntheses* are intended for use only by persons with proper training in experimental organic chemistry. All hazardous materials should be handled using the standard procedures for work with chemicals described in references such as "Prudent Practices in the Laboratory" (The National Academies Press, Washington, D.C., 2011; the full text can be accessed free of charge at http://www.nap.edu/catalog.php?record_id=12654). All chemical waste should be disposed of in accordance with local regulations. For general guidelines for the management of chemical waste, see Chapter 8 of Prudent Practices.

In some articles in *Organic Syntheses*, chemical-specific hazards are highlighted in red "Caution Notes" within a procedure. It is important to recognize that the absence of a caution note does not imply that no significant hazards are associated with the chemicals involved in that procedure. Prior to performing a reaction, a thorough risk assessment should be carried out that includes a review of the potential hazards associated with each chemical and experimental operation on the scale that is planned for the procedure. Guidelines for carrying out a risk assessment and for analyzing the hazards associated with chemicals can be found in Chapter 4 of Prudent Practices.

Discussion

Ketene-alkene [2+2] cycloadditions represent a useful method to construct cyclobutanones, a valuable functionality in organic synthesis.[2] These reactions have predominately relied on the inherent reactivity of each coupling partner. Recently, we disclosed Lewis acid-promoted [2+2] ketene-alkene cycloadditions.[3] These reactions demonstrated improved reactivity, stereo- and regioselectivity compared to the thermal cycloadditions. This method is exemplified by the above procedure and is applicable for a variety of alkenes and ketenes (Table 1 **3-7**).

Table 1. Representative Scope of Lewis acid-promoted ketene-alkene [2+2] cycloadditions

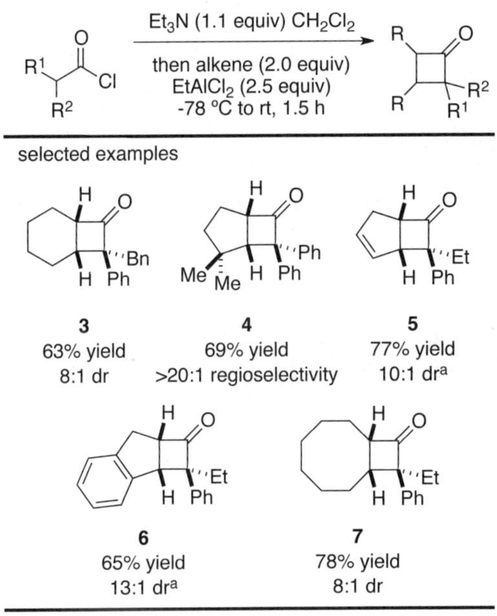

[a] Reaction was performed at –78 °C for 3 h.

Several points regarding the Lewis acid-promoted cycloadditions are noteworthy (Table 2). (1) When compared to thermally-induced reactions, Lewis acid-promoted variants give rise to increased yields for unactivated alkenes. For example, Lewis acid-promoted conditions yield cycloadduct **8** in 84% yield and 13:1 dr. The corresponding thermally-induced cycloaddition requires forcing conditions that yield cyclobutanone **8** in only 5% yield and 1:1 dr. (2) Increased selectivity is also observed. Lewis acid-promoted cycloaddition yields **9** in 18:1 dr while the thermal cycloaddition yields **9** in 40% in 2:1 dr. (3) Perhaps the most intriguing feature, is the reversal in diastereoselectivity for aryl/alkyl ketene cycloadditions under Lewis acid conditions. Lewis acid-promoted cycloaddition yields **10a** in 59% yield and 7:1 dr. The corresponding thermal cycloaddition, however, yields the opposite diastereomer, **10b**, as major in 71% yield and 6:1 dr.

Table 2. Comparison of thermal and Lewis acid-promoted ketene-alkene [2+2] cycloadditions

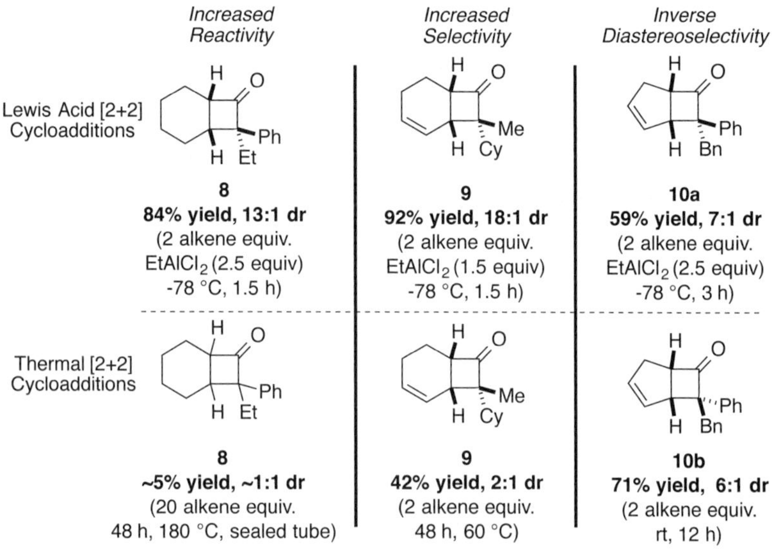

References

1. Department of Chemistry, Indiana University, 800 E. Kirkwood Ave. Bloomington, IN 47401. We thank the NIH (RO1GM110131) and Indiana University for support of this work.
2. (a) Lee-Ruff, E.; Mladenova, G. *Chem. Rev.* **2003**, *103*, 1449. (b) Namyslo, J. C.; Kaufmann, D. E. *Chem. Rev.* **2003**, *103*, 1485.
3. (a) Rasik, C. M.; Brown, M. K. *J. Am. Chem. Soc.* **2013**, *135*, 1673. (b) Rasik, C.; Brown, M. K. *Synlett* **2014**, *25*, 760. (c) Rasik, C. M.; Hong, Y. J.; Tantillo, D. J.; Brown, M. K. *Org. Lett.* **2014**, *16*, 5168.

Appendix
Chemical Abstracts Nomenclature (Registry Number)

Diphenylacetyl chloride: Benzeneacetyl chloride, α-phenyl- (1871-76-7)
Dichloromethane: Methane, dichloro-; (75-09-2)
Triethylamine: Ethanamine, N,N-diethyl-; (121-44-8)
Cyclopentene: cyclopentene; (142-29-0)
Ethylaluminum dichloride: Aluminum, dichloroethyl- ; (563-43-9)

Christopher M. Rasik was born in Swanton, Ohio. He obtained his B.A. at The Ohio State University and began his graduate work under the tutelage of Prof. M. Kevin Brown at Indiana University in 2011.

Eleni M. Salyers was born in Indianapolis, Indiana. She is currently an undergraduate student at Indiana University studying biochemistry.

Kevin Brown grew up in the suburbs of Chicago. Kevin received his B.A. degree from Hamilton College in 2002 and then moved to Boston College to pursue graduate studies under the mentorship of Professor Amir Hoveyda. Upon completion of his graduate studies he began a postdoctoral fellowship in the laboratories of E. J. Corey at Harvard University. In 2011 he began his independent career at Indiana University as an assistant professor.

Charel Prost graduated wth a degree in Chemistry from the University of Zurich in 2015. He is currently a Masters student in the group of Prof. Cristina Nevado at the University of Zurich.

Estíbaliz Merino obtained her Ph.D. degree from the Autónoma University (Madrid-Spain). After a postdoctoral stay with Prof. Magnus Rueping at Goethe University Frankfurt and RWTH-Aachen University in Germany, she worked with Prof. Avelino Corma in Instituto de Tecnología Química-CSIC (Valencia) and Prof. Félix Sánchez in Instituto de Química Orgánica General-CSIC (Madrid) in Spain. At present, she is research associate in Prof. Cristina Nevado´s group in University of Zürich. She is interested in the synthesis of natural products using catalytic tools and in the development of new materials with application in heterogeneous catalysis.

Organic Syntheses

Synthesis of a 2,2-Dichloroimidazolidine-4,5-dione and its Application in a Chlorodehydroxylation

David Schilter*[a,b] and Christopher W. Bielawski[1*a,b]

[a]Center for Multidimensional Carbon Materials (CMCM), Institute for Basic Science (IBS), Ulsan 44919, Republic of Korea; [b]Department of Chemistry, Ulsan National Institute of Science and Technology (UNIST), Ulsan 44919, Republic of Korea

Checked by Jo Balsamo, Christopher Borths, and Margaret Faul

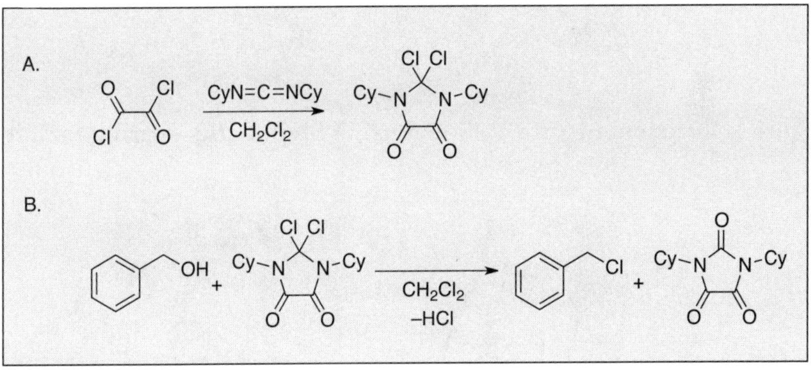

Procedure

A. *2,2-Dichloro-1,3-dicyclohexylimidazolidine-4,5-dione* (**1**Cl_2). An oven-dried 250 mL 24/40 round-bottomed flask (Note 1), equipped with a 3 cm Teflon-coated magnetic stir bar, is sealed with a rubber septum. A Teflon-coated thermocouple is inserted through the septum and the flask is cooled under a stream of N_2 delivered via a Schlenk line through a needle inserted into the septum. The septum is briefly removed, *N,N'*-dicyclohexylcarbodiimide (12 g, 58 mmol) (Note 2) is added to the flask, and the septum with the thermocouple is reattached to the flask. Dichloromethane (125 mL) (Note 3) is added via cannula and the resulting

DOI: 10.15227/orgsyn.093.0413

Figure 1. Reaction mixture obtained using Step A after stirring for 2 h

Figure 2. Cannula filtration setup used in Step A

mixture is stirred to effect dissolution. The mixture is cooled to *ca.* –5 °C in an ice/NaCl bath, after which oxalyl chloride (7.9 g, 5.4 mL, 61 mmol, 1.05 equiv) (Note 4) is added dropwise over 10 min using a 5 mL gastight syringe (Note 5). The solution is allowed to stir for 2 h, during which most of the ice melts and the reaction mixture develops a pale yellow color (Note 6) (Figure 1). The septum is removed, the flask is quickly attached to a rotary evaporator, and the solution is evaporated to dryness. Upon completion of the evaporation, the flask is blanketed with nitrogen while still on the rotary evaporator. The flask is then quickly removed and capped with a septum. A septum is attached to the flask and the remaining off-white powder is washed with hexane (4 × 50 mL) that is added via syringe (Note 7). The source flask is kept under nitrogen and house vacuum (ca. 25-40 mmHg) is used as needed to remove the solvent through the filter cannula (Figure 2). The washings sequentially become less yellow in color. The resulting solids are dried under vacuum (<2 mmHg) to give product **1**Cl_2 as a white powder (16.5 g, 49.3 mmol, 85%) (Notes 8, 9, and 10) of 99% purity (Note 11).

B. *Benzyl chloride.* A 500 mL 24/40 round-bottomed flask equipped with a 3 cm Teflon-coated magnetic stir bar (Note 1) is sealed with a rubber septum and cooled under a stream of N_2 at a Schlenk line. Dichloromethane (250 mL) (Note 2) is added via cannula and benzyl alcohol (5.0 g, 4.8 mL, 46 mmol) (Note 12) is added, with stirring, using a 5 mL gastight syringe. 2,2-Dichloro-1,3-dicyclohexylimidazolidine-4,5-dione (**1**Cl_2) (16.4 g, 49.3 mmol, 1.1 equiv) is quickly weighed out (Note 13) and then promptly added using a powder funnel, which is then replaced with a septum immediately after the addition. The resulting solution is stirred for an additional 3 h (Note 14), after which it is transferred to a 500 mL 24/40 round-bottomed flask and the volatiles are removed on a rotary evaporator (Notes 15 and 16). The crude material is purified using column chromatography on SiO_2 (column dimensions: 8 cm high × 3.5 cm diameter) (Note 17) packed in pentane (Note 18). While eluting with pentane, the first UV-active band is collected and the volatiles are removed on a rotary evaporator (Note 15) to give benzyl chloride as a colorless oil (5.1 g, 40 mmol, 86% yield, 99% purity as determined by quantitative NMR spectroscopy) (Notes 19 and 20).

Notes

1. Due to the sensitivity of the reagents, the Submitters recommend the use of a Schlenk flask equipped with a vacuum pump to facilitate evaporation through the Schlenk line. Higher yields (94-96%) were reported by the Submitters when using a Schlenk flask and vacuum line for solvent removal.
2. *N,N'*-Dicyclohexylcarbodiimide (Alfa Aesar, 99%), as received, may be rather clumpy in nature. Large crystals can be broken up with a metal spatula under a cone of N_2.
3. New bottles of anhydrous HPLC-grade CH_2Cl_2 (Sigma-Aldrich) were sparged with N_2 prior to use. Karl-Fischer titration determined that the water content in the solvent was <0.001%. The Submitters used CH_2Cl_2 that was dried over alumina in a solvent purification system, such that the final moisture content is 1.1 ppm according to Karl-Fischer titration.
4. Oxalyl chloride (Alfa Aesar, 98%) is used as received. Exposure to air should be limited and the liquid should be taken up in the syringe as quickly as is safe and practical.
5. The Submitters oven-dried (110 °C) the syringe and the 20 gauge Luer-lock needle for an hour and then cooled the instruments under N_2 before use. The Checkers did not perform this step.
6. The punctured septum is replaced with a new septum and secured with a copper wire. The same is done after the cannula filtration. After 2 h, an aliquot of the reaction mixture is subjected to LC-MS analysis, which confirms formation of 2,2-dichloro-1,3-dicyclohexylimidazolidine-4,5-dione.
7. Hexane (Sigma-Aldrich, HPLC grade) was sparged with N_2 prior to use. Karl-Fischer titration indicated that the water content in the solvent was <0.01%. The fourth wash was almost colorless.
8. The solids are dispersed by stirring, after which they are allowed to settle. The filter cannula barrel is then immersed in the supernatant and the pale yellow solution is removed under a positive pressure of N_2.
9. A separate reaction performed on a similar scale yielded 18.05 g (88%) of the same product.
10. Analytical data for **1**Cl_2: ^1H NMR (400 MHz, $CDCl_3$) δ: 1.33 (m, 6H), 1.69 (m, 2H), 1.93 (m, 8H), 2.25 (m, 4H), 3.74 (tt, *J* = 12.3, 3.8 Hz, 2H). ^{13}C NMR (400 MHz, $CDCl_3$) δ: 24.7, 25.9, 28.8, 57.0, 103.2, 154.4. ^1H and ^{13}C NMR data referenced to residual solvent signals at 7.26 and

77.2 ppm, respectively. IR (film): 2936, 2858, 1732, 1412, 1372, 1319, 1260, 1186, 1117, 894, 871 cm^{-1}. mp 176–178 °C (dec.).

11. The product was determined by quantitative NMR spectroscopy to be 98.7% pure when benzyl benzoate was used as an internal standard.
12. Benzyl alcohol (Alfa Aesar, 99%) is used as received.
13. Solid $\mathbf{1}Cl_2$ can be briefly handled in air (e.g., for weighing), but is best stored in a Schlenk flask under dry N_2 or in a vial inside a desiccator. Moisture induces the formation of its colorless hydrolysis product according to the reaction: $\mathbf{1}Cl_2$ + H_2O → 1,3-dicyclohexylimidazolidine-1,4,5-trione (**1O**) + 2 HCl. Analytical data for **1O**: ^{1}H NMR ($CDCl_3$, 400 MHz) δ: 1.27 (m, 6H), 1.68 (m, 2H), 1.74 (m, 4H), 1.86 (dt, J = 13.4, 3.0 Hz, 4H), 2.07 (m, 4H), 4.01 (tt, J = 12.4, 3.9 Hz, 2H).
14. LC-MS was used to monitor the reaction until its completion.
15. If the reaction has gone to completion, any losses in yield may be due to evaporation of the product during workup.
16. P = 75 mmHg and an ice/water bath is used. Extended application can result in excessive evaporation of benzyl chloride. In the Checker's experimental run, residual solids, presumably of the chlorinating reagent, remained in the round-bottomed flask (Figure 2). These were removed using a syringe filter after mixing benzyl chloride with pentane.
17. SiO_2 (Merck, Silica Gel 60, 60 – 200 μ m) is used as received. Although the authors used silica gel chromatographs, the Checkers performed the purification using a Biotage chromatography system equipped with a Biotage 50 g Snap Ultra KP-Sil flash chromatography cartridge.
18. Pentane (Sigma-Aldrich, HPLC grade) is used as received. The crude product consists of an oil and solids. Rather than directly pipetting the oil onto the column, it is best taken up in pentane, transferred to the column, and then washed with additional pentane to ensure that all of the benzyl chloride is loaded.
19. Analytical data for benzyl chloride: ^{1}H NMR (400 MHz, $CDCl_3$) δ: 4.61 (s, 2H) 7.30–7.42 (m, 5H). ^{13}C NMR (400 MHz, $CDCl_3$) δ: 45.9, 128.6, 137.5. IR (film): 676 cm^{-1}. R_f (SiO_2/pentane) = 0.7. Quantitative NMR spectroscopy was used to determine the purity by mass of the isolated material against benzyl benzoate (Sigma-Aldrich) as a standard.
20. A separate reaction on the same scale afforded 5.84 g (86%) of benzyl chloride.

Working with Hazardous Chemicals

The procedures in *Organic Syntheses* are intended for use only by persons with proper training in experimental organic chemistry. All hazardous materials should be handled using the standard procedures for work with chemicals described in references such as "Prudent Practices in the Laboratory" (The National Academies Press, Washington, D.C., 2011; the full text can be accessed free of charge at http://www.nap.edu/catalog.php?record_id=12654). All chemical waste should be disposed of in accordance with local regulations. For general guidelines for the management of chemical waste, see Chapter 8 of Prudent Practices.

In some articles in *Organic Syntheses*, chemical-specific hazards are highlighted in red "Caution Notes" within a procedure. It is important to recognize that the absence of a caution note does not imply that no significant hazards are associated with the chemicals involved in that procedure. Prior to performing a reaction, a thorough risk assessment should be carried out that includes a review of the potential hazards associated with each chemical and experimental operation on the scale that is planned for the procedure. Guidelines for carrying out a risk assessment and for analyzing the hazards associated with chemicals can be found in Chapter 4 of Prudent Practices.

The procedures described in *Organic Syntheses* are provided as published and are conducted at one's own risk. *Organic Syntheses, Inc.*, its Editors, and its Board of Directors do not warrant or guarantee the safety of individuals using these procedures and hereby disclaim any liability for any injuries or damages claimed to have resulted from or related in any way to the procedures herein.

Discussion

While dihaloimidazolidinediones have been known for more than a half a century,[2] their utility has only recently been demonstrated. Simple to prepare in almost quantitative yield,[3] the dihaloimidazolidinediones exhibit electrophilic characteristics applicable to the synthesis of, *inter alia*, dioxoguanides[4] or imidazolidinetriones upon treatment with primary amines or water, respectively. Additionally, such reagents can facilitate efficient and mild chlorodehydroxylations of alcohols. While the conversion of benzyl alcohol to benzyl chloride is demonstrated here, dihaloimidazolidinediones may also be used to chlorodehydroxylate secondary and some tertiary alcohols.

The utility of dihaloimidazolidinediones is, however, not limited to chlorodehydroxylations. Such species can be viewed as Cl_2 adducts of diamidocarbenes (DACs), to which they can be converted by reduction with elemental K,[5] or NaI.[6] The resulting *N,N′*-disubstituted imidazolyl-idinediones, isolable when ^{t}Bu or 1-adamantyl *N*-substituents are present, offer less crowded steric profiles than their 6-membered analogues.[7] The 5-membered DACs also distinguish themselves from imidazolylidenes in view of their enhanced π-basicity. The dihaloimidazolidinediones and their carbene derivatives thus may be considered as privileged reagents envisioned for use in a range of transition metal-mediated and organocatalyzed transformations.

References

1. Center for Multidimensional Carbon Materials (CMCM), Institute for Basic Science (IBS), Ulsan 44919, Republic of Korea; Department of Chemistry, Ulsan National Institute of Science and Technology (UNIST), Ulsan 44919, Republic of Korea; *E-mail: d.schilter@gmail.com or bielawski@unist.ac.kr; Financial support from the Institute for Basic Science (IBS-R019-D1) is gratefully acknowledged.
2. Ulrich, H.; Sayigh, A. A. R. *J. Org. Chem.* **1965**, *30*, 2781–2783.
3. Moerdyk, J. P.; Bielawski, C. W. *Chem. Eur. J.* **2014**, *20*, 13487–13490.
4. Zhao, F.; Wang, Y.; Zhang, W.-X.; Xi, Z. *Org. Biomol. Chem.* **2012**, *10*, 6266–6270.
5. Moerdyk, J. P.; Bielawski, C. W. *Chem. Commun.* **2014**, *50*, 4551–4553.
6. Zhao, F.; Wang, Y.; Xu, L.; Zhang, W.-X.; Xi, Z. *Tetrahedron Lett.* **2014**, *55*, 4597–4600.
7. Hudnall, T. W.; Moerdyk, J. P.; Bielawski, C. W. *Chem. Commun.* **2010**, *46*, 4288–4290.

Appendix
Chemical Abstracts Nomenclature (Registry Number)

Oxalyl chloride: Ethanedioyl chloride; (79-37-8)
N,N'-Dicyclohexylcarbodiimide: *N,N'*-Methanetetraylbiscyclohexanamine (538-75-0)
2,2-Dichloro-1,3-dicyclohexylimidazolidine-4,5-dione: 4,5-Imidazolidinedione, 2,2-dichloro-1,3-dicyclohexyl (3621-77-0)
Benzyl alcohol: Benzene methanol; (100-51-6)
Benzyl chloride: Benzene, (chloromethyl)-; (100-44-7)

David Schilter was educated at The University of Sydney, from which he received B.Sc. (Hons.) and Ph.D. degrees in 2005 and 2009, respectively, the latter under the auspices of Louis M. Rendina. These studies were followed by postdoctoral research at the University of Illinois at Urbana-Champaign (Thomas B. Rauchfuss) and at the IBS Center for Multidimensional Carbon Materials at UNIST (Christopher W. Bielawski), where he has served as a research scientist since 2014. His interests include coordination, organometallic and materials chemistry.

Christopher W. Bielawski received a B.S. degree in chemistry from the University of Illinois, Urbana-Champaign and a Ph.D. degree in chemistry from the California Institute of Technology (Caltech). After a postdoctoral appointment (also at Caltech), he assumed an independent position at the University of Texas at Austin where he directed efforts in a broad range of synthetic projects. Recently, he moved his research program to the Ulsan National Institute of Science and Technology (UNIST) and is currently participating in a new initiative focused on the synthesis and study of novel macromolecular materials.

Joanna Balsamo earned a B.S. in chemistry from Florida State University in 2003 and conducted undergraduate research in Dr. Michael Blaber's laboratory. In 2006 she joined Amgen in the Small Molecule Preformulation group, specializing in formulation development and solid state characterization. In 2015 she joined the Synthetic Technologies and Engineering group within Pivotal Drug Substance Technology department, where she currently works on the development of small molecule continuous manufacturing processes.

Christopher J. Borths earned a Ph.D. in synthetic organic chemistry from the California Institute of Technology in 2004 for developing novel organocatalytic methods with Prof. David MacMillan. After completing his graduate studies, he joined the Chemical Process Research and Development Group at Amgen where he is currently a Principal Scientist. He is a group leader in the Synthetic and Hybrid Technologies group within Pivotal Drug Substance Technology department, where he currently works on the development of robust and safe manufacturing processes for the production of active pharmaceutical ingredients, including traditional synthetic small molecule drugs, antibody-drug conjugates, and bioconjugation technologies.

Author Index Volume 93

Author Index Volume 93